Specialist Techniques and Materials for Concrete Construction

Proceedings of the International Conference
held at the University of Dundee, Scotland, UK
on 8-10 September 1999

Edited by

Ravindra K. Dhir
Director, Concrete Technology Unit
University of Dundee

and

Neil A. Henderson
Materials Engineer, Mott MacDonald Ltd

Published by Thomas Telford Publishing, Thomas Telford Limited, 1 Heron Quay, London E14 4JD.

URL: http://www.t-telford.co.uk

Distributors for Thomas Telford books are
USA: ASCE Press, 1801 Alexander Bell Drive, Reston, VA 20191-4400, USA
Japan: Maruzen Co. Ltd, Book Department, 3–10 Nihonbashi 2-chome, Chuo-ku, Tokyo 103
Australia: DA Books and Journals, 648 Whitehorse Road, Mitcham 3132, Victoria

First published 1999

The full list of titles from the 1999 International Congress 'Creating with Concrete' and available from Thomas Telford is as follows

- *Creating with concrete*
- *Radical design and concrete practices*
- *Role of interfaces in concrete*
- *Controlling concrete degradation*
- *Extending performance of concrete structures*
- *Exploiting wastes in concrete*
- *Modern concrete materials: binders, additions and admixtures*
- *Utilizing ready-mixed concrete and mortar*
- *Innovation in concrete structures: design and construction*
- *Specialist techniques and materials in concrete construction*
- *Concrete durability and repair technology*

A catalogue record for this book is available from the British Library

ISBN: 0 7277 2825 3

© The authors, except where otherwise stated

All rights, including translation, reserved. Except for fair copying, no part of this publication may be reproduced, stored in a retrieval system or transmitted in any form or by any means, electronic, mechanical, photocopying or otherwise, without the prior written permission of the Books Publisher, Thomas Telford Publishing, Thomas Telford Ltd, 1 Heron Quay, London E14 4JD.

This book is published on the understanding that the authors are solely responsible for the statements made and opinions expressed in it and that its publication does not necessarily imply that such statements and/or opinions are or reflect the views or opinions of the publishers or of the conference organizers.

Printed and bound in Great Britain by MPG Books, Bodmin, Cornwall

PREFACE

Concrete is the key material for Mankind to create the built environment, the requirements for which are both demanding in terms of technical performance and economy and yet greatly varied from architectural masterpieces to the simplest of utilities. This presents the greatest challenge and the question is how best to advance concrete and create imaginatively.

In response, the Concrete Technology Unit (CTU) of the University of Dundee organised this Congress following on from its established series of events, namely, Concrete in the Service of Mankind in 1996, Concrete 2000: Economic and Durable Concrete Construction Through Excellence in 1993 and Protection of Concrete in 1990.

Under the theme of Creating with Concrete, the Congress consisted of five Seminars: (i) Radical Design and Concrete Practices, (ii) Role of Interfaces in Concrete, (iii) Controlling Concrete Degradation, (iv) Extending Performance of Concrete Structures and (v) Exploiting Wastes in Concrete, and five Conferences: (i) Modern Concrete Materials: Binders, Additions and Admixtures, (ii) Utilising Ready-Mixed Concrete and Mortar, (iii) Innovation in Concrete Structures: Design and Construction, (iv) Specialist Techniques and Materials for Concrete and Construction and (v) Concrete Durability and Repair Technology. In all, a total of 421 papers were presented from 67 countries.

The Opening Addresses were given by Mr Henry McLeish, MP, MSP, Minister for Enterprise and Lifelong Learning, Scotland, Dr Ian Graham-Bryce, Principal and Vice Chancellor of Dundee University, Mrs Helen Wright, Lord Provost, City of Dundee and Professor Peter Hewlett, President of the Concrete Society. This was followed by four Opening Papers by leading international experts; Dr Bryant Mather, US Army Corps of Engineers, Professor Charles F Hendriks, Delft University of Technology, Netherlands, Dr Bjørn Jensen, Danish Technological Institute, Dr Oliver Kornadt, Philipp Holzmann AG, Germany, Professor Jurek Tolloczko, Concrete Society, UK, Mr Michael Téménidès, CIMBÉTON, France and Professor Yves Malier, Ecole Normale Superieure de Cachan, France. The Closing Address was given by Professor John Morris, University of the Witwatersrand, South Africa.

The support of 20 International Professional Institutions and 31 sponsors was a major contribution to the success of the Congress. An extensive Trade Fair, participated in by 50 organisations, formed an integral part of the Congress. The work of the Congress was an immense undertaking and all of those involved are gratefully acknowledged, in particular, the members of the Organising Committee for managing the event from start to finish; members of the International Advisory and National Technical Committees for advising on the selection and reviewing of papers; the Authors and the Chairmen of Technical Sessions for their invaluable contributions to the proceedings.

All of the proceedings have been prepared directly from the camera-ready manuscripts submitted by the authors and editing has been restricted to minor changes where it was considered absolutely necessary.

Dundee
September 1999

Ravindra K Dhir
Chairman, Congress Organising Committee

INTRODUCTION

Concrete is a versatile construction material that is finding use in ever more demanding environments and situations. In order to meet these challenges, the technology involved in designing concrete has become more sophisticated and specialised. This has led to the development of concrete mixes incorporating a variety of new materials designed to meet the demands of specific applications.

Hand-in-hand with any advancement in the use of different materials in concrete must come the development of techniques that enable these mixes to be used in practice. Since many of these techniques are unique to a particular concrete or material type they are by necessity specialised. The concept of using specialised construction techniques to take account of new materials represents a major advancement in concrete construction. In many situations the use of specialist materials and techniques leads to improvements in terms of cost, environmental benefits and/or durability. This is exemplified by the increase in the number of specialist suppliers who provide construction materials and techniques that achieve the desired result in a cheaper, faster and more environmentally friendly way. However, there is a need to transfer these technologies to designers and specifiers so that they are aware of the benefits and can make best use of the materials and techniques currently available. In order to facilitate innovation in this area the benefits of using specialist materials and techniques for concrete construction need to be explained in the context of conventional concrete practice.

This Conference reviewed progress in the development of six new materials and the associated techniques used for their effective use in concrete construction. These represent relatively established technologies such as sprayed and foamed concrete which are being developed further to allow their use in ever more demanding situations.

The Proceedings for the Conference; *Specialist Techniques and Materials for Concrete Construction* dealt with issues in six themes, namely (i) Grouting, (ii) Sprayed Concrete, (iii) Foamed Concrete, (iv) Underwater Concrete, (v) Floor Toppings and Overlays and vi) Non-Ferrous Reinforcement. The Conference was opened by a Leader Paper presented by the foremost exponent in the field. Each of themes was opened by a Keynote Paper presented by internationally acclaimed experts. There were a total of 43 papers presented during the Conference which have been complied into these Proceedings.

Dundee
September 1999

Ravindra K Dhir
Neil A Henderson

ORGANISING COMMITTEE
Concrete Technology Unit

Professor R K Dhir, OBE (Chairman)

Dr M R Jones (Secretary)

Mr M D Newlands (Joint Secretary)

Professor P C Hewlett
British Board of Agrément

Dr N A Henderson
Mott MacDonald Ltd

Professor V K Rigopoulou
National Technical University of Athens, Greece

Dr S Y N Chan
Hong Kong Polytechnic University

Dr N Y Ho
L & M Structural Systems, Singapore

Dr M J McCarthy

Dr M C Limbachiya

Dr T D Dyer

Dr K A Paine

Dr T G Jappy

Mr P A J Tittle

Mr J C Knights

Mr S R Scott (Unit Assistant)

Miss A M Duncan (Unit Secretary)

INTERNATIONAL ADVISORY COMMITTEE

Dr H M Z-Al-Abideen
Deputy Minister
Ministry of Public Works and Housing, Saudi Arabia

Professor M S Akman
Emeritus Professor of Civil Engineering
Istanbul Technical University, Turkey

Dr R Amtsbüchler
Manager-Technical Services
Blue Circle Ltd (South Africa), South Africa

Professor C Andradé
Director
Institute of Construction Sciences, Spain

Professor J M J M Bijen
Director
INTRON B.V., The Netherlands

Professor A M Brandt
Head of Section
Polish Academy of Sciences, Poland

Dr J-M Chandelle
Managing Director
CEMBUREAU, Belgium

Professor P Helene
Head of Civil Construction Engineering Department
University of Sao Paulo, Brazil

Dr G C Hoff
Senior Engineering Consultant
Mobil Technology Company, USA

Professor I Holand
Senior Research Engineer
SINTEF, Norway

Professor B C Jensen
Director
Carl Bro, Denmark

Professor S Mirza
Professor of Civil Engineering and Applied Mechanics
McGill University, Canada

Professor S Nagataki
Professor of Civil Engineering and Architecture
Niigata University, Japan

Professor H Okamura
Vice President
Kochi University of Technology, Japan

Professor E A e Oliveira
Director
Laboratório Nacional de Eng Civil, Portugal

Professor J-P Ollivier
Director of LMDC-INSA
LMDC, France

INTERNATIONAL ADVISORY COMMITTEE (CONTINUED)

Professor R Park
Professor of Civil Engineering
University of Canterbury, New Zealand

Mr S A Reddi
Managing Director
Gammon India Limited, India

Professor H-W Reinhardt
Head of Construction Materials Institute
University of Stuttgart, Germany

Professor R Rivera-Villarreal
Chief of Concrete Technology Department
Ciudad Universitaria, Mexico

Professor A Samarin
Consultant
Sustainable Development Technological Sciences and Engineering, Australia

Professor A E Sarja
Research Professor
Technical Research Centre of Finland, Finland

Professor S P Shah
Walter P Murphy Professor or Civil Engineering
Northwestern University, USA

Professor H Sommer
Head of Research Institute
VÖZ, Austria

Professor I Soroka
Professor of Civil Engineering
National Building Research Institute, Israel

Professor M Tang
Research Professor
Nanjing University of Chemical Technology, China

Professor T Tassios
Professor
National Technical University of Athens, Greece

Professor K Tuutti
Vice President
Skanska Technik AB, Sweden

Professor T Vogel
Professor of Structural Engineering
Swiss Federal Institute of Technology ETH, Switzerland

Professor F H Wittmann
Head of Building Materials Laboratory
Swiss Federal Institute of Technology ETH, Switzerland

Professor A V Zabegayev
Head of Department RC Structures
Moscow State University of Civil Engineering, Russia

NATIONAL TECHNICAL COMMITTEE

Mr P Barber
Manager of the Scheme, The Quality Scheme for Ready Mixed Concrete

Professor A W Beeby
Professor of Structural Design, University of Leeds

Mr B V Brown
Divisional Technical Executive, Readymix (UK) Ltd.

Dr T W Broyd
Technology Development Director, W S Atkins Ltd.

Professor J H Bungey
Professor of Civil Engineering, University of Liverpool

Dr P S Chana
Director, CRIC, Imperial College of Science, Technology & Medicine

Professor J L Clarke
Principal Engineer, The Concrete Society

Dr P C Das
Group Manager, Structures Management, Highways Agency

Dr S B Desai, OBE
Principal Civil Engineer, Department of the Environment, Transport and the Regions

Professor R K Dhir, OBE (Chairman)
Director, Concrete Technology Unit, University of Dundee

Mr C R Ecob
Director Special Services Division, Mott MacDonald Ltd.

Professor F P Glasser
University of Aberdeen

Professor T A Harrison
Technical Consultant, Quarry Products Association

Professor P C Hewlett
Director, British Board of Agrément

Professor J Innes
Director of Roads, Scottish Office

NATIONAL TECHNICAL COMMITTEE (CONTINUED)

Mr K A L Johnson
Director, AMEC Civil Engineering Ltd.

Dr M R Jones
Senior Lecturer, Concrete Technology Unit, University of Dundee

Mr P Livesey
National Technical Services Manager, Castle Cement Ltd.

Professor A E Long
Director of School, Queens University of Belfast

Professor P S Mangat
Head of Research, Sheffield Hallam University

Mr G Masterton
Director, Babtie Group Ltd.

Professor G C Mays
Director of Civil Engineering, Cranfield University

Mr L H McCurrich
Technology Development Consultant, Fosroc Construction

Professor R S Narayanan
Partner, SB Tietz & Partners Consulting Engineers

Dr P J Nixon
Head, Centre for Concrete Construction, Building Research Establishment Ltd.

Dr W F Price
Senior Associate, Messrs Sandberg

Professor G Somerville, OBE
Director of Engineering, British Cement Association

Professor D C Spooner
Director, Materials and Standards, British Cement Association

Dr H P J Taylor
Director, Tarmac Precast Concrete Ltd.

Mr M Walker
Technical Manager, The Concrete Society

Dr R J Woodward
Senior Project Manager, Transport Research Laboratory

SUPPORTING INSTITUTIONS

American Concrete Institute, USA

American Society of Civil Engineers, USA

Australian Concrete Institute

Concrete Association of Finland

Concrete Society of Southern Africa

Concrete Society, UK

Danish Concrete Society, Denmark

Fédération de l'Industrie du Beton, France

German Concrete Association (DBV)

Hong Kong Institution of Engineers

Indian Concrete Institute

Institute of Concrete Technology, UK

Institution of Civil Engineers, UK

Instituto Brasileiro Do Concreto, Brazil

Japan Concrete Institute

Netherlands Concrete Society

New Zealand Concrete Society

Norwegian Concrete Association

Singapore Concrete Institute

Spanish Association for Structural Concrete

Swedish Concrete Association

SPONSORING ORGANISATIONS WITH EXHIBITION

AMEC Civil Engineering Ltd.

Babtie Group Ltd.

Bardon Aggregates

Blue Circle Cement

Blyth & Blyth

British Board of Agrément

British Cement Association

Building Research Establishment

Castle Cement Ltd.

Cementitious Slag Makers Association

CIMBÉTON, France

Du Pont de Nemours International S.A., Switzerland

ECC International Ltd.

Elkem Ltd. (Materials)

Fosroc International Ltd.

Grace Construction Products

HERACLES General Cement Co., Greece

John Doyle Group

Lafarge Aluminates

L M Scofield Europe Ltd.

Minelco Ltd.

Mott MacDonald Ltd.

O'Rourke Group

Ove Arup and Partners

SPONSORING ORGANISATIONS WITH EXHIBITION (CONTINUED)

Readymix (UK) Ltd.

Rugby Cement

Scottish Enterprise Tayside

Sika Ltd.

SKW - MBT Construction Chemicals

Thomas Telford Publishing Ltd.

United Kingdom Quality Ash Association

W A Fairhurst & Partners

ADDITIONAL EXHIBITORS

Christison Scientific Equipment Ltd.

CMS Pozament Limited

The Concrete Society

David Ball Group plc.

E & FN Spon

Flexcrete Ltd.

Germann Instruments A/S, Denmark

Natural Cement Distribution Limited

Palladian Publications Ltd.

Quality Scheme for Ready Mixed Concrete

UK Certification Authority for Reinforcing Steel

Wacker-Chemie GmbH, Germany

Wexham Developments

CONTENTS

Preface	iii
Introduction	iv
Organising Committee	v
International Advisory Committee	vi
National Technical Committee	viii
Supporting Institutions	x
Sponsoring Organisations With Exhibition	xi
Additional Exhibitors	xii

Leader Paper
Concrete research in practice - the fourth dimension — 1
B J G van der Pot, HBG Group Technology, Netherlands

THEME 1 GROUTING

Keynote Paper
Grouting - Materials and Techniques on the approach to the millennium — 11
A M Haimoni, AMEC Piling, United Kingdom

Triaxial creep behaviour of grouted sand — 31
E Ribay, R Cabrillac and D Gouvenot

Cementitious grouts - 1999 update — 41
D J Johnson

Studies on the injecting properties of crack repair materials — 49
T Iisaka, H Umehara and T Sumi

Improving the properties of cement grout using organic fiber and superplasticizer — 57
W-H Huang and C-C Tseng

Cement grouting technology for consolidating soils — 69
M S Akman and M Mutlu

Multi-blend cementitious injection grouts for repair and strengthening of masonry structures — 79
E E Toumbakari, D van Gemert and T P Tassios

THEME 2 SPRAYED CONCRETE

Keynote Paper
Sprayed concrete for artistic interior shells   93
P Teichert, Laich SA, Switzerland

Hygrothermal stress induced problems in large scale   103
sprayed concrete structures
O Hrstka, R Černý and P Rovnaníková

Wet-mixed spray concrete: Improvement via GGBS   111
M F Nuruddin, S A M Yunus and A B Diah

The influence of accelerator on the long-term strength of   121
steel fibre reinforced shotcrete
G Baker, R Hockings and J Blanck

Numerical model for accelerating admixture in DEM modelling of shotcrete   131
U C Puri and T Uomoto

Workability, shear strength and build of wet-process sprayed mortars   141
S A Austin, P J Robins and C I Goodier

Innovations in the field of shotcrete repairs   153
D Beaupre, P Lacombe, N Dumais, S Mercier and M Jolin

Shotcrete : International practices and trends   163
K F Garshol

THEME 3 FOAMED CONCRETE

Keynote Paper
Micro-properties of foamed concrete   173
E P Kearsley and M Visagie, University of Pretoria, South Africa

The influence of the mix design on the properties   185
of micro-cellular concrete
L De Rose and J Morris

Moisture fixation and transfer in clayey cellular concrete-   199
Relation with thermal characteristics
L Marmoret, A Bouguerra, A t'Kint de Roodenbecke, O Douzanet and M Quenuedec

Configuration of pores on the mechanical and thermal   209
characteristics of cellular concretes through a homogenization method
Z Malou and R Cabrillac

Mixture design optimisation of cellular concrete A Rodriguez, M Pedraza, J Luciano and D Constantiner	219
Just foamed concrete - An overview E P Kearsley	227

THEME 4 UNDERWATER CONCRETE

Keynote Paper

Underwater concrete placement: Materials, methods and case studies J E McDonald and B Neeley, US Army Engineer WES, United States of America	239
Underwater concrete construction and repair T J Collins	257
Role and responsibility of the engineer-diver in the assessment of underwater concrete T M Browne	269
Underwater concrete construction L McLennan	281

THEME 5 FLOOR TOPPINGS AND OVERLAYS

Keynote Paper

Thickness design methods for slabs on ground D Beckett, University College of London, United Kingdom	289
Delaminations of bonded concrete overlays S J Sopko	303
Use of construction remainder soil in pavement construction M Ohno and K Fukai	311
Techniques for the early-life in-situ monitoring of concrete industrial ground floor slabs S A Austin, P J Robins and J W Bishop	317
Measuring the near-surface moisture condition of concrete slabs and gypsum screeds R P West, M L O'Neill and A D Rynhart	331
Design of steel fibre reinforced floors on foundation piles H Thooft	343
Floors - The working surface of industry A Dennis	355

THEME 6   NON-FERROUS REINFORCEMENT

Keynote Paper
Non-ferrous reinforcement 365
R H Scott, University of Durham, United Kingdom

Local strength reduction at boundaries due to 377
non-uniformity of steel fibre distribution
P Stroeven

Experimental investigation of polypropylene fibre reinforced concrete 389
A Al-Robaidi and M R Resheidat

Mechanical properties of carbon fibre reinforced concrete (CFRC) 395
Y Sato, C Kiyohara, K Ueda, H Sakai and M Nakamura

Structure and mechanical behaviour of concrete reinforced 405
with hydbrid steel-carbon fibres
Z Shui, Y R Cheng, P Stroeven and D H Dalhuisen

Behaviour of carbon fibre sheet as flexural reinforcement 413
in RC beams with aramid FRP rods
Y Takahashi, C Hata, Y Sato and T Maeda

Study on mechanical behaviour of mortar reinforced 425
with discontinuous carbon fibres
N Koshiishi

Improvement of bending load-bearing capacity by externally bonded plates 433
R Žarnić, S Gostič, V Bosiljkov and V B Bosiljkov

Effects of carbon fibre sheets on shear strength of 443
reinforced concrete columns adjoining walls or sashes
O Joh and Y Goto

Seismic performance of RC coumns laterally confined by 455
carbon fiber reinforcing plastic tube
T Yamakawa and P Zhong

Index of Authors 465

Subject Index 467

LEADER PAPER

CONCRETE RESEARCH IN PRACTICE – THE FOURTH DIMENSION

B J G van der Pot
HBG Group Technology
Netherlands

ABSTRACT. In the Netherlands the focus in concrete research is on "defined performance concrete". The aim is to put emphasis on developing knowledge to obtain defined performance, full control and no surprises rather than on ultra high strength and extreme performances. A brief synthesis is given of a research project called "Tolerant Concrete" which is focussed on developing an engineering concept which enhances the specification and control of performance properties in concrete structures.

Keywords: Defined performance, Particle packing, Self-levelling, Self-compacting, Rheology.

Ir Bart J G van der Pot is Director of HBG Group Technology in Rijswijk, the Netherlands. HBG is a large European construction group with the Netherlands, the UK and Germany as its home markets. Group Technology is the central department where technology initiatives are taken and facilitated.

INTRODUCTION

For a building contractor there are good reasons to be interested in concrete research. Most of the buildings and houses in the Netherlands are concrete based. Sometimes this is directly visible, but in most cases it is not because masonry facades or curtain walls of glass, steel, aluminium or natural stone hide the concrete columns, floors walls staircases etc. These houses and buildings are designed and built by building contractors who have to make some profit to enable them to stay in business. In order to do so, they have to be innovative, efficient and effective.

The relation between research at one side and performance, effectiveness, efficiency and ultimately profit at the other side is the combination I want to emphasise here. It is the link between concrete research and profit, or rather time, because in construction there is a strong relation between time and profit and an even stronger relation between lost time and lost profit. In most cases it are the building contractors who carry the cost for everything that goes wrong and requires extra time in construction.

In the title I have called this the fourth dimension, it is the dimension of time, money and usefulness, or in other words performance.

This dimension is reflected in the current way of thinking in concrete research in the Netherlands. In the last decade concrete research has, to a large extent, either been a sub-item of more integral research efforts into items such as underground construction, environmentally friendly construction etc., or it has been focussed on finding ever higher value applications in terms of higher strength or other properties. This research has obviously value, but it relates only to a low percentage of our business, it is directed to very special purposes and applications. For the majority of our daily work, all the houses, buildings and structures that we continuously produce, we need another focus; at least if we believe that there are still things to learn and to improve in these "normal" areas. In the Netherlands we have this belief and we have called this focus "Beton op maat", probably best translated by "defined performance concrete", as a contrast to the "high performance" approach.

Last year CUR, which is the Dutch organisation which co-ordinates and programs the concrete research activities in the Netherlands, has installed a programme advisory committee on concrete research. This committee has come to a programme approach with a focus on "defined performance concrete".

What we want is probably less spectacular, but more useful; not ultra high strength or extreme properties, but full control and defined performance. We want to be able to properly plan our work and have no surprises whatsoever; we want to provide our customers with a guaranteed product in full conformance with their specifications and expectations as well as a lot of other things which are not spectacular but can improve the speed, the control and the quality of our products and working circumstances.

Ideally concrete, as a material, has to be able to comply with an endless amount of requirements which vary from application to application and from project to project. As a result of all the research and development over the past decades, concrete has become a very complicated and knowledge intensive material. It can be given an endless variety of properties by varying its composition, additions or treatment.

Knowledge has become a vital ingredient for getting the most out of the material, for being commercially successful and for enabling creative thinking. The average user of the material, either in the design, specification, production, procurement or execution phase, only knows a fraction and only has access to a fraction of the available knowledge which could help him to prevent problems or to bring him to innovative solutions or improvements. As part of our "defined performance" approach we also have identified knowledge and knowledge technology as key attention areas.

Hereafter I give a short briefing of a Brite Euram research project named Tolerant Concrete in which we have participated and which has this focus on defined performance, full control and no surprises.

TOLERANT CONCRETE

Scope

During the last decade a large number of civil engineering structures were built using concretes which were designed for excellent properties with regard to production (placing and compaction), structural behaviour and durability. In several cases however these objectives have not been realised. Part of the reasons can be traced back to the consolidation phase of the concrete. They showed up in e.g. segregation in macro and microstructure of the concrete, intrinsic cracking and bad air void structure.

This resulted in unforeseen repair and increased maintenance costs and reduced service life as well as extra costs during construction resulting from rejections, corrective actions and delays.

These deficiencies are basically caused by the high sensitivity of the concretes in the consolidation phase during placing and compaction and in the early phase where the concretes, due to low strain capacity are sensitive to thermally, chemically and mechanically induced stresses and displacements. One of the observations was that in laboratory circumstances, with small specimens, a level of control over the properties could be achieved which in real circumstances could not be met at all times

The low tolerance of concrete for all kind of variations and occurrences has resulted in costly problems and repairs.

To solve this problem the BRITE EURAM proposal 'Tolerant Concrete designed for high performance during production and structural service life' was granted by the EC. The Tolerant Concrete research project was executed from 1994 till 1998.

The objective of the project was to develop an engineering concept to enable the designer of concrete to specify concretes with clearly defined performances being:

- tolerant to minor errors during production and curing and to the type of mineral aggregates; the choice of aggregate is largely dependent of local availability because large amounts are needed and transport costs may therefore be a substantial part of the material costs

- easy to compact with low or no energy consumption; related types of concrete are the so called flowing concretes, self levelling and self compacting.
- low sensitivity to intrinsic shrinkage
- improving the working condition of labourers
- reducing noise production during compaction

Gains were expected to be:

- higher production reliability
- improved durability due to the homogeneous and dense structure
- increase of the production rate
- reduction of the impact on labourers
- general improvement in environmental conditions and consumption of resources.

It was aimed to develop a concrete which requires no or highly reduced energy for compaction, which is structurally stable during transportation, pumping, placing and consolidation, which does not become inhomogeneous in the micro structure during setting and which lends itself to a predictable and gentle curing. In short: risks in execution should be avoided and there should be full control on the properties during all phases.

This new type of concrete had to be based on actual theories concerning the effects of combining minerals and silica fume spheres in the cementitious binders. To achieve this objective it is necessary to establish an understanding of the interaction between these micro particles and to develop dispersing agents, which at the same time work on all constituents.

Project

The research project was executed by a consortium consisting of 8 partners:

- Dansk Beton Technik A/S [DK]
- Geological Survey of Denmark and Greenland[DK]
- Demex A/S [DK]
- English China Clay Int. Ltd [UK]
- Hollandsche Beton Groep NV (HBG) [Nl]
- Intron BV [Nl]
- Eggerding & Co Groep [Nl]
- Perstorp Chemitec AB [SE]

The project started in 1994 and finished in 1998. The 12 project tasks can be clustered in 3 groups:

- identification of the research topics
- definition of the 'high performance concept'
- identification of related research and patents
- identification of commercially available clays in Europe and description of their characteristics
- identification of surface active agents which may act as dispersers for micro particles in multi powder combinations at high pH-levels

- development of techniques and understanding
  - optimisation of binders
  - ultra fine grinding technologies
  - verification of the basic theories of the research
  - development of 6 different types of concrete based on the verified principles
  - development of low compaction techniques

- verification
  - verification of rheological and functional properties of the new type of concrete through trial castings
  - assessment of the impact on the entire production process on working conditions and external environment compared to traditional production
  - documentation of properties in full scale structures

Principles

The aimed performance is strongly related to an optimal grain packing on all length scales. This optimum must be obtained in the production process. Important are therefore:

- the quality and proportions of the constituents
- storage, mixing and transportation of the fresh mix
- moulding, placing and compaction

Regarding the constituents three principles may be considered:

- Extending the principle of underwater concrete (wash out concrete or non-dispersive concrete) which is self-compacting and non-segregating. The technique is based on addition of chemical admixtures to improve the viscosity and thereby preventing dissolution of particles in the surrounding water. The result however is a sticky mix, which may result in inhomogeneous concrete, especially near rebars etc. Also the high cement content may result in unfavourable heat development and high material costs and furthermore the use of large amounts of superplasticiser may lead to fast loss of workability, retarded setting time, low early strength and further increase of material costs.

- The second concept, which is mainly due to Japanese research, is based on optimisation of the sieve curves of traditional materials. The relative volume of aggregates is reduced to prevent agglomeration of aggregates when the concrete passes obstacles. The powder to water ratio was adjusted to approximately 1 to produce an adequate viscosity to avoid segregation and to obtain self-levelling properties. The powder implies large amounts of mineral particles like slag, fly-ash and mineral fillers of e.g. limestone and quarts. This concept is risky. Noted are a high sensitivity to overdose of superplasticiser, variations in content of water adhering to the relative fine mineral aggregates and a high internal material disorder.

- The concept followed in this Brite Euram project is to add mineral constituents to realise the desired properties without having the above mentioned disadvantages. Theories are hardly available so the research bears an empirical character.

All three concepts are based on optimisation of the rheological behaviour of the species in order to avoid separation and to reduce compaction effort and finishing works. The use of super plasticisers is inevitable.

The usual defects in concrete are of the following nature:

- bonding defects at coarse aggregate and reinforcement, mainly evolved as highly porous zones of varying thickness
- uneven homogeneity of the cement paste, i.e. varied porosity
- cracks, both macroscopic in exposed surfaces and microscopic in the volume
- uneven distributions of constituents
- poor quality of the air void system (low specific surface)
- local lack of hydration of cement grains

For these defects two main causes may be noted:

- instability at micro – and macro level in the fresh concrete
- inappropriate handling, primarily excessive vibration

Particle Packing

It has been recognised for many years that, besides grading of aggregates, the particle size distribution on micro level is important for the properties of the slurry and the concrete. Therefore calculations have been made to get insight in the optimal composition, which is defined as the one with the highest packing density. As the numerical effort is considerable the calculations were restricted to 2 dimensions. Two models have been developed. In a kinematic model only geometrical compatibility of circular elements, which are the 2 dimensional equivalent of spheres, is demanded. A more advanced dynamic model was subsequently built to take into account forces between particles, like gravity, attraction- and elastic forces, and constitutive properties, like elasticity and viscosity. These models were made available for computerised use in the software program FEMMASSE. The steps in the calculations are:

- generation of a particle structure (suspension, slurry or solid) according to the applied material laws
- link mechanisms to the generated structure
- optimisation with regard to the criterion of optimised density as a consequence of particle packing.

It appeared to be possible to optimise the composition by means of the developed calculation scheme and the founding assumptions. This mix was subsequently made in the laboratory and appeared to behave well.

Dispersion

One of the conditions to arrive at an optimal packing of the particles is an optimal dispersion of the fines during the fluid stage of the mix. A dispersion agent may furthermore act like a plasticiser for the cementitious part of the binder. Demands posed on the mix from practice may be formulated as:

- workability of the mix for more than 3 hours
- pumpability for more than 2 hours
- initial setting time less than 6 hours.

The theoretical knowledge is rather well known as well as the action of candidate dispersing agents:

- disperse materials in water
- stabilise the dispersion

Due to the stabilisation workability is increased and setting is retarded, so demands on these aspects may be contradictory. These agents usually are of inorganic nature and are usually based on for example naphthalene or melamine. They lower the viscosity and thereby allow reduction of the water content and thus result in strength increase of the concrete. A major obstacle for more complex mixes becomes that the dispersion affinity of a type of disperser differs for different powders.

Good flow properties and high levelling and filling capacity are key properties for high performance concrete.

Grinding

As grinding of slag, cement and minerals is of great importance for the success of the promoted addition, methods, fineness and accuracy of grinding processes have been investigated. A typical classification according to resulting size is:

- crushing $100\% > 200\ \mu m$
- grinding $100\% < 200\ \mu m$
- micronising $100\% > 7\ \mu m$
- sub-micronising $100\% < 7\ \mu m$

Another classification is to grinding principles:

- crushing
- impact
- attrition
- scraping / shearing

Furthermore a large number of grinding mills have been discussed and tried out and a large number of selection criteria have been considered. The effect of micronising the additives is to obtain a better build-up of the paste i.e. optimal filling of the pores. Since these additives are not hydraulically active, wet grinding is the most suitable technology. To obtain a fineness of $1\ \mu m$ an eccentric vibration mill or a pearl mill is necessary. Otherwise the higher capacity dry ball mill can be used. For micronising hydraulic components dry ball milling process is recommended.

Next to size also shape may influence concrete strength and rheological behaviour. Experimental determination of these effects is obliged for optimisation.

Selection of Mixes

After a first selection of mixes, candidate mixes have been identified by using laboratory tests. The chosen mixes have been evaluated at this stage of the research by semi full-scale tests.

- On beforehand the use of 3 types of cement (CEM I 42.5 from Aalborg, CEM I 32.5 R from ENCI and CEM III/B 42.5 from CEMIJ), 4 ranges in water-cement ratio (0.35 – 0.55) and use of 2 types of superplasticisers (melamine and naphthalene based) was foreseen. From the possible combinations a selection was made. The aim of the first part of the research was to select a small number of candidate operational mixes with locally available mineral aggregates
- The optimal mixing procedure was found to be dry mixing of all powder components and aggregates, addition of water and finally step-wise addition of superplasticiser. A practical viscosity-measurement apparatus, the concrete tester, was found to be indispensable for the determination of the amount of superplasticiser to be added.

A second series of tests has been carried out in Denmark with locally available mineral aggregates. The objectives were to test mix designs and mixing procedures with regard to the demands for tolerant mixes:

- workable under difficult conditions
- transportable under vibration without loss of homogeneity
- low compaction effort needed

The mixes were prepared with great care in order to eliminate differences arising from the preparation procedure. The resulting mixes and concretes have been evaluated with respect to:

- workability (slump and flow test)
- air content
- density
- cohesiveness, homogeneity, bleeding (subjective by inspection by bare eye and by microscopic evaluation of epoxy impregnated plane sections)
- compressive strength and Young's modulus
- frost resistance
- chloride diffusion and permeability
- petrographic and air void analysis

It was concluded that all three mixes are promising for developing an operational tolerant concrete.

Full Scale Trials

The aim of the full-scale tests was to verify and to demonstrate the practical applicability of the mixes, which acted well in the laboratory tests with regard to:

- less or strongly reduced vibration for compaction
- avoidance of aggregate pockets
- smooth surfaces
- homogeneous product

As a realistic but hard to fabricate building, a L-shaped mould with a volume of 7 m^3, a height of 3.3 m and the inclusion of a casting joint was chosen. The test was executed on side of a mixing plant with two batch capacities of 2.5 m^3 and a production capacity of 160 m^3 / h. Local gravel aggregate from the North Sea and the blast furnace slag and Portland fly-ash cements available at the plant were used.

The full-scale trials have shown that self-compacting concrete based on the developed theory can be made on a practical scale. Both blast furnace slag and Portland cement may be used, as well as aggregate of different environmental origin. Critical aspects are:

- the batch size should be at least 1 m^3 as smaller batch resulted in a relatively large amount of the sticky binder on the batch walls
- the type of aggregates has a large impact on the mix properties
- provisions for avoiding air entrapment near the mould; entrapped air may probably be a reason to apply reduced (mould) vibration
- the pumpability.

Actual Test

Having concluded from the full-scale test that the concept of tolerant concrete is practical and workable –at least regarding the flow and compaction properties- an actual building site was sought. In Uelzen a local river underpasses a canal and for several reasons the siphon had to be reinforced. So a new concrete wall had to be built against the existing one. This new wall was to be built in segments, which were to be casted one at a time. It was foreseen to make a number of them in tolerant concrete.

Before applying the concrete in the segments of the siphon, two test segments have been cast just outside the siphon. These segments had been reinforced heavily. Thus the full industrial process has been simulated, including the mixing, the transport and the use of raw materials. The casting was done by pumping over a distance of 50 m. The mixes used in both casting were almost identical.

After demoulding it appeared that the findings of the full scale test were confirmed. The bulk material was (very) satisfying, but the surface was of less quality. In the casting of the first mould the pump pressure was low, in order to apply the concrete very gentle. This in contrast with the full scale test. In the second trial the pump pressure was normal.

It is observed that the smoothness of the surface was very good in case the Zemdrain was used. At the surfaces were Zemdrain is not applied, smooth areas and areas with air and water inclusions could be recognised. A correlation with (vertical) movements of the casting pipe has been recognised.

Temperature measurements all over the mould have shown a maximum increase of the temperature during hardening of 20 °C.

From the tests it was concluded that the tolerant concrete principle works and is able to yield good mixes, although the application technique may require some development. The mix itself has very good flow properties and is self-compacting. Slight mould vibration could be helpful to obtain smooth surfaces

CONCLUSIONS

The results of the Tolerant Concrete project supports the "defined performance" approach with respect to concrete research.

The overall conclusions of the Tolerant Concrete research project are:

1. the developed product is tolerant with regard to handling, though some aspects may not be disregarded (minimal batch size, form work provisions) but not to changes and variations in composition

2. no compaction energy is needed, except in case light mould vibration is chosen as the form work provision for obtaining smooth surfaces

3. mechanical properties, including shrinkage, can be accurately realised

4. as a consequence of the avoidance of vibration labour conditions and noise production during execution are seriously improved

5. as a consequence of the well controllable properties durability of structures can be improved by using the developed mixes

REFERENCES

BRITE EURAM. T BRE2-CT94-0972, 1993. Tolerant Concrete designed for High Performance during production and structural service.

THEME ONE:
GROUTING

Keynote Paper

GROUTING - MATERIALS AND TECHNIQUES ON THE APPROACH TO THE MILLENNIUM

A M Haimoni

AMEC Piling

United Kingdom

ABSTRACT. Grouting, essentially to modify the properties of the ground or to fill voids, has been used for centuries. Usually the objective is to make the ground/structure less permeable and/or stronger and stiffer. Traditionally, permeation grouting was used to achieve this objective via the use of fluid grout, which sets and stiffens. In the last few decades a number of new grouting techniques have evolved such as compaction grouting, jet grouting and compensation grouting. With these techniques, new materials have evolving while others have become obsolete. These techniques and materials have allowed grouting in a wider range of soils and led to a wider range of applications to achieve a variety of objectives.

Grouting contracts can vary in value between few hundreds of pounds too more than 100 millions. The variation in ground properties and the degree of expertise and knowledge required to achieve success has a similarly large range. Therefore, despite the many cases of success, it is perhaps not surprising that failure can sometimes occur. This paper briefly describes the available techniques and the materials currently in use. It describes two case histories to highlight the important factors that lead to success and failure. In addition, a brief description is given of the equipment employed in the industry.

Keywords: Grouting techniques, Grout materials, Cement grouts, Chemical grouts, Ground treatment.

Dr A M Haimoni is a design manager, AMEC Piling U.K. He is responsible for the technical and engineering aspects of the 'specialists' ground engineering activities undertaken by the company, such as grouting, ground anchorages and soil nailing, and minipiling. He has undertaken a number of research programmes on grouting and published several papers on the subject.

INTRODUCTION

Grouting has been used in civil and mining engineering for centuries to solve real problems and facilitate construction activities. Over the years, practitioners have evolved new techniques and materials, mainly by trial and error. The development of scientific bases and models for the grouting processes generally lagged behind and indeed this state is still prevalent in today's scientific era. This state, however, has not prevented grouting from evolving and establishing itself as a recognised engineering process. More and more techniques are being pioneered and a host of new materials are becoming available. There exists a huge range of different grouting techniques and applications with widely variable natures that make them wholly separate and unique. Nonetheless, they have one universal common feature that is the grout itself. All grouting techniques make use of grout material to achieve their objectives. The grout material can vary greatly in constituents, properties and behaviour. However, they all have the common feature of exhibiting a fluid like behaviour during placement, which will then gel and solidify with time. This simple feature encompasses and defines all the grouting techniques and grouting materials. The various grouting techniques are simply a description of the process of placing the grout. Indeed, concrete and its placement can be considered as a form of grouting. In mine infilling, for example, concrete 'like' material is often used to fill large cavities and limits grout spread.

Another unique feature of all the grouting processes is void filling. These voids are either present prior to the start of grouting or made during the grouting process itself.

GROUTING TECHNIQUES

Permeation grouting has been used for centuries to treat soils and rocks to improve their engineering behaviour or make them less permeable. In this instance, the grout is injected into the ground structure to invade and fill the voids present within the mass. Thin fluid grouts are essential to achieve adequate penetration into the small voids present between the soil particles or thin rock fissures. The inability of permeation grouting to deal with the very fine voids prompted the development of new grouting techniques such as compaction grouting and jet grouting. Compaction grouting, for example, uses a thick mortar paste to displace and compact the soil particles to increase the soil stiffness and strength. Jet grouting, on the other hand, breaks the soil structure by the action of pressurised water jets, to create a cavity, which is then filled with grouts.

A more recent development is the compensation grouting technique, which deals with the particular objective of minimising ground movement associated with tunnelling and other forms of ground excavation.

This section is not intended to give a complete description of the grouting techniques but simply as a list of the available techniques and their uses. For more information, the reader is referred to [1&2] and to the references stated herewith in each section.

Permeation Grouting/Rock Grouting/Crack Repair

Although these three processes differ in some details, they all describe the process of careful and slow grout injection into voids present within the mass of the soil, rock or structure.

Pressure and flow rate are limited so as not to disturb the structure of the body being grouted by the initiation of new fractures or cracks. The fundamental aspects of this process are given in [3-5]. In this process, the existing voids are fully or partially invaded and filled by the grout material, which when solidified, will result in modification of the properties of the whole body. The soil/rocks which can be permeated are limited by the permeability of the mass and the size of the pores/fractures existing. It is also dependent on the viscosity of the grout material used and the size of the particles present within the grout mix. For soils, Figure 1 indicates the limiting permeability for penetration for commonly used grouts. It also indicates the anticipated improvement as measured in the field.

The resulting strength/stiffness of the grouted mass is dependent on the strength/stiffness of the grout/gel, degree of penetration and the original strength of the mass prior to grouting. Although extremely high strength can be achieved, practically, this is limited since high strength gels are usually highly viscous which limits their penetration. Therefore, in normal practice, permeation grouting is normally limited to achieve strength of between 0.1 MN/ m^2 to 3 MN/ m^2.

Frequently the permeation grouting process is used to:

- reduce the mass permeability of the ground

- stiffen and strengthen the ground

- structural repair

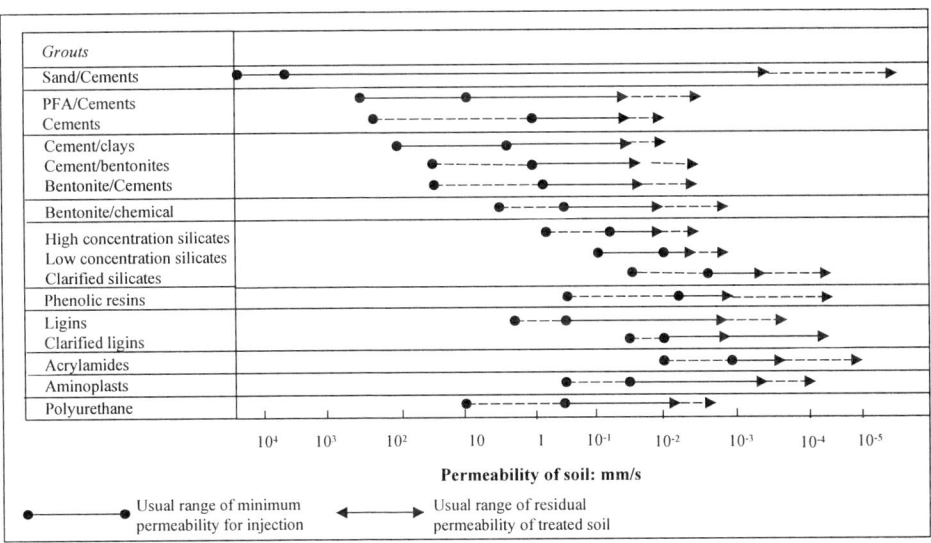

Figure 1 Indicative range of grouting treatments: if initial permeability lies at the lower end of its acceptable range, the resulting residual permeability tends to the high end of its range because of increased difficulty of treatment; permeability ranges are indicated are for superficial soils only [6]

Void Filling

This process is similar to permeation grouting in that it aims to fill existing voids by careful injection of grout materials without opening new cracks or fissures. The difference between the two processes, lie in the size of the voids. The most notable application of this process is the mine infilling process, which has been recognised as a unique process [7]. However, this process encompasses a wide range of applications. Each could be classified as a unique process such as cementation of the casing in the oil well industry, grouting around tunnel lining, grouting of anchors, rock bolts, soil nails and piles, etc.

Hydrofracture Grouting/Squeeze Grouting

This process lies on the opposite end of permeation grouting and void filling in that it deliberately disturbs the ground structure and initiates fracturing of the ground, see [5&8] for the conditions required to initiate hydrofracturing. It can be used in conjection with permeation grouting to increase accessibility, while compensation grouting in the main relies on ground fracturing to induce heave in order to compensate for excavation induced settlement.

In some practices particularly European, the technique known as claquage (hydrofracture) is used with permeation grouting to increase accessibility and the range of treatable soils [9-11]

Squeeze grouting (thick fracture) is generally used to achieve ground compaction, [12].

Compaction Grouting

Similar to hydrofracturing and squeeze grouting, compaction grouting relies on altering the structure of the ground by moving the particles closer to each other, hence achieving compaction, more information can be found in [13&14]. In contrast to hydrofracturing, thick mortar grout is used to prevent or limit hydrofracture. Compaction grouting is usually terminated on the onset of ground heave since at this point ground compaction will cease. However, this ground heave can be utilised as a means to induce controlled heave in order to affect structural jacking, [15] or compensation grouting [16].

Jet Grouting

In this process a high-pressure water jet (an air annulus can be added to increase efficiency) is used to cut and erode the structure of the soil with grout filling the hole being created. The fine soil particle such as clay and silt will be washed out by the flushing water while the larger particles tend to sink and mix with the grout. A review of the process and the factor affecting it is given in [17]. Case histories for horizontal and vertical barriers formed by this system can be found in [18&19].

Since this process uses extremely high pressure, extra care is needed to limit the risk of severe damage to the structure of the soil and any structures/services lying in the area being grouted.

Strength of between 0.5 to 30MN/m^2 can be achieved with this system with the strength dependent on the strength of the grout used and the degree of soil/grout mixing.

Mix-in-place

Here a boring tool is inserted into the soil, which is designed to break the structure of the soil. Grout is injected through the stem of the tool to the tip to achieve soil and grout mixing, hence mixing the soil in place.

The diameter of treatment is usually limited to the diameter of the mixing tool [20]. In some cases improvements in quality and size of treatment is achieved by increasing the number of boring tools to 2, 3 or 4, [21]

The strength/stiffness of the treated mass is dependent, similar to permeation grouting, on the degree of mixing and properties of the grout mix. This is normally limited to 1 to 10MN/ m^2.

Compensation Grouting

This is a relatively new technique which evolved solely to deal with excavation induced settlement. It is in some ways similar to structural jacking where grouts are used to induce heave by hydrofracturing or using the compaction grouting technique. The two techniques differ in the philosophy and timing of injection. Ideally, compensation grouting should take place before or during settlement development (concurrent compensation grouting) [22], while structural jacking is carried out afterwards. Nonetheless, the less attractive observational compensation grouting has been used successfully to compensate for the induced settlement. Here the settlement is allowed to develop and only when the 'trigger' limit is reached, a grout injection regime is implemented to jack up the structure/ground and reduce the settlement [23]. In this way, the settlement is prevented from being fully developed.

GROUT MATERIAL

Cement Grout

Ordinary Portland Cement (OPC) based slurries are the most widely used grouts. When mixed with other additives and admixtures, a wide range of mix properties, both in the fresh state and after curing, can be produced. For example, strength of between 0.2 MN/m^2 to 80MN/m^2 is usually attainable with cement grouts. Figure 2 shows typical cement grouts properties. The range indicated on this figure can be greatly modified by the use of additives and admixtures. Consequently, cement based grouts are used in all the grouting techniques described above. With permeation grouting, their use is limited only by the particulate nature of the cement. The cement particles tend to arch and filter which rapidly precludes penetration in the fine pores. This arching ability is enhanced in cement slurries by particle coagulation which enlarges the effective particle size and reduces the amount of available water by trapping water within the coagulates.

Admixtures can be used to reduce the risk of coagulation and enhance penetration. Nonetheless, this will be limited to pores of size greater than 200 to 300 microns and a soil permeability of greater than 10^{-3} m/s. To increase grout penetration microfine cements are sometimes used [25] which could tackle finer pores, but care is needed to ensure that particle coagulation does not offset the beneficial effects of the smaller particle sizes. This cement is relatively expensive and its use needs to be justified in terms of increased grout penetration.

Ultrafine cement grouts [26], and mineral grouts [27], have been used successfully to permeate the fine pores which are not accessible by microfine cement, but again cost plays a major factor in limiting their use.

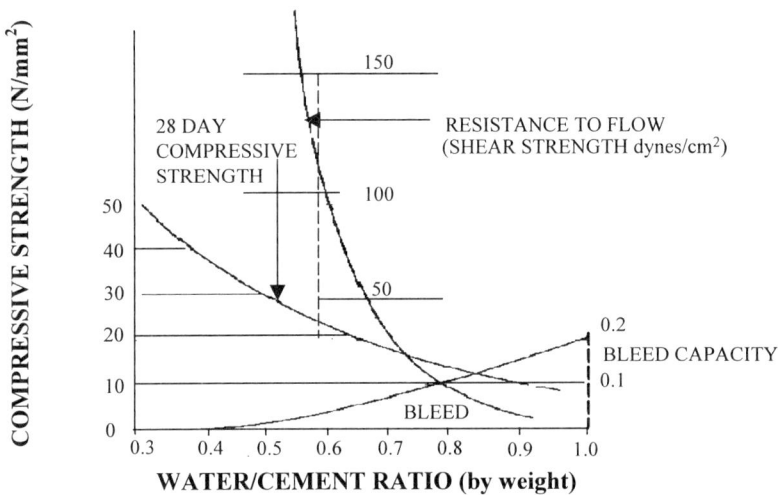

Figure 2 Effect of water content on grout properties [24]

OPC grouts are invariably not expected to achieve permeability less than 10^{-5} to 10^{-6}. To ensure a permeability of less then 10^{-5} it is common practice to supplement the OPC grouts with other grouts such as sodium silicate grouts.

Chemical Grout

Large numbers of what are termed chemical grouts have been used in the grouting industry. These grouts are usually more expensive than the Ordinary Portland cement based grout and in many cases suffer from being toxic. Therefore, their use is limited to areas where cement grouts are unsuited. The most common situation for the use of chemical grout is with fine grain soils and fine rock fissures where cement based grout is unable to penetrate. A wide range of strengths, gel times and viscosities can be achieved by the use of such grouts and the selection of the appropriate grout will depend on the particular problem. An interesting review of chemical grout is given in [28].

By far, sodium silicate based grout is the most widely used chemical grout since a wide range of viscosities, strengths and gel times can be provided by changing the proportions of the chemical constituents. It is a relatively inexpensive chemical grout and is generally considered to be non toxic. Gel times of between a few seconds and 180 minutes can be obtained by varying the grout constituent and concentration. Grouted soil strength of between 100 kN/m² to 2000 kN/m², [29&30] is generally achievable, which makes it a very versatile grout.

Silicate grout, however, is not able to treat soils with permeability of less than 10^{-6} m/s and resins are used in these circumstances. These grouts are very expensive and usually mildly to significantly toxic and their use is limited to special cases.
Types of resins include acrylic, phenol formaldehyde, polyurethane, etc. Some of these grouts have been used extensively in the past but became obsolete such as AM9 and Rocagil BT (Acrylic resins), due to their toxicity.

The intrinsic strength of the chemical gels used in ground consolidation and structured repair can vary from a few kPa to thousands of kPa. Soft gels are typically used in ground consolidation and waterproofing. The strong gels in the main have been reserved for structural uses, such as structured repairs, owing to their high cost and high viscosity. These gels normally exhibit elastic properties and excellent adhesion to concrete which makes them well suited for sealing cracks and joints present in structures.

It is important to note that all chemical gels are visco-elastic and their strength is dependent on the rate of application of stress. Unlike cement grouts, they exhibit considerable amount of creep under sustained load. Continuing creep can lead to rupture at much lower stress level than the ultimate which can be achieved in the short term. In [31] it has been demonstrated that failure due to continued creep can occur within few hours even at a stress ratio ($^{applied}/_{ultimate}$) as low as 0.15 in the unconfined stress state. This behaviour should be studied and allowed for, if the soil is to be structually loaded. The grouted soil is usually many folds stronger than that of the pure gel, [28&32]. In the field, allowance should be made for partial and incomplete treatment, particularly in mixed and layered soils. Here, low permeability zones or layers may receive little or no strength improvement.

For most waterproofing application the strength of the grouted soils is of little significance, since in fine pores a grout strength as low as 50-100 N/M m² is able to withstand a hydraulic gradient in excess of 100 without displacement [3].

Durability

Cement grouts are durable under most normal conditions. Only in abnormal environmental conditions where special measures are necessary to ensure long term performance of cement based grouts. Such conditions many include, soil with high concentration of sulphates or acids and large-scale temperature fluctuation. Normal rules and guidelines established for concrete are applicable in these instances.

Chemical gels suffer from diffusion as a result of concentration gradient and also suffer from dissolution caused by flowing water. This is likely to take place over surface than through the body of the gel since the permeability of the gel is very low (typically 10^{-10} m/s). Therefore, the best defence against this phenomena is good grouting practice to achieve maximum grout penetration. The choice of the correct grout formulation to reduce shrinkage and syneresis (bleed) helps to reduce decay. In normal circumstances, stable chemical grouts are expected to live more than 25 years [33]. Care needs to be taken in particularly aggressive and contaminated grounds which may affect the quality of the gel both in the short and the long term.

Syneresis can present a problem if large body of gel is present which is not the case in fine-grained soil and fine rock fissures. Soils and rocks containing large voids is best treated with cement grout first before chemical grouts injection. This two stage grouting is often economical owing to the cheapness of the cement grouts.

Depending on the mix proportions and degree of alkalinity cement based grouts and silicate-based grouts can give corrosion protection to steel. Other chemical grouts are not so alkaline and the protection they offer is limited to the formative of a barrier which limits the steel contact with water, salts, gasses and other oxidisers. Some constituent parts of the chemical gels are highly corrosive in their pure state but the resulting gels are usually, but not always, neutral. Consequently it is very important in these circumstances to ensure good mixing and no phase separation during injection.

PLANT AND EQUIPMENT

Drilling

The first stage in any grouting works is the placement of the delivery system or achieving access to the zone of ground to be grouted. This is carried out in many ways depending on the type of grouting works and the geometrical details of the zone to be treated and the area around it. In the majority of cases, this involves drilling or driving a pipe into the ground, which then can be used for grout injection. Drilling and driving equipment can come in all sorts of shapes and sizes and vary from the hand held jackhammer type to the very large and powerful drilling rigs capable of drilling holes hundreds of metres in length. Rigs can have state-of-the-art computer control and monitoring equipment which includes depth, inclination and position monitoring together with drilling parameter monitoring, such as penetration rate, torque, flush pressures, etc. These could be used to ensure correct positioning and alignment of the hole and aid in understanding the quality of the material being drilled. It could also be used to detect voids and cavities deep in the ground that may need grouting.

During or after the completion of drilling, hole alignment could be checked using systems such as the Maxibore or the EMS, and the quality of the rock surveyed using CCTV survey techniques.

Grout Placement

Once access to the ground to be treated is achieved by drilling of holes or driving of pipes, grouting can proceed by pumping the grout to that zone. Virtually all types of pumps have been used in the grouting works, such as piston pumps, moving cavity pumps, centrifuge pumps and rotary pumps. The best pump for any particular grouting operation is dependent on the grout type and its consistency, and the pressure and flow rate required. In many cases, there is more than one type of pump available to choose from. In recent years, a major advancement has been made in the monitoring and control of the grout pumps, which has led to major improvements in the quality and speed of grouting works.

CONTROL AND RECORDING

The recent advances in computer technology, electronic control and data acquisition and recording have led to a vast improvement in the control and analysis of the grouting works. The delivery pump, in many cases, is controlled by a computer that is able to regulate pressure, flow rate and volume leaving little room for human error. Data can be continuously captured and recorded. It can then be stored on disk for subsequent use and analysis.

On Contract 102 of the Jubilee Line Extension, London, a chemical grouting manifold was developed by AMEC-Geocisa allowing up to four different lines to be grouted simultaneously. Each line is independently controlled and monitored by a Programmable Logic Controller mounted on the manifold which is in turn under the supervision of a remote Supervisory Control And Data Acquisition (SCADA) computer. Up to three such manifolds can be operated simultaneously by the SCADA.

Grout is delivered to the manifolds from a pumpset comprising three individually controlled pumps each supplying a separate constituent of the chemical grout in metered quantities. The SCADA monitors the demand from each manifold line, and controls the pumpset accordingly, in order to provide the required total volume.

Control is such that the grout constituent ratios can be maintained within 0.5% at all times. Real time graphical and numerical displays are available throughout the operation at both the manifold and the SCADA. All data is also stored on the SCADA for subsequent analysis and reporting.

The individual grout constituents are delivered to inline mixers situated close to the manifolds. This ensures that grout is freshly mixed on demand and wastage of material and flushing of lines is limited to the short distance between the mixer and the manifold. The raw material storage and pumping area is static and can be situated hundreds of metres from the manifolds.

Pictorial representation of data is often used to speed up, simplify and improve the quality of the analysis process. Analyses and graphs can be drawn at the touch of a button which can show a detailed presentation of one injection to a summary of all injections in the whole grouting area.

Ground Monitoring

Monitoring of the ground and structures in the vicinity of the grouting works is an integral part of the grouting process. It can play an important role in the control of the grouting works and leads to a better understanding of ground response. It can also help to avoid problems that could have disastrous consequences if they go unchecked. These could take the form of ground heave leading to distortion of the adjacent structures/services, or over pressurisation of adjacent underground structure such as tunnels, sewers, etc.

Ground movement monitoring can include precise levelling, position indicator, verticality prism, extensometer, electro-levels, etc. The type, number and frequency of monitoring is dependent on the sensitivity of the ground and the structures to ground movement, and the type of the grouting work being undertaken.

GROUTING – SUCCESSES AND FAILURES

The widespread use of grouting to achieve a host of differing objectives coupled with the huge range of available techniques and materials and the variable and unknown nature of the ground, yield a recipe for many cases of success and perhaps not surprisingly, a few cases of partial or complete failure. Obviously the grouting industry strives for success and aims to limit failure.

Cases of success are quite often reported but due to commercial liabilities and pressures, cases of failures frequently go unreported and a great deal of useful information is lost in the process. Failures can be related to reasons other than human error or lack of professionalism on the part of the 'specialist' grouting contractor.

The reasons which lead to failure can be related, in the main, to the lack of knowledge of the ground being grouted and lack of understanding of grout/ground interaction.

The ground normally represents the biggest unknown in the grouting process. The site investigations are frequently inadequate, rarely carried out for the specific objective and directed to the needs of the grouting operation. Boreholes represent a tiny fraction of the ground being grouted and the sampling and testing are frequently inappropriate for obtaining the parameters required for designing the grouting system.

One way forward could be by the implementation of a two-tier site investigation. In the first stage, the site investigation would aim at obtaining a general ground characterisation and classification with general sampling and testing similar to the majority of site investigations carried out nowadays. If grouting is judged necessary, then a second stage should be implemented in which the site investigation will be targeted at obtaining the parameters required for the design of the grouting process. Liaison with a specialist-grouting contractor, at this stage, would be advantageous. On large contracts, a full-scale trial could represent an integral part of this detailed site investigation. Obviously, clients are reluctant to spend additional funds on a second site investigation and it is the consultants, or the client adviser's, responsibility to ensure that the client is aware of the considerable benefits and rewards of such an investigation. This should lead to a huge reduction in risks and associated costs.

An improvement in some of the SI technique is necessary to obtain the high quality soil parameters required. The most notable techniques which require further consideration are:-

(i)   permeability testing in soils and rock
(ii) sampling granular soil with fines under water
(iii) sampling of layered soils.

In addition to the unknown represented by the ground itself, there exists a lack of knowledge of the behaviour of the grout in the ground, in other words, the ground/grout interaction. There exists a dearth of information and theoretical modelling of the grouting processes. Those few models that exist are at best conceptual and generally unable to predict accurately the grouting parameters. Consequently, grouting is often based on empirical rules and past experience. This state is the product of the many complex processes that take place during even the simplest of grouting operations. Take for example the permeation grouting of soil/rock using cement grouts. This operation is dictated by many processes for which the parameters are frequently unavailable and the theoretical models are insufficient.

Cement grout is a complex suspension that can be thixotropic and time and shear dependent. Any flow parameters measured are dependent on the measuring system and often are transient and changing with time, [34]. When cement grout flows through small pores, the cement particles tend to arch and filter leading to blockage of these pores. This process is dependent on the size of the pores, which are variable in the ground, the physical and chemical nature of the particles in the suspension, and rheology of the grout. Cement particles in suspension tend to coagulate.

This leads to an enlargement of effective particulate size and a reduction of available free water within the suspension, since water is trapped between the coagulated particles. Coagulation is dependent on many factors, such as size of particles, type of cement, water cement ratio, type of additives and admixtures, etc.

All these variables lead to very complex behaviour, which is virtually impossible to predict even in the simplest geometries. In the ground with variable pore sizes, which are very difficult to measure and quantify, the picture is even more complex. Despite this we have come a long way in understanding what can and cannot be done. Owing to these difficulties, much attention has been focused on alternative processes to permeation grouting.

Jet grouting, for example, avoids some of the complication associated with permeation grouting by cutting and replacing/mixing the soil. Jet grouting is not free from difficulties, however, the extremely high pressure used in this process and the large volume of slurry produced create their own problems, e.g. ground leave, and adds to the cost of the system. The most important parameter in this process is the size of the hole created which cannot be predicted with certainty.

From the above short review, it is apparent that accurate prediction of the grouting parameters and grout/ground interaction is not possible. Despite this, a large proportion of grouting works are satisfactory and only in the minority of jobs some kind of difficulties occur. In order to minimise these occurrences, it is recommended that the following is implemented on grouting jobs:-

(1) Undertaking good site investigation, which addresses the needs and requirements of the chosen grouting system. A site trial may prove to be advantageous on large jobs. This ensures that the appropriate grouting process is selected.

(2) Keeping an open mind and being ready for alterations to methods of work and the grout mixes used. To avoid contractual difficulties, all parties need to recognise the limits of predictions and allow, in the contract, for changes in the method of work and grout type, using the observational method.

(3) Undertaking continuous monitoring, recording, observing and analyses of results in order to optimise the process and introduce the appropriate changes if necessary. This is the best way to ensure that the desired results are being achieved.

(4) Ensuring that the personnel in charge of the site work have adequate experience in the type of work being carried out and are able to analyse and understand the meaning of the parameters being monitored.

With the advances in electronic monitoring, recording and analysis, it is quite simple nowadays to monitor all sorts of parameters which should ensure adequate control of the grouting works. Nonetheless, it is important to be selective and ensure that the most vital of the parameters are not swamped by the huge data collected and that the engineer controlling the work is not overloaded with numbers.

CASE HISTORIES

Grouting literature is full of case histories on all types of grouting techniques. The reader only needs to look in the recent conferences on the subject to find all sorts of case histories detailing various grouting techniques, materials, projects and objectives [35-41]. Ciria reports [42&43], contain good sources of references. This brief section is not meant to give yet another case history but rather a reflection on two published case histories to attempt, by example, to explain some of the reasons for success and failure in grouting.

It has been stated earlier that grouting relies mainly on empirical methods and the observational technique. Consequently, works carried out in such a way stand a good chance of succeeding. It is imperative that the contractual arrangement and specifications reflect the need for such method and that adequate time and flexibility are built in for the truly observational technique to be employed. Time limitations, strict unsuited specifications and contractual arrangements quite often stand in the way of achieving the best results. The specialist contractors can also sometimes be blamed for unsatisfactory results, which can be the result of the choice of unsuitable grouting techniques or the use of inexperienced personnel. Quite often commercial pressure forces the specialist contractors to accept undue risks and contractual terms.

To overcome these difficulties, all parties involved in the grouting works including clients, consultants, main contractors and specialist sub contractors need to be educated in the process and limitations of the grouting techniques. All should be made aware of the advantages and disadvantages, for proper evaluation of risks and expectations. CIRIA's attempt [43] is encouraging. A grouting job should not be seen in a similar way to the construction of a structure or a road. Her, virtually all the materials used are well understood, covered by a code and the results of the work are clearly visible to the operator, in clear contrast to grouting in the ground.

In the first example, the grouting work on contract 102 of the Jubilee Line Extension, London, is briefly considered. This work represents one of the biggest and most complex contracts ever undertaken. The Jubilee Line Extension project itself represents one of the biggest civil engineering contracts undertaken in the UK in recent years. It involves the construction of twin running tunnels and a number of stations and service tunnels constructed mainly in London Clay The grouting work on this contract (worth approximately £20m) made use of the oldest grouting technique in the form of permeation grouting and the newest in the form of compensation grouting.

The permeation grouting was used to enable the safe construction of the shafts and the tunnels in and below the water bearing permeable Terrace Gravel. 11,000 cubic metres of cement/bentonite and sodium silicate/ester grouts were injected in the most prestigious and heavily congested parts of London. Grouting works were undertaken routinely and all permeation grouting was completed successfully and achieved the desired results.

The size of the grouting works, the long contract time, and access and space difficulties lead to the development of the manifold grouting system described briefly earlier.

Compensation grouting is a relatively recent technique and many of the methods employed and materials used were developed and evolved during the works. More than 60 prestigious buildings including the Big Ben clock tower were protected using this technique.

To achieve this, 35,000 metres of large diameter tube "a" manchettes (TAMs) were installed and more than 4,000m³ of grout were injected.

The large diameter steel TAMs (76mm) were developed to enable thick mortar paste to be repeatedly injected from the same port. Extensive laboratory testing and full-scale field trials were undertaken by AMEC-Geocisa in association with the Department of Civil Engineering, University of Bradford, to develop and check the system before field application.

This mortar paste was used to produce thick grout wedges rather than thin fractures that associated with fluid grouts traditionally injected through TAMs. This allowed a high degree of control over ground movement both during the tunnel excavation (concurrent compensation grouting) or afterwards using observational grouting techniques [22].

The success of this system can be clearly seen in the results shown in Figure 3 and 4. Figure 3 shows typical observational grouting episode carried out to reduce the deflection of Big Ben clock tower. More than twenty of such grouting episodes were carried out effecting a total deflection reversal of approximately 200mm at the top of the tower.

Figure 4 demonstrates the ability of the concurrent compensation grouting in limiting the settlement of the Royal Automobile Club as the eastbound running tunnel (4.85m diameter) passed 7 to 9m underneath the foundation of the building, Figure 5. Without compensation grouting, the settlement was anticipated to exceed 25mm. These two examples represent a tiny fraction of the grouting work carried out on contract 102.

Part of this success can be related to the good understanding of the behaviour of the grout mix which was heavily researched and the good understanding of the behaviour of the over consolidated London Clay.

A large part of the success can also be related to the contractual arrangement and teamwork which was employed on this contract. Due to the complexity of the work, it became apparent at an early stage that the best way to overcome difficulties is through the use of partnering and one team working.

This allowed the flexibility required for the introduction of a number of novel ideas and methods that led to achievement of such results.

In the second example, the compaction grouting technique was employed in an attempt to form a series of grout bulbs within a very soft peaty soil in order to form continuous columns. The aim of these columns is to transmit the load from the granular soils above the peat to the over consolidated Kimmeridge clay underneath

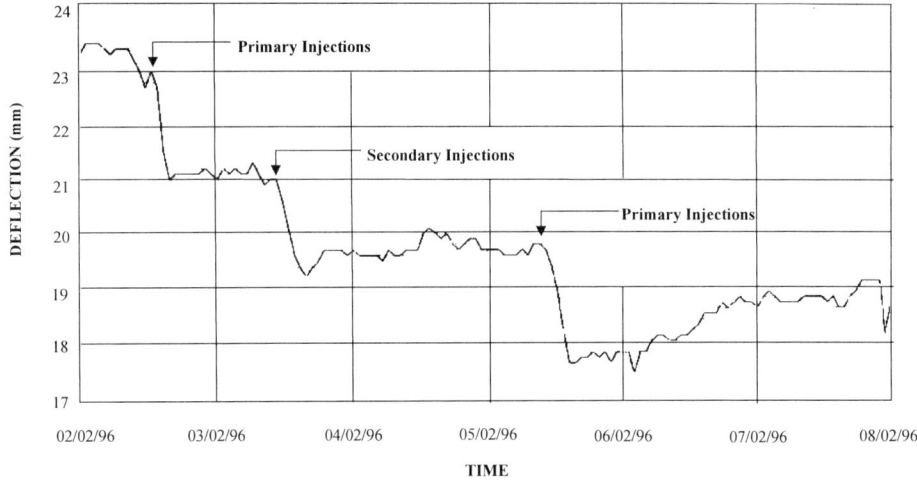

Figure 3 Reduction off deflection on Big Ben clock tower using observational grouting deflection measured by electrical plumb bob mounted 55m above base

For more information on this case history the reader is referred to [44]. Although the compaction grouting technique is well established, two new details were introduced on this contract. These are (i) the use of compaction mortar to form bulbs in peaty soils, (ii) the addition of plasticiser to the grout mix, to reduce blockages during pumping. As the work progressed, it became apparent that ground heave was being generated much earlier than can be predicted for a spherical bulb.

Consequently, the heave was related to the development of horizontal fractures, which, in turn, may be related to the low shear strength of the peaty soils or the influence of the plasticiser on the grout mix.

Time limitation did not allow adequate investigation to be undertaken nor did the contractual arrangement. Consequently, the exact reason for this anomalous result is still a mystery and further research works is needed to improve our understanding of the behaviour of mortar grouts in soft clayey and peaty soils. Although the result is a dubious success, in the words of the authors, this case highlights a number of typical factors that normally lead to problematic grouting contracts. These could be summarised as follows:

(i)   Lack of time – Although similar to most grouting works the observational technique was employed, lack of time did not allow proper evaluation to be made in order to introduce suitable changes and alterations.
(ii) Contractual arrangements – The contractual arrangements did not recognise the limitation of the technique and the need to review and instigate the appropriate changes.
(iii) Lack of understanding of material and ground behaviour. Although the compaction grouting technique is well understood in permeable soils using frictional grout mixes, our knowledge is limited in other soils particularly if plasticiser is added to the grout mix.

Grouting : Materials and Techniques 25

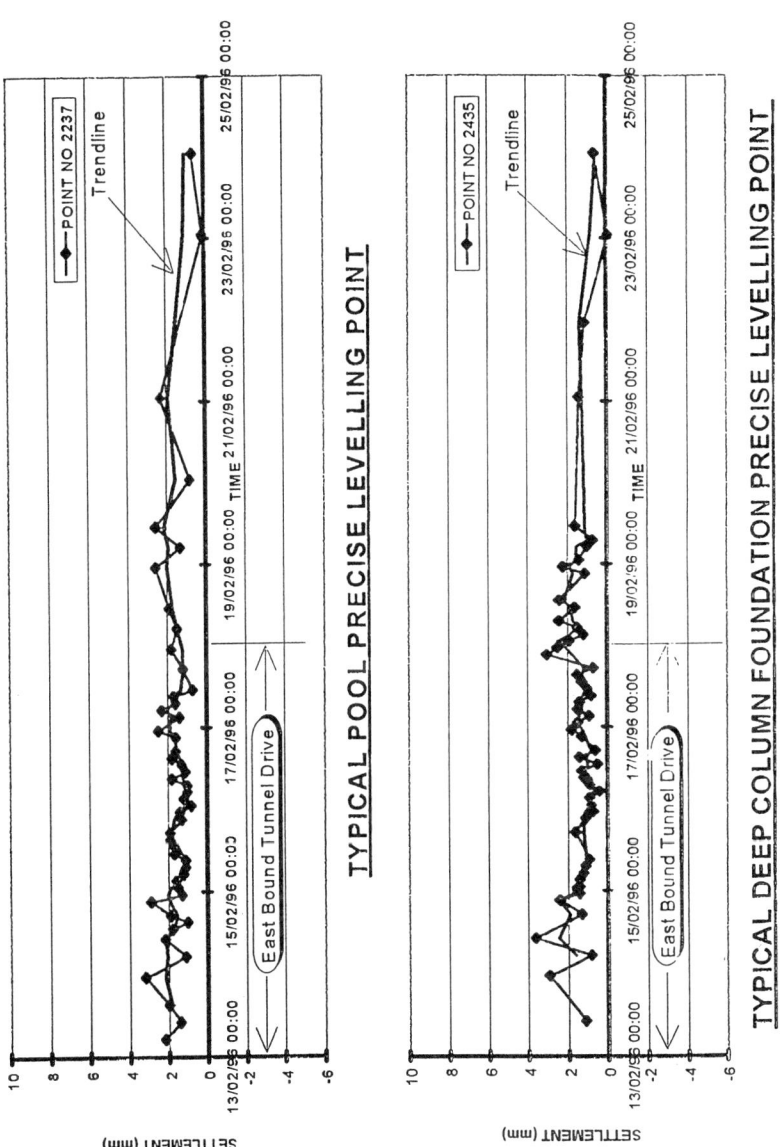

Figure 4 Time plot showing settlement at the Royal Automobile Club after the passage of the eastbound running tunnel

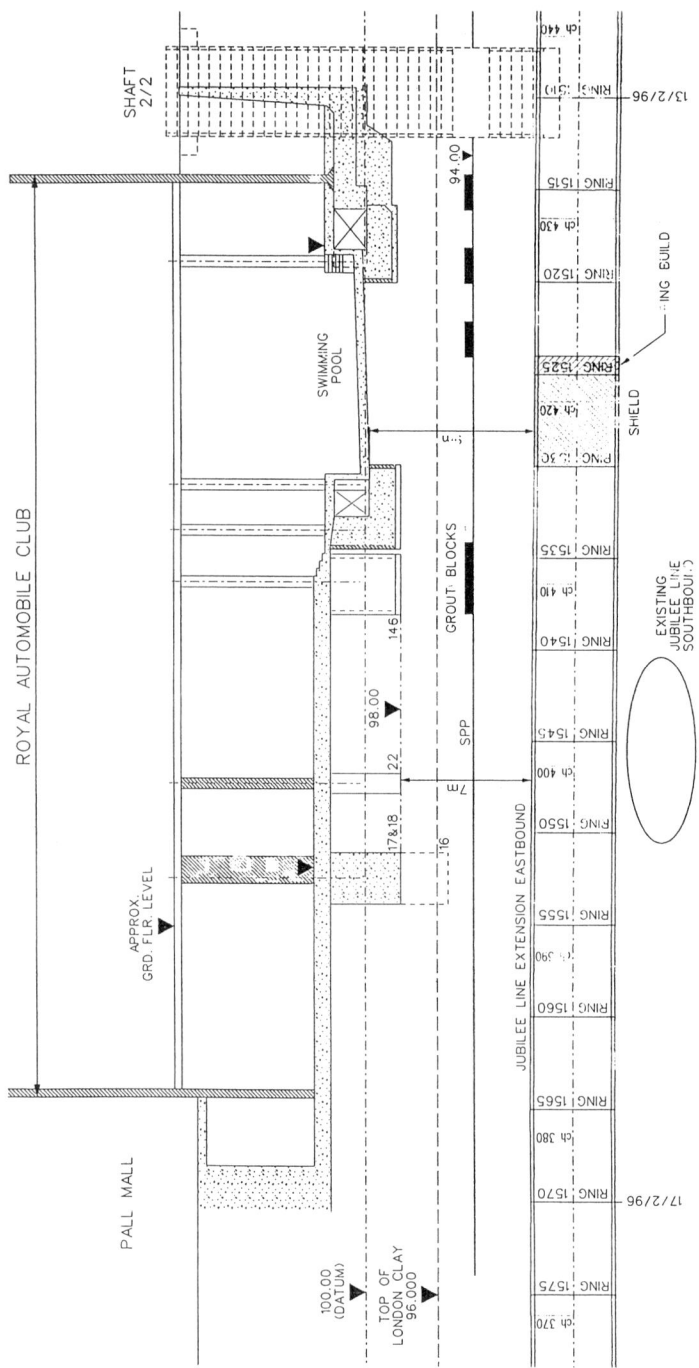

Figure 5 Section along the tunnel at the Royal Automobile Club showing the concurrent grouting blocks relative to tunnel heading

FINAL REMARKS

From the above short review, it is clear that given the appropriate conditions, grouting works can deliver the goods and produce the desired results even with the most complex and highly demanding projects. Nonetheless, grouting should not be seen as another routine construction contract. We should recognise the limitations of each technique and the limitation in our knowledge of the grout behaviour in the ground, which itself is highly variable and difficult to quantify. Therefore, it is important that the appropriate contractual arrangements are formulated which allow for these limitations and recognise the need to apply the observational technique based on adequate monitoring and evaluation of the results. This should be combined with the appropriate site investigation and the use of experienced personnel to ensure that the most suitable grouting process is selected at an early stage.

We have come a long way in the past few decades particularly on the monitoring and control. These advances should lead to a better control and evaluation of the works. This should ultimately lead to better results and more confidence in the grouting work. Nonetheless, they should be used in conjunction with experienced personnel, not to replace them.

The speed in data acquisition and analysis should allow the grouting operator to follow the progress of the work and instigate the appropriate changes at the appropriate time. Hopefully, this will not be hindered by unsuitable and restrictive specifications and contractual arrangements.

By far, cement based grouts are the most widely used material in the grouting industry. Microfine cement grouts or even ultrafine cement/mineral grouts are gaining popularity but they are relatively expensive and further research is required in order to evaluate the improvements they offer in comparison to the additional cost. On the chemical grouts, sodium silicate grouts have emerged as the most widely used and accepted chemical grouts due to its relatively low cost and since it is generally considered to be non toxic. Many chemical grouts have appeared in the past few decades and many have disappeared due to their toxicity.

It is most certain in the next century that toxicity will play a vital part in the development and introduction of new grouts. In the current era of environmental awareness and legislation's, there is little scope for any toxic grout to be accepted.

REFERENCES

1. BOWEN R, Grouting in engineering practice. Second edition, Applied Science Publishers, Essex, UK, 1981.

2. NONVEILLER E, Grouting theory and practice. Developments in Geotechnical Engineering, 57, Elsevier Science Publishers B.V., Netherland, 1989.

3. SCOTT R A, Fundamental considerations governing the penetrability of grouts and their ultimate resistance to displacement. ICE Conf. on grouts and drilling muds in engineering practice. Butterworths, London, 1963, pp 10-14.

4. RAFFLE J F & GREENWOOD D A, The relation between the rheological characteristics of grouts and their capacity to permeate soil. Proc. 5th Int. Conf. SMFE, Paris, 2, 1961, pp 789-795.

5. MORGENSTERN N R & VAUGHAN P R, Some observations on allowable grouting pressures. ICE. Conference on grouts and drilling muds in engineering practice. Butterworths, London, 1963, pp 36-43.

6. GREENWOOD D A & THOMSON G H, Ground stabilisation – Deep Compaction & Grouting, ICE Works Construction Guides, Thomas Telford Ltd, London. 1984.

7. HEALY P R & HEAD J M, Construction on abandoned mine workings, CIRIA special publication 32. Construction Industry Research and Information Association, London. 1984.

8. WONG H Y & FARMER I W, Hydrofracture mechanisms in rock during pressure grouting. Rock Mech. 5, 1973, pp 21-41.

9. CAMBEFORT H, The principles and applications of grouting. Q.J. Engng Geol, 10(2), 1977, pp57-95..

10. CAMBEFORT H, Injection des sols. Editions Eyrolles, Paris. 1964.

11. CARRON C, The state of grouting in the 1980's. Proc. Conf. On Grouting in Geotechnical Engineering, New Orleans, American Society Civil Engineers, New York 1982.

12. GREENWOOD D A & HUTCHINSON M T, Squeeze grouting unstable ground in deep tunnels, Grouting in Geotechnical Engineering, Proc. of conference, ASCE Geotechnical Engineering Division, New Orleans, LA, USA. 1982, pp 631-651.

13. WARNER J, Compaction grout: rheology vs effectiveness, Grouting, soil improvement and geosynthetics, Proc. of conference, Geotechnical Special Publication 30,ASCE Geotechnical Engineering Division, New Orleans, LA, USA. 1992, pp 229-239.

14. WARNER J F, SCHMIDT N, REED J, SHEPARDSON D, LAMB R AND WONG S, Recent advances in compaction grouting technology. Grouting, soil improvement and geosynthetics, Proc. of conference, Geotechnical Special Publication 30,ASCE Geotechnical Engineering Division, New Orleans, LA, USA. 1992, pp 252-264.

15. GREENWOOD D A, Re-levelling a gasholder at Rhyl. Quarterly Journal Engineering Geology, London. Vol 17, 1984, pp 319-326.

16. BAKER W H, CRADING E J AND MACPHERSON H H, Compaction Grouting to Control Ground Movements during Tunnelling, Underground space, Vol. 7, 1983, pp 205-212.

17. COVIL C S AND SKINNER A E, Jet grouting – a review of some of the operating parameters that form the basis of the jet grouting process. Grouting in the Ground, Proc. Of conference, ICE, London, UK. 1992, pp 605-629.

18. CEPPI G, MAGGIONI F, DE PAOIL B, STELLA C, LOTTI A AND PEDEMONTE S, Horizontal jet grouting as a temporary support for the "Montelimpino 2" tunnel, International congress on progress and innovation in tunneling, Toronto, Canada. 1989.

19. ASCHIERI F, JAMIOLKOWSKI M AND TORNAGHI R. Case history of cut-wall executed by jet grouting. Proc. Eighth European Conf. On Soil Mechanics and Foundation Engng., Helsinki, Vol.1. 1983, pp 121-126.

20. BLACKWELL J. A case history of soil stabilisation using the mix-in-place technique for the construction of deep manhole shafts at Rochdale. Grouting in the Ground, Proc. Of conference, ICE, London, UK. 1992.

21. HARNAN C N AND IAGOLNITZER Y, COLMIX: The process and its applications. Grouting in the Ground, Proc. Of conference, ICE, London, UK. 1992, pp 511-524.

22. HAIMONI A, Tunnelling with compensation grouting – Jubilee Line Extension, ICE National Conference, Going Underground 97, Birmingham. 1997.

23. MAIR R J, HARRIS D I, LOVE J P, BLAKEY D AND KETTLE C. Compensation grouting to limit settlements during tunneling at Waterloo station, Tunneling Conf. UK, 1994.

24. LITTLEJOHN G S, Design of cement based grouts, Proceeding of the Conference on Grouting in Geotechnical Engineering. ASCE, New Orleans, Louisiana. 1982.

25. JENKINS S J, GARTSHORE G AND LEVITT K, The rheology and bleed properties of Blue Circle H900 microfine cement for grouting purposes. Grouting in the Ground, Proc. Of conference, ICE, London, UK, 1992, pp 591-604.

26. CLARKE W J, BOYD M D AND HELAL M, Ultrafine cement tests and dam test grouting. Grouting, soil improvement and geosynthetics, Proc. of conference, Geotechnical Special Publication 30,ASCE Geotechnical Engineering Division, New Orleans, LA, USA., 1992, pp 626-638.

27. SHERWOOD D E AND GANDAIS M, Neutral mineral micro particle grout and its application in recovery of a collapsed tunnel in Switzerland. Grouting in the Ground, Proc. Of conference, ICE, London, UK, 1992, pp 631-648.

28. LITTLEJOHN G S, Chemical Grouting, the South African Institution of Civil Engineers (Geotechnical Division) Grouting Course, University of Witwatersrand, Johannesburg. 1983.

29. LITTLEJOHN G S AND HAJI-BAKER I, Engineering Properties and Behavious of Silicate Grouted sand, Grouting in the Ground, Proc. Of conference, ICE, London, UK, 1992, pp25-35.

30. LITTLEJOHN G S, Chemical Grouting, Ground Engineering, 18(2), 13-16, (3), 23-28, (4), 29-34. 1985.

31. LITTLEJOHN G S, AND MOLLAMAHMUTOGLU M. Time dependant behaviour of silicate grouted sand, Grouting in the Ground, Proc. Of conference, ICE, London, UK. 1992, pp 37-51.

32. SHEIKH BAHAI A AND JEFFERIS S A. Prediction of the strength of chemically grouted sand, Grouting in the Ground, Proc. Of conference, ICE, London, UK, 1992, pp 53-70

33. HEWLETT P C AND HUTCHINSON M T, Quantifying Chemical Grout Performance and Potential Toxicity. Proc. 8th European Conference on soil. Mech & Founf, Engng. Helsinki,Vol. 1, 1983, pp. 361-366

34. HAIMONI A M, Rheology of a specific oilwell cement. PhD Thesis, Department of Civil Engineering, University of Surrey, Guildford, U.K. . 1987.

35. BAKER W H, Grouting in Geotechnical Engineering, Proc. of conference, ASCE Geotechnical Engineering Division, New Orleans, LA, USA. 1982.

36. BAKER W H, Issues in dam grouting, Proc. of conference, ASCE Geotechnical Engineering Division, Denver, CO., USA. 1985.

37. FELSH J P, Soil improvement – a ten-year update, ASCE Committee on Placement and Improvement of Soils, Special Publication 12, Proc. of symposium at ASCE convention, Atlantic City, NJ, USA. 1987.

38. BORDEN R H, HOLZ R D AND JURAN I, Grouting, soil improvement and geosynthetics, Proc. of conference, Geotechnical Special Publication 30,ASCE Geotechnical Engineering Division, New Orleans, LA, USA. 1992.

39. BELL A L, Grouting in the Ground, Proc. Of conference held in 1992, ICE, London, UK, 1994.

40. Proc. Eighth European Conf. On Soil Mechanics and Foundation Engng., Helsinki, Vol.1. 1983.

41. Proceedings of IS-Tokyo . Second International Conference on Ground Improvement Geosystems, Grouting and Deep Mixing. Tokyo, Balkema, 1996.

42. CIRIA REPORT PR60, Geotechnical Grouting: a bibliography, Construction Industry Research and Information Association. 1997.

43. CIRIA FUNDERS REPORT/CP/56, Grouting for ground engineering, Construction Industry Research and Information Association. 1998.

44. GREENWOOD D A, LORD J A AND HAIMONI A M,. Settlement correction by intrusion grouting in organic clays: a dubious success. Grouting in the Ground, Proc. Of conference, ICE, London, UK, 1992, pp 361-374.

TRIAXIAL CREEP BEHAVIOUR OF GROUTED SAND

E Ribay

R Cabrillac

University of Cergy Pontoise

D Gouvenot

Soletanche-Bachy Society

France

ABSTRACT. Silicate grouts used for a long time in soil grouting are chemically unstable. Triaxial creep tests were performed on sand grouted with two new grouts, a cement base grout (Microsol) and a mineral base grout (Silacsol) in comparison with the silicate grouted sand. Results show that sand grouted with new grouts presents few creep strains whereas silicate grouts make sand viscous. Creep limit of silicate grouted sand is below 30 % of its unconfined strength. The parameter m of the phenomenological equation postulated by Singh and Mitchell was determined for each soil studied. It is between 0.7 and 1 and it is independent of the loading level whatever the type of soil.

Keywords: Creep, Triaxial tests, Sand, Grouted sand, Silicate grout, Cement grout, Mineral grout, Creep limit, Phenomenological relationship.

Estelle Ribay is preparing a thesis on the mechanical behaviour of grouted sand at the University of Cergy Pontoise.

Professor Richard Cabrillac is Head of the laboratory of Civil Engineering at the University of Cergy Pontoise. He is also Director of the University Institute of Civil Engineering Science of the same University.

Dr Daniel Gouvenot is the Technical Manager of Soletanche-Bachy Society.

INTRODUCTION

Some soils have to be treated before it can be built on them. Grouting is a treatment method which can increase mechanical strength and/or decrease soil permeability [1].

Silicate grouts used for a long time in soil grouting are chemically unstable. Many studies show that grouted sand with silicate grouts have large time-dependent accumulations of strain when they are subjected to a constant stress [2], [3], [4]. Therefore we have been searching for new types of grouts which in addition to increasing strength and decreasing permeability have stable mechanical properties in time.

In this study we have been particularly interested in creep characteristics of a sand grouted with different grouts, a silicate grout and two new grouts, a cement-base grout (Microsol) and a mineral-base grout (Silacsol). Results display the importance of the type of grout on the stress-strain-time effects of grouted sand. The creep behaviour of grouted sand was studied from creep curves by determining a creep limit and parameters of the phenomenological relationship postulated by Singh and Mitchell [5].

NOTION OF CREEP BEHAVIOUR OF SOILS

Creep is the strain induced when soil is subjected to a constant stress and a constant temperature. Three phases of creep can be distinguished depending on the load level. For small loading primary creep (damped creep) can be observed. This creep phase is characterized by a strain rate which decreases with time. Strains cease after some period of time. Under larger loading secondary creep (stationary creep) may appear. The strain rate is constant. Under high loading tertiary creep (no damped creep) can be reached. Strain rate increases continuously until failure occurs.

Determination Of Creep Limit

The following method is based on that used by different authors [6], [7], [8]. In a graph presenting the evolution of strain versus time for a given load (Q), we note (α) the slope of the linear part of the curve. The variation of the slope α_i with the load level Q_i shows three separate fields. For each field the parameter α increases linearly with the load Q. The value of the strength corresponding to the second bend is admitted as creep limit (Q_f). For loading below this limit soil is stable. Beyond this limit soil is in an unstable phase. Creep failure will occur after a finite time. This method is illustrated in Figure 1.

Creep Relationship

Singh and Mitchell [5] have postulated a three parameter phenomenological relationship which has often been used to describe the creep behaviour of soils.

$$\frac{d\varepsilon}{dt} = A e^{\alpha q} \left(\frac{t_1}{t}\right)^m$$

with : dε/dt : Strain rate at any time.
      t_1 : Unit time (1 minute, 1 hour).
      q : A function of stress intensity.
      m : Absolute value of slope of the straight line on the log(dε/dt) - log(t) plot.
      α : Slope of the linear part on the log(dε/dt) - D plot.
      A : Logarithm of the fictive value of dε/dt for D = 0 and for t = t_1.

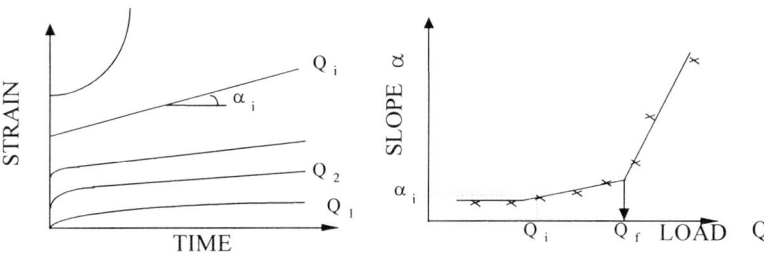

Figure 1 Determination of the creep limit

MATERIALS

Tests were carried out on four natures of soil, sand without any treatment and sand grouted with three different grouts.

Granular Framework

The natural sand was the normalised NE 34 Fontainebleau sand with the following characteristics :

$$\gamma_s = 2.65 \; ; \; e_{min} = 0.51 \; ; \; e_{max} = 0.9 \; ; \; D_{60} = 0.23 \text{ mm} \; ; \; D_{10} = 0.14 \text{ mm}$$

Grouts

The first grout is a sodium silicate grout with a molecular ratio SiO_2/Na_2O = 3.3. The ester hardener solution is Rhône Poulenc 600C.

The second grout studied is MICROSOL, patented by the society Soletanche-Bachy. It is a suspension of very fine cement in water.

The last grout, SILACSOL, is patented by the Society Soletanche-Bachy. It is a chemical-liquid-mineral base grout. The reaction between the different components is a hydraulic type one. Stable crystals of hydrated calcium silicates are formed during this reaction.

EXPERIMENTAL PROGRAM

Grouted Soil Samples Preparation

Sand columns of 80 mm in diameter and 75 cm in length were first saturated with water. Columns are then grouted by the base. Grouting is stopped when excess grout (equivalent to approximately 120 % of the pore volume) has passed through the sample.

The column ends are sealed. Columns are maintained in humid atmosphere. Just before the test, columns are sawed and cored to obtain samples of 54 mm in diameter and 108 mm in length corresponding to a slenderness of 2.

Grouts were formulated to have unconfined compression stress of 2 MPa at 28 days at a rate of 20 mm/mn.

Creep Test Apparatus

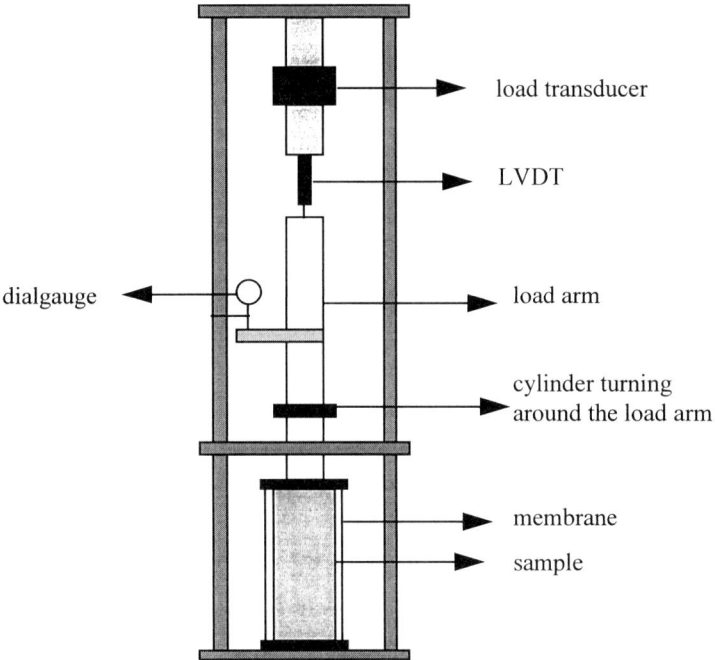

Figure 2 Creep test apparatus

Creep tests were carried out in triaxial cells (Figure 2). A kind of cylinder was turned around the load arm to reduce the friction effects. Strength and deformations were recorded with time. Deformations were measured at once by a linear voltage displacement transducer (LVDT) and by a dialgauge with an accuracy of 1/1000 mm.

Test Description

During creep tests, samples were drained by the base and the top. Samples were surrounded with two membranes, a latex one coated with silicone oil and a neoprene one to avoid osmosis between the sample and fluid used to apply the confining pressure. A confining pressure of 0.1 MPa was applied.

For each sample loads are applied following a stepwise system (Figure 3). Loading levels for grouted soils are expressed as a percentage of the unconfined compression strength at 28 days. They are 5 % for the silicate grouted sand, and 10 % for the Microsol and Silacsol grouted sand. For the natural sand, load level corresponds to 10 % of its confined compression test under 0.1 MPa. The transition between two loading levels was carried out at a low rate to avoid the development of pore pressure into the sample.

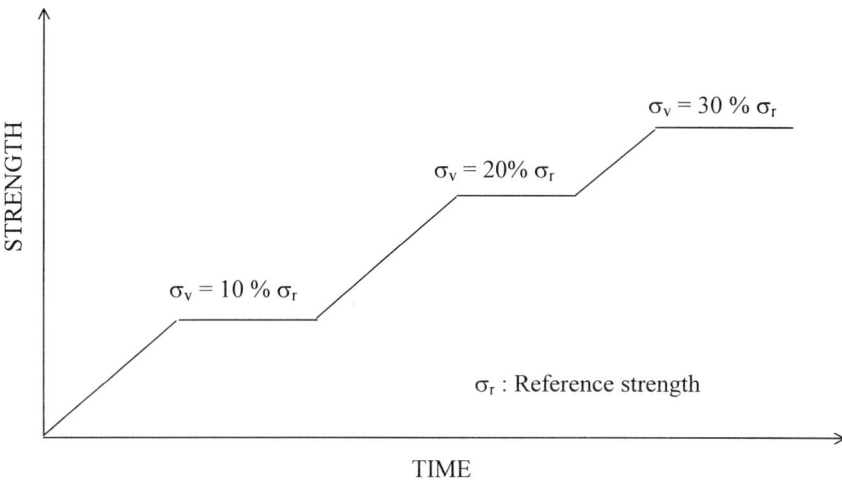

Figure 3 Loading program

The duration of a loading step can vary between 1 and 3 weeks depending on the evolution of strains. Tests lasted between 3 and 6 months.

EXPERIMENTAL RESULTS

We obtained classical creep curves for the different soils. Natural Fontainebleau sand sample and silicate grouted Fontainebleau sand sample reached the tertiary creep whereas only the primary creep could be observed for grouted sand with Microsol and Silacsol. Figure 4 shows the results of the silicate grouted sand.

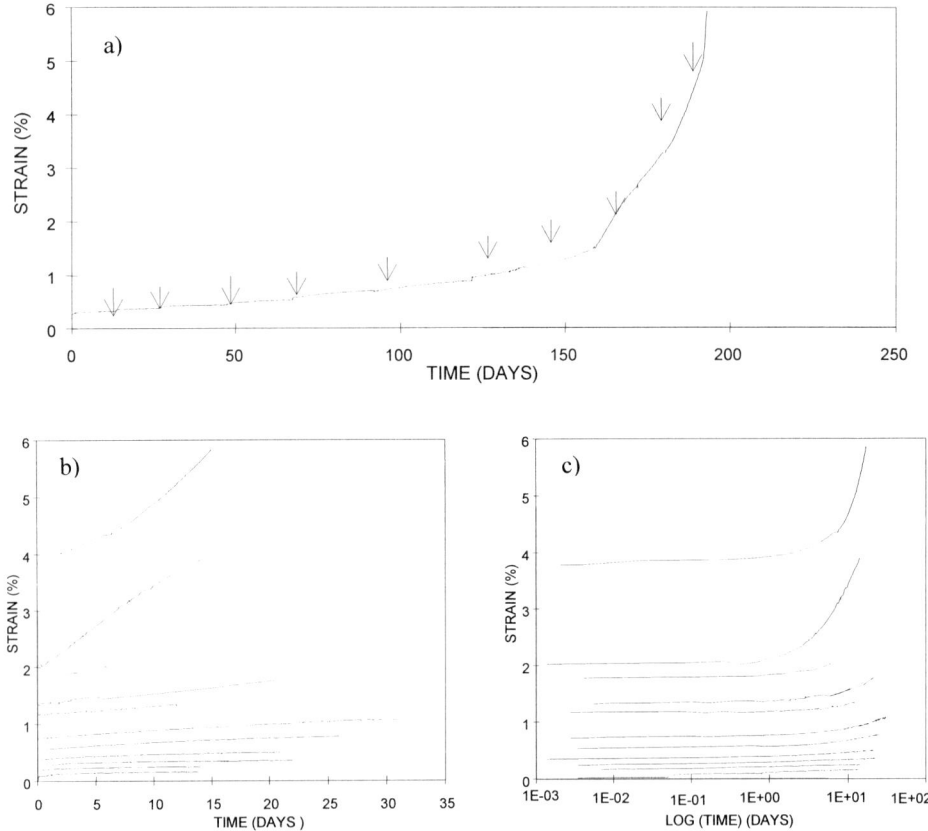

Figure 4 Creep curves of the silicate grouted sand a) Successive steps plot. Arrows indicate the transition between two steps. b) Curves with each step beginning at the time t=0. c) Strains versus logarithm of time.

CREEP LIMIT

For a same relatively low load level (20 % of the confined compression strength under 0.1 MPa), we can already note the tendency of the silicate grouted sand to have creep strains larger than the natural sand. On the other hand Microsol and Silacsol grouted sand samples have very low creep slopes (Figure 5).

The values of creep limits determined by the previously expressed method are in percentage of the confined compression strength under 0.1 MPa respectively 50 % and 25 % for the natural Fontainebleau sand and the sodium silicate grouted sand (Figure 6). For the latter the limit corresponds to 30 % of its unconfined compression strength. Microsol and Silacsol grouted sand samples show low strains. We can't determine a possible creep limit.

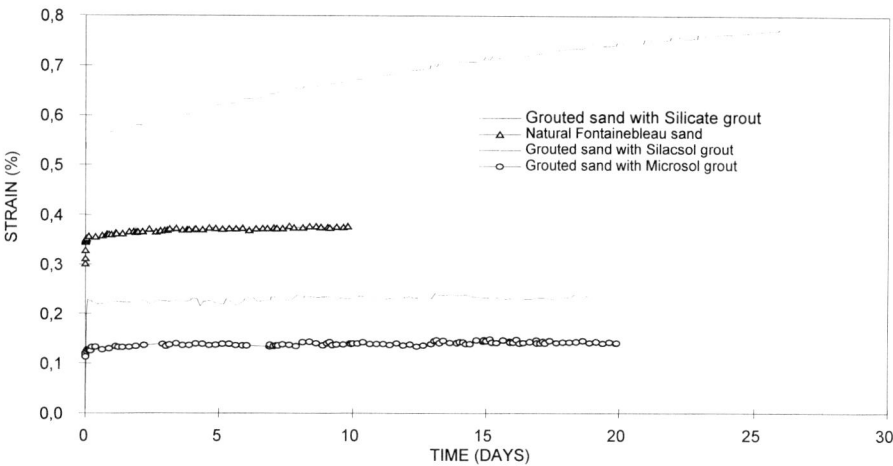

Figure 5   Creep curves for a load level of 20 % of the compression strength under 0.1 MPa of confining pressure

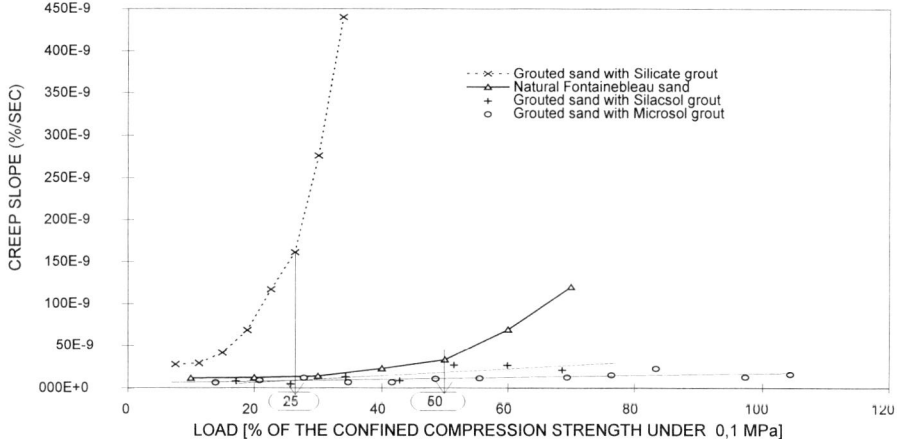

Figure 6 Creep limit

These tests show the importance of the type of grout on the creep behaviour of the grouted sand. Sodium silicate grout gives sand a viscous character whereas the two other grouts Microsol and Silacsol have a tendency to reduce the possible creep strains.

CREEP RELATIONSHIP

Figure 7 shows the evolution of strain rate versus time for each type of soil studied. The parameter m from the Singh and Mitchell relationship can be determined through this Figure.

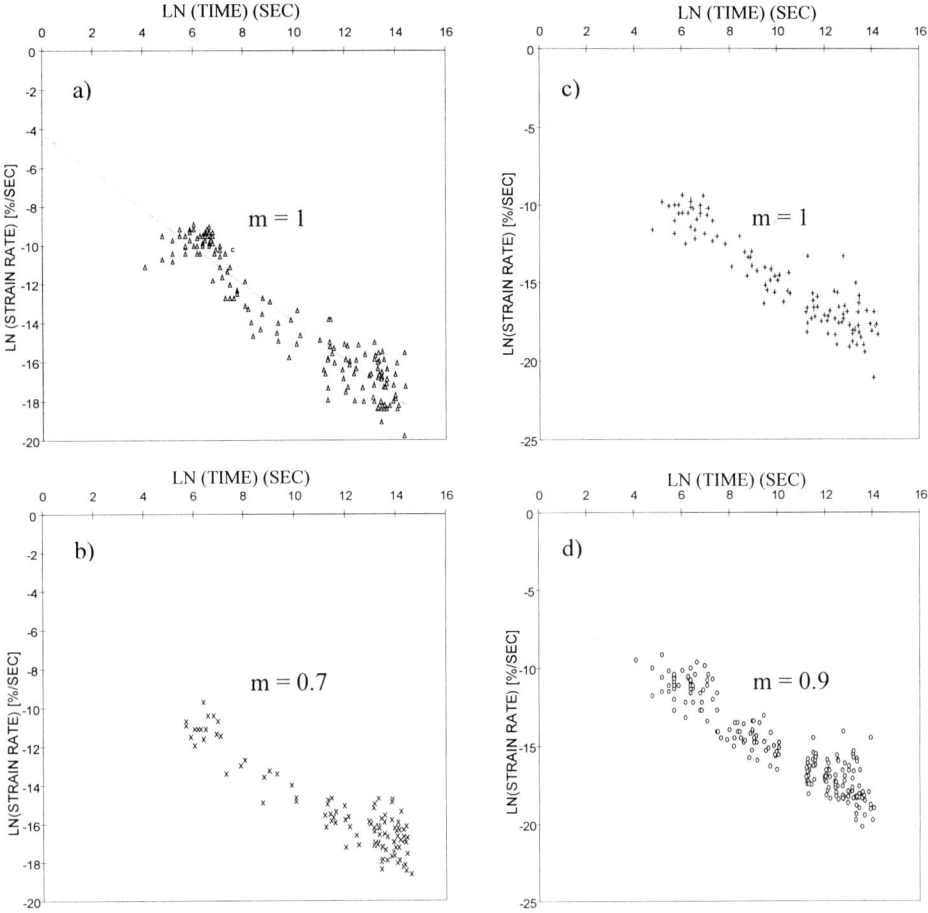

Figure 7 Values of m for every load level a) Natural Fontainebleau sand b) Fontainebleau sand grouted with sodium silicate grout c) Fontainebleau sand grouted with Microsol d) Fontainebleau sand grouted with Silacsol

In accordance with the results [5], [9] we note that the slope m is independent of the load level (Figure 7).

Results obtained are in a same range and have a tendency to show that the slope m is relatively independent of the type of soil (Figure 8). Values of m determined for the natural Fontainebleau sand (1) and the Microsol and Silacsol grouted sand samples (respectively 1 and 0.9) are rather in the upper range of values given by Singh and Mitchell [5] (between 0.7 and 1). On the other hand the value of m for the silicate grouted sand (0.7) is lower but it is in agreement with the value obtained by Ata [8] on the same type of soil. The parameter m controls the rate at which the strain rate decreases with time. The fact that m is lower for the silicate grouted sand confirms that this type of soil has a more viscous behaviour.

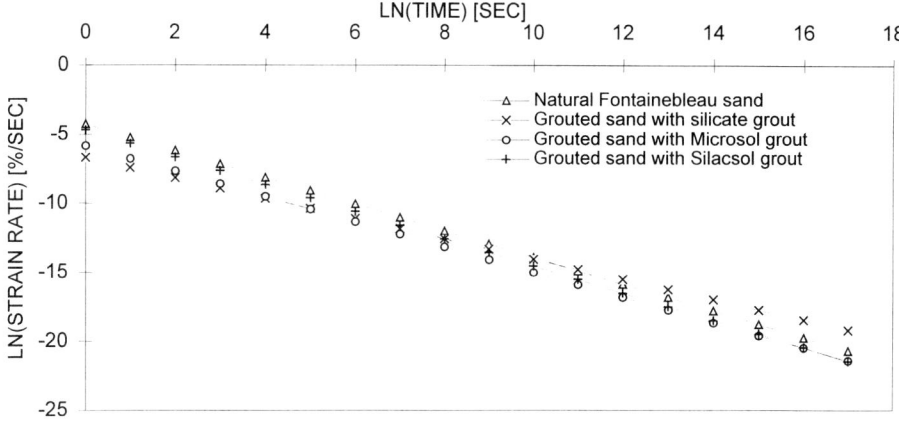

Figure 8 Slope m for each kind of soil

CONCLUSION

Triaxial creep tests carried out on natural sand and on sand grouted with different types of grout show a significant influence of the grout nature on the stress-strain-time effects. Cement base (microsol) grouted sand sample and mineral base (Silacsol) grouted sand sample present low creep strains whereas silicate grout gives sand an important viscous behaviour with a creep limit equal to 30 % of its unconfined compression strength.

Regardless the grouted sand the parameter m of the Singh and Mitchell relationship is independent of the stress level with a value of 0.7 – 1.0 for soils with large creep strains.

The current study deals with unconfined compression creep tests. This work has two objectives. The first one is to study the influence of the confining pressure on the stress-strain-time effects and particularly on the value of the creep limit. The second one is to determine the parameters (α and A) of the Singh and Mitchell relationship and to propose a creep relationship in accordance with the type of grout used.

REFERENCES

1. CAMBEFORT, H. Injection des sols, Tome I, principes et méthodes. Editions Eyrolles, 1967.

2. BORDEN, H, KRIZEK, R J, BAKER, W H. Creep behavior of silicate-grouted sand. Proceedings of the conference on Grouting in Geotechnical Engineering, New Orleans, Louisiana, february 10-12, 1982, pp. 450-469.

3. LUONG, M P, GANDAIS, M, ALLEMAND, P. Comportement mécanique des sols injectés aux produits chimiques. Annales de l'Institut Technique du Batiment et des Travaux Publics, octobre 1967, n°354, SF 145, pp. 14-35.

4. MOLLAMAHMUTOGLU, M, LITTLEJOHN, S. Varying temperature and creep of silicate grouted sand. Ground improvement, 1997, pp. 59-64.

5. SINGH, A, MITCHELL, J K. General stress-strain-time function for soils. Journal of the soil mechanics and foundations division, janvier 1968, pp. 21-46.

6. MURAYAMA, S, SHIBATA, T. Flow and stress relaxation of clays. IUTAM Symposium Grenoble, pp. 99-129, 1964.

7. TAN TJONG-KIE. Proceedings of the 5th international Conference on soil mechanics and Foundation engineering, Paris 1961, vol III discussions, pp. 141-143.

8. VIALOV, S S, SKIBITSKI, M. Problems of the rheology of soils. Proceedings of the 5th international Conference on soil mechanics and Foundation engineering, Paris 1961.

9. FODIL, A, ALOULOU, W, HICHER, P Y. Comportement viscoplastique d'une argile naturelle. Ouvrages géomatériaux et interactions (modélisations multi-échelles), Editions Hermes, 1998.

10. ATA, A. Effect of cohesive and adhesive properties of grouts on the static and the cyclic behavior of grouted sand. Dissertation, University of Houston, décembre 1993.

CEMENTITIOUS GROUTS - STANDARDS UPDATE 1999

D Johnson
Construction Materials Consultant
United Kingdom

ABSTRACT. Cementitious grouts have a wide variety of uses and their compositions depend on the required performance and behaviour. The majority of the current grout standards refer to the high performance grouts which exhibit structural type strengths and are used, for example, in tendon grouting or ground anchorages. These are based on Portland cement as the main binder which provides the strength and durability required for the application. A variety of test methods exist which characterise the more important performance properties, such as strength, fluidity, bleeding and volume change. These, however, constitute only a small proportion of the UK grouting industry whilst the majority are used for void filling and ground stabilisation. New standards are expected in the next century which will provide a comprehensive guide to geotechnical grouting.

Keywords: Cement, Grout, Standard.

Dr David Johnson is a materials consultant with over 15 years experience in the construction industry. He specialises in the use of new hydraulic cements and pozzolans in concrete, mortar and grout.

INTRODUCTION

Grout is defined [1] as a flowing material, that hardens after application and is used for filling fissures and cavities. This naturally raises the questions of how fluid must the grout be to be termed a 'grout' and how hard to have 'hardened'. There are a wide variety of grouts available and a more general definition is provided by a draft geotechnical standard [2]:

'A pumpable material (suspension, solution, emulsion or mortar) injected into a soil or rock formation which stiffens and sets with time and thereby changes the physical characteristics of the formation'.

This definition clearly addresses the problem that the range of applications for grouts require a wide range of properties, from the non-hardening gels to the high strength structural cementitious grouts. This wide range results from the many practical problems such as size of the cavity, the strength requirement, the durability, cost, etc. A grout which may be suitable for filling abandoned mine workings is unlikely to be suitable for grouting fine fissures in rock and masonry.

For many years, the grout supplier has had to provide materials to meet obscure standards with specifications which had requirements which were not relevant to the intended application. Different test methods and requirements were difficult to correlate, resulting in the use of the wrong grout in certain situations. Over the last 5 years, a number of standards have been introduced which concentrate on the chief requirements for a grout i.e. strength and fluidity. This paper deals specifically with cementitious grouts designed for use in the Civil Engineering and Mining Industries, the grouts being primarily mixtures of Portland cement, water, admixtures and pozzolans. Standards in Europe and the USA are reviewed as well as the proposed standards which seek to remove inconsistencies in the test methods and requirements.

CEMENTITIOUS GROUTS

The majority of cementitious grouts used in the industry are those based on Portland cement; the exceptions include the weak pozzolanic grouts used in the stabilisation of ancient structures and calcium aluminate cement in offshore applications. Portland cement is used either on its own with water or in combination with sand or pozzolans such as pulverised fuel ash. Examples of the former include the anchor grouts, oil-well cement slurries and the ultra-fine Portland cements for sealing fine fissures in rock; examples of the latter are the tunnelling grouts and the commercial compositions for specific applications.

Irrespective of composition, the relevant grout properties are :

- strength (usually compressive)
- rheology or fluidity
- bleeding
- set time
- volume change
- water retentivity
- chemical resistance
- permeability
- heat of hydration

The order of importance will vary with the grout composition and intended application; however, existing grout standards do not cover all the properties.

Structural Grouts

Structural cementitious grouts are used for a wide range of applications including the installation of steel anchors or bolts into rock and concrete. In addition to providing a high degree of compressive strength, the alkaline nature of the cement protects the metal from corrosion thus ensuring long term stability. In the majority of cases, the grout is required to have a certain ultimate strength and rate of strength development and in some instances, the grout-anchor interface is additionally tested [3].

The grout is composed of Portland cement alone or in combination with sands or pozzolans at a low water to cement ratio. Examples of the former are the grouts used for installing ground anchors in rock or in oil-well drilling for sealing the annulus between the walls of a borehole and the steel casing. This latter usage requires the Portland cement to possess specific properties and therefore, composition, and is reviewed elsewhere [4]. Proprietary mixtures, composed of cements, sands, pozzolans and admixtures are used for specific civil engineering applications such as the bolting of heavy machinery to concrete floors and the grouting of post-tensioned concrete bridges.

Non-Structural Grouts

These grouts usually are of low compressive strength, the most important property being the ability to completely fill the cavity. Thus, good flow properties and reduced bleed are required, the first to ensure that the grout reaches the cavity and the second to reduce the creation of secondary voids. Few standards directly apply to these grouts although there are standardised test methods for the relevant properties. The successful use of the non-structural grouts underground minimises surface disruption, such as mining subsidence, and provides gas seals.

In tunnelling and mining operations, the terms 'structural' and 'non-structural' are vague and can apply to the same situation. An example is the grouting of the void between the tunnel lining and the cut rock; this is both structural, as the hardened grout spreads the load acting on the tunnel roof around the concrete ring, and non-structural as the grout fills the empty space. This has hindered the standardisation of grouts although work is progressing within Europe [2] to address the matter.

EXISTING STANDARDS

Five years ago, the majority of the Western European standards for cementitious grouts referred to those used for prestressing tendons and admixtures in concrete, mortar and grout (Table 1). In the UK, advice was provided on grouts for prestressing tendons and ground anchorages. In the USA, a group of standards covered non-shrink grouts and the use of hydraulic cement grouts for preplaced aggregate concrete, a construction method seldom used in the UK. The harmonisation of standards throughout Europe saw the recent introduction of three cementitious grout standards covering the test methods, limits and grouting procedure for prestressing tendons. The British Standard BS 3892 was recently revised to control the properties of pulverised fuel ash for use in cementitious grouts whilst experiences in the Mining Industry led to a British Standard [3] for cable bolting grouts.

Table 1 Current cementitious grout standards

COUNTRY	NUMBER	TITLE
Austria	B/2270:1983	Grouting works, contracts for material and labour
Finland	SFS 4636:1991	Foundation. Instruction for grouting.
France	NG P18	Admixtures for concrete, mortar and grout
	NF P26-401:1942	Hardware. Grouting pads
Germany	DIN 18309:1988	Ground treatment by grouting
	DIN 18999:1991	Admixtures for concrete, mortar and grout
Italy	UNI 8993 to 8998	Preblend expansive mortars for grouting
Spain	UNE 83-200	Admixtures for concrete, mortar and grout
UK	BS EN 445 to 447:1996	Grout for prestressing tendons
	BS 3892:Part 3:1997	Specification for pulverised-fuel ash for use in cementitious grouts.
	BS 7861:Part 2:1997	Specification for birdcaged cablebolting.
	BS 8081:Clause 7.1:1989	Code of practice for ground anchorages Cementitious grouts
	BS 8110:Part 1: Clause 8.9 1985	Structural use of concrete. Grouting for prestressing tendons
USA	C937 to 942	Preplaced aggregate concrete. Standard test methods
	C1090-93	Standard test method for measuring height of cylindrical specimens from hydraulic cement grout
	C1107-91	Specification for packaged dry hydraulic cement grout (non-shrink).

Grout Composition

The USA ASTM standards are rather vague regarding composition, using such terms as 'includes' and 'prepared from'. Thus, non-shrink grout 'includes hydraulic cement, fine aggregates and other ingredients' whilst the grout for preplaced aggregate concrete is 'a mixture of Portland cement, pozzolan and fine aggregate in water'. BS EN 447 and BS 8110 specify Portland cement as the binder whilst BS 8081, BS 7861 and BS 3892 refer to cement, binder and cementitious binding agent, respectively This reflects the age of the standard e.g. the latter two recognise the use of binders other than Portland cement as long as the performance requirements are met.

Standard Test Methods

The specific test methods for each grout property often differ between the standards and where the general principle is the same, the dimensions of the test apparatus may be different, making test data incompatible. An example is the cone test used for measuring the fluidity of the grout. The ASTM standards uses a cone with an aperture of 12.7mm (i.e.0.5 inch) whereas BS EN 445 uses an aperture size of 10mm. The various standards have test methods for compressive strength, fluidity, bleeding, setting time, volume change and water retentivity. In some standards, more than one test method exists for the same property.

Strength

The shape or dimensions are the chief difference; BS EN 445 allows the use of cylinders and the prisms employed in BS EN 196 [5]. The ASTM test method is a 2 inch cube whilst BS 8110 and BS 3892 use the 100mm cube. It is known that specimen size can affect the strength of concrete [6] and similar effects would be expected for grout. Tunnelling grouts, especially those containing large proportions of pulverised fuel ash, would be expected to give relatively low strengths when tested as small specimens due to the leaching out of the lime from the specimen thereby inhibiting the pozzolanic reaction [7].

Fluidity

The rheological properties of the grout may be more important in some applications than the strength of the grout. Very careful use of admixtures and raw materials can result in grouts with various degrees of thixotropy and fluidity. There are a number of different test methods for assessing the fluidity or flow of the grout, each being the one most commonly used in that industry or country. Thus, grouts used for tendon grouting use the cone as this test characterises the very fluid grouts. Tunnelling grouts are relatively more viscous and may, additionally, be thixotropic such that the cone is not so reliable; the flow channel specified in BS 3892 is commonly used instead. The German immersion method is an alternative test in BS EN 445 whilst the USA standards use the cone but recommend the use of a flow table when the efflux or outflow time is over a certain value. The mining standard, BS 7861, advises that the grout be mixed and pumped through typical mining equipment in order to assess suitability.

Bleeding

The ability of a grout to bleed is, in general, a negative effect leading to the partial filling of voids; where this occurs in a tendon sleeve or in a tunnel, the resulting gap can act as a path for water flow with potential structural failure. Bleeding is calculated either as a height change or as a volume change although BS EN 445 allows the use of both methods. The ASTM test specifies the use of a 1 litre graduated cylinder of unspecified dimensions. However, vertical shrinkage does not show a linear relationship to the height of the grout column [8] so great care is still required when predicting effects based on laboratory test data. Thus, when bleeding is an important property, tests based on actual site conditions should be used.

Setting Time

Few of the standards consider setting time as an important property, presumably because it is affected by many site parameters e.g. size of void, temperature, etc. ASTM C937 specifies the Vicat apparatus but does not specify the water content, rather that the grout be of a fluid consistency for use with preplaced aggregate.

Volume Change

The amount of shrinkage that occurs after setting is generally small compared to the shrinkage due to bleeding, however, this small amount can be important for some critical applications e.g. grouting prestressing tendons. The ASTM standards recognise that two stages exist, the first being the volume change before hardening and, secondly, the changes after hardening. Some grouts are designed to provide positive expansion in the former with or without expansion in the set state. The pre-hardening method relies on the movement of a ball seated on the grout whilst the post-hardening method measures the height of a column of grout which has been stored under certain conditions over a specified time period. BS 7861 uses ASTM C490 test apparatus to measure expansion and shrinkage, the small specimen size being more appropriate to the site application.

Current Performance Requirements

Most standards which specify test methods also include performance requirements (Table 2) which advise the maximum and minimum values for the various properties so that the grout performs satisfactorily on site. The absolute values, however, are specific to the test methods and the reader is referred to the appropriate standard for further information.

Table 2 Grout requirements

STANDARD	REQUIREMENT
BS EN 446	Assessment of the contract, equipment, procedure and quality control.
BS EN 447	Physical and chemical properties of the constituents; fluidity, bleeding, volume change, strength and mixing of the grout
BS 3892	Physical and chemical properties of the pfa; fluidity and strength of OPC/pfa blend for certain applications.
BS 7861	Fluidity, strength, elastic modulus and expansion of the grout; system performance
BS 8081	Physical and chemical properties of the grout constituents; strength and bleed of the grout
BS 8110	Physical and chemical properties of the grout constituents; fluidity, bleeding, volume change and strength of the grout.
ASTM C937	Effect of fluidifier on grout with respect to set time, expansion, bleeding and strength
ASTMC1107	Strength and volume change of the grout.

PROPOSED STANDARDS

Despite the number of standards relating to cementitious grouts, the majority refer to one type of application i.e. grouting of steel cables into rock or masonry. Whilst such standardisation is vital to ensure the integrity of a structure, most grouting is carried out in the ground for geotechnical purposes. Standardisation of these materials has always been difficult as ground conditions vary so much over a relatively short distance, necessitating the use of various types of grout. This is being addressed by CEN with a standard [2] which will control the test methods and grout properties with particular emphasis on grouting procedure.

Grout Composition

The draft standard allows the use of any pumpable material which can be injected into soil or rock formation. This includes:

- suspensions e.g. cementitious grouts
- solutions e.g. silicates
- emulsions
- mortars e.g. as used in compensation grouting.

The grout composition should be capable of setting or stiffening in such a way as to change the physical characteristics of the grouted area. Thus, a composite material is formed with properties that are different to those of the separate components (the set grout and the soil).

Test Methods

With the wide variety of materials available for ground treatment, it is not surprising that the number of proposed test methods (Table 3) are high although these may be reduced in the final document. The test methods are those which are mainly used in the geotechnical industry.

Table 3 Draft test methods

PARAMETER	TEST APPARATUS
Fluidity	Cone or viscometer
Density	Mud balance
Water retention	Filter press
Bleeding	Measuring cylinder
Setting time	Overturned glass beaker
Strength	Unconfined compression
Durability	Mechanical flow test
Thixotropy	Viscometer
Penetrability	Grouting test

Performance Requirements

Great emphasis is placed on the execution of the grouting i.e.workmanship and requires full geological, geochemical and geophysical surveys in order to choose the correct type of grout and grouting method. The selected cementitious grout must satisfy three levels of performance requirements :

- physical and chemical properties of the constituents, to comply with their respective National Standards
- rheology, setting time, permanence, strength and toxicity of the grout
- strength, permanence and water tightness of the grouted soil.

CONCLUSION

The standardisation of cementitious grouts in Europe and the USA is almost complete and can be attributed to the acceptance that grout is a material in its own right rather than being concrete minus the large stones. Thus, the way in which the constituents interact to affect performance and durability cannot be predicted solely on their performance in a concrete. The next millennium should see the completion of the task with the new geotechnical grout standards.

REFERENCES

1. BRITISH STANDARDS INSTITUTION. Glossary of Building and Civil Engineering Terms. Blackwell Scientific Publications, 1993, Section 1, 100 4410.

2. BRITISH STANDARDS INSTITUTION. prEN 12715 (1997). Execution of special geotechnical works - grouting.

3. BRITISH STANDARDS INSTITUTION. BS 7861:Part 2:1997. Strata reinforcement support system components used in coal mines. Part 2. Specification for birdcaged cablebolting.

4. BENSTED, J. Oilwell cements. World Cement, October 1989, pp 346-357.

5. BRITISH STANDARDS INSTITUTION. BS EN 196-3:1995. Methods of testing cement. Determination of setting time and soundness..

6. NEVILLE, A M. Properties of Concrete, Longman Scientific and Technical, 1987, pp 557-565.

7. LEA, F M. The Chemistry of Cement and Concrete, Arnold, 1970, pp 414-435.

8. JEFFERIS, S A. Application of Settlement and Bleed Theory to Problems of Offshore Grouting, Grouts and Grouting for Construction and Repair of Offshore Structures, HMSO, 1988, pp 72-83.

STUDIES ON THE INJECTING PROPERTIES OF CRACK REPAIR MATERIALS

T Iisaka

Meijo University

H Umehara

Nagoya Institute of Technology

T Sumi

Nagoya Express Way Public Corporation

Japan

ABSTRACT. Cracks can be observed in general on concrete structures as its grow old. These cracks may give bad influences on durability, water tightness and beauty etc. of structures. It is necessary to repair them. Injecting is a method of repair the cracks by injecting material. In this study, as a repairing material instead of epoxy resin, we tried to inject ultra fine powder, using the same cement as used in the concrete structures. The cracks of test pieces used for experiment are made by setting up two plates of glass in V form. Into these test pieces, injecting epoxy resin and ultra fine powder cement as repair material, we investigated the injecting and filling characteristics. As a result, we confirmed that ultra fine powder cement, as injecting material of cracks, could be injected as equally as organic material of epoxy resin etc. Moreover we could presume the finishing time of injecting from the characteristics value of injecting material and the depths of cracks.

Keywords: Concrete cracks, Repair materials, Ultra fine powder cement, Epoxy resin, Arrival time, Finishing time.

Dr T Iisaka is an Associate professor of Civil Engineering at Meijo University, Nagoya, Japan. He is a member of Japan Society of Civil Engineers(JSCE) and Japan Concrete Institute(JCI). His research interests include the resin concrete and repair materials for cracks in concrete.

H Umehara PhD is Professor of civil engineering at Nagoya Institute of Technology, Nagoya, Japan. He is a member of JSCE and JCI. His research interests include thermal effects at concrete structures.

Mr T Sumi is Chief Engineer of Nagoya Express Way Public Corporation, Nagoya, Japan. He is a member of JSCE. His research interests include repair materials for cracks in concrete.

INTRODUCTION

Crack promotes a deterioration of concrete structures and invites its depression, loss of beauty and psychological reliability. There are in the repairing method for these matters some injecting methods using various repairing materials. An organic epoxy resin is generally used in these repairing materials, however, it is said that this has weak points in thermal properties, durability, etc., and an inorganic cement is inferior in the injecting efficiency due to its large particle size.

This study compared injecting and filling efficiencies using inorganic and organic injecting materials and taking their inherent conditions into consideration, and investigated the injecting efficiency of inorganic materials.

In Japan actual injecting works, injecting finish time is not yet specified and is judged by a skilled person from his workmanship. From these circumstances, this study quantified the injecting finish time and the like by carrying out the regression analysis of experimental results.

MATERIALS

A cement made of an inorganic material is unsuitable for injecting because of its large particle size, however, its fine powder cement has been produced thanks to the pulverizing and processing technologies.

In this experiment, a super fine powder cement produced by mixing and pulverizing a normal Portland cement with a blast furnace slag was used, and organic epoxy resin and acrylic resin were also used to compare.

EXPERIMENTAL METHOD

Viscosity and Fluidity of Injecting Materials

The viscosity of injecting material was measured with a single cylinder rotational viscometer, and the measurement of fluidity was carried out in conformity with " PC injecting test method" in the standard of JSCE.

Preparation of Crack Specimens Made of Concrete and Injecting Method

As shown in Figure 1, concrete crack specimens are used dividing it into two in the direction of its long side using a mold for concrete bending strength (15×15×53cm). The divided long side is set vertically on the compression test machine. A wedge is placed in the center of the upper part of the set specimen and loaded.

When confirmed crack in the side face of the specimen over a range more than 50%, loading is stopped and the surface crack is sealed with a sealing material after measuring its crack width, and it was made to be a crack specimen for injecting.

Injecting is carried out to the crack specimen by a natural casting method. After the injecting, the specimen is divided into two along the crack with the same method as that in the preparation of crack. The degree of injecting and filling was examined by photograph and sketching the one side of its section.

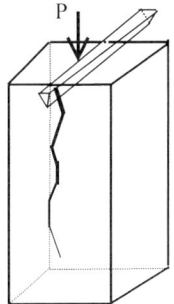

Figure 1 Cracking the concrete specimen

Preparation of Crack Specimens Made of Glass Plates and Injecting Method

A glass plate cracked specimen was prepared to observe directly the state of injecting and filling. This was done by overlapping two transparent glass plates (height 100cm×width 50cm× thickness 5mm) in a V shape and by inserting a metal spacer to make the upper width 1.2mm and the lower width 0mm and by sealing with tape to prevent leakage from both sides. Injecting and filling efficiencies were examined by placing this specimen vertically and casting a injecting material from the center of the upper part. With the dimensions of glass plate crack specimen prepared in 5 kinds (50×20~100cm), a injecting material was injected corresponding to the amount of each crack volume. The quantification of injecting efficiency was determined by measuring the time that a injecting material reaches to a crack width of 0.2 mm and the injecting finishing time that the total injecting volume in injected.

RESULTS AND CONSIDERATIONS

Injecting Materials

Figure 2 shows the viscosity test result of inorganic materials. It is a matter of course that the greater the water cement ratio is, the smaller the viscosity. When a water cement ratio is 60-70%, viscosity changes widely, and after passing 10 minutes, a great change occurs in it. When placed a injecting material in still standing after mixing it and agitated it, viscosity decreases a little due to the agitating effect in small water cement ratio. However, when water cement ratio is 80~120%, the effect of agitation on viscosity is small.

Figure 3 shows the test result of fluidity. When a fineness of an inorganic material is large, its flow value becomes small. A fine powder cement has a fineness three times larger than that in a normal port land cement, but its flow value is almost the same as that in a normal port land cement. When a water cement ratio is 120%, it approximates to the flow time of water.

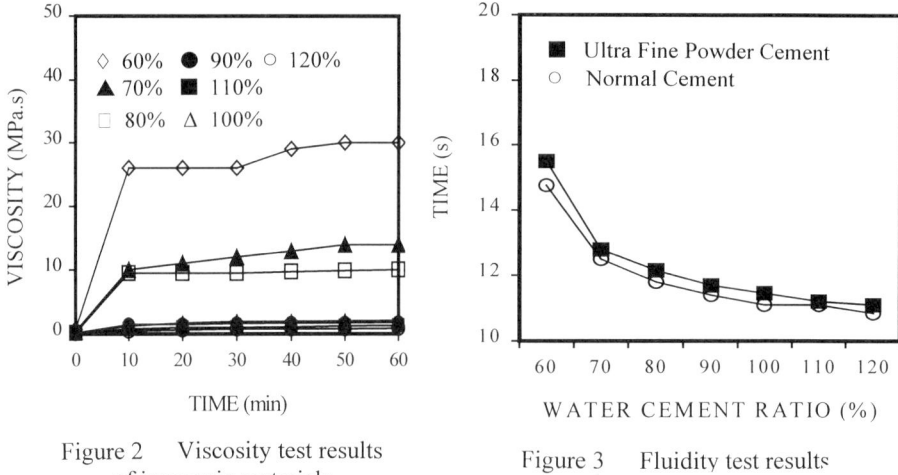

Figure 2   Viscosity test results of inorganic materials

Figure 3   Fluidity test results of inorganic materials

Figure 4 shows the relationship between the added amount of a dilute and viscosity using an organic epoxy resin. When added a 10% of dilute, viscosity decreases about 45%. When added a 20% of dilute, viscosity decreases 64%. Added a dilute more than 60%, the change in viscosity becomes small and almost constant.

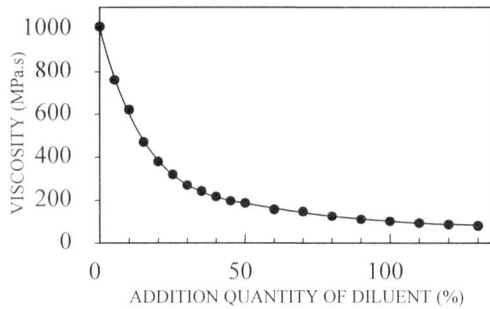

Figure 4   The result of addition quantity and viscosity by epoxy resin

From these facts, each material of inorganic and organic groups can be used as a injecting material based on the test results of viscosity and fluidity.

Injecting Efficiency of Concrete Crack Specimens

Figure 1 show the result of profile spitted into two after injecting the injecting materials into the crack specimens. It has been said up to now that the particle size of cement is so large that injecting and filling can not be carried out if a crack width 10~30 times larger than the particle size does not exist there.

However, when used a very finely pulverized cement, its injecting efficiency can be remarkably improved. The smaller the viscosity of injecting material is, the better the injecting efficiency, and injecting can be carried out even to a fine crack. In Japan, allowable crack width is designated as 0.2 mm, however, the author judges that the material used in this experiment clearly demonstrated its injecting efficiency. Furthermore, the author infers that the size of crack injecting port affects greatly its injecting efficiency.

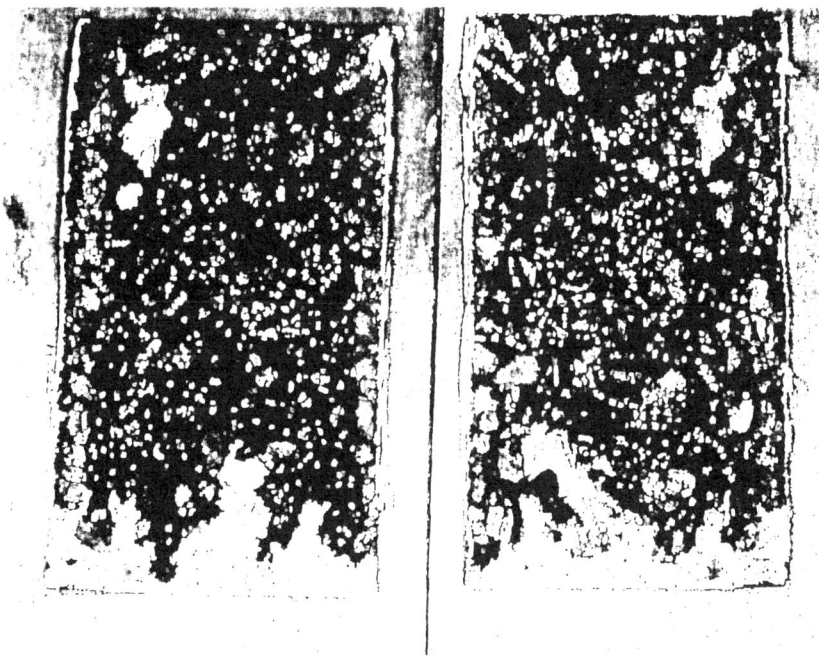

Figure 5 The injecting condition after injecting inorganic materials into crack specimen

Figure 6 shows the result of injecting efficiency using epoxy resin. It is understandable from this result as well why a large amount of epoxy resin are applied to the injecting of fine crack. The viscosity of epoxy resin is desirable to be used below 220 MPa.s when injected in the natural casting method, and it revealed that the width of injecting port affects greatly its injecting efficiency.

There can be seen in these results some differences between viscosity and density of injection materials.

Injection materials are extended laterally according to the degree of cracking, and after that, injected and filled toward the lower part. Particularly when a water cement ratio is great, its density becomes small and its effect works widely.

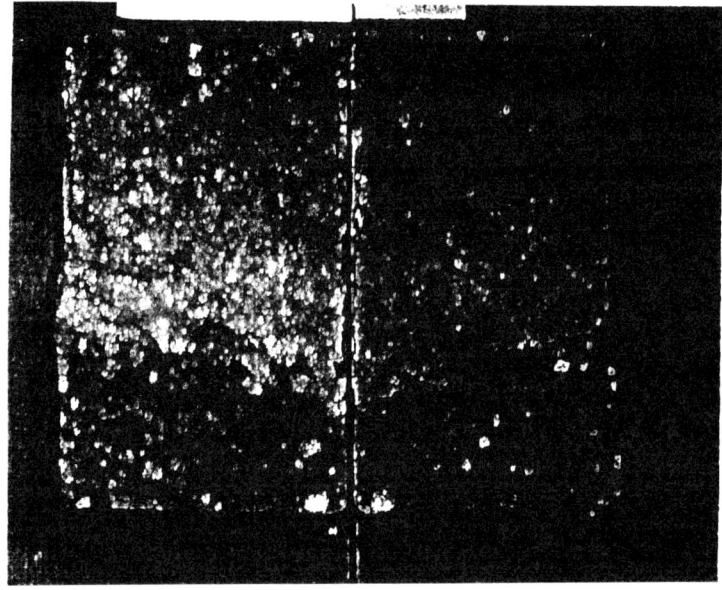

Figure 6 The injecting condition after injecting epoxy resin into crack specimen

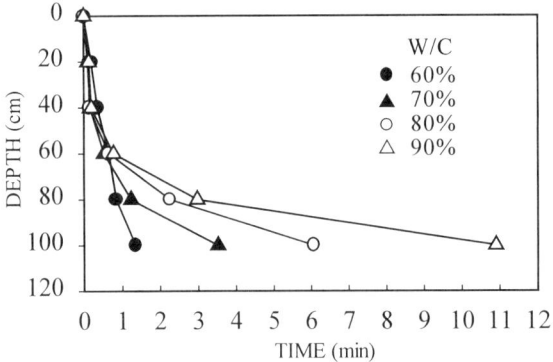

Figure 7 The test result for arrival time in case of glass plate

Compared only its injecting time with its injecting depth, the smaller a water cement ratio is, the shorter the injecting time and the better the injection efficiency. However, compared the filled areas in water cement ratios of 60 and 80%, the latter filled area is 4 times larger than the former and its injecting efficiency is high.

Figure 8 and Figure 9 shows the result of regression analysis making arrival time and injecting finishing time as an objective variable and making specific gravity and viscosity of injecting materials and crack depth as an explanatory variable in these results.

Coefficient of determination shows a high value. Each factor is significant at the level of significance of 1% in the result of test and is useful. From the above results, injecting efficiency can be estimated. These estimating equations are shown below.

In equations (1) and (2),

$$T = \frac{0.061\, \eta^{0.807}\, h^{1.022}}{d^{4.548}} \quad \ldots (1) \quad T = \frac{7.551\, \eta^{0.481}\, h^{0.749}}{d^{5.543}} \quad \ldots (2)$$

where,

Ta : Estimated value of injecting arrival time
Tf : Estimated value of injecting finishing time
η : Viscosity of material
h : Simulator crack depth
d : Specific gravity of material

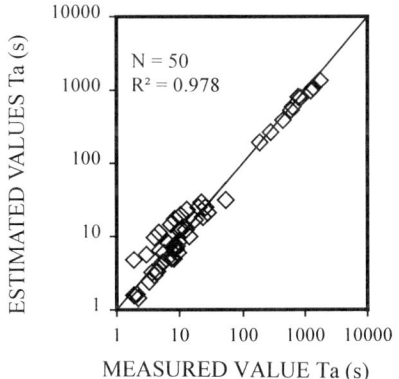

Figure 8 Correlation diagram of arrival time

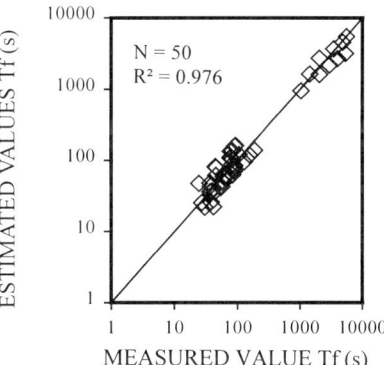

Figure 9 Correlation diagram of finish time

CONCLUSIONS

As the result of experimenting injecting and filling efficiencies to crack using inorganic materials, etc., this paper concludes as follows:

1. A fine powder cement having a water cement ratio of 60-80% is good to finish its injecting with one hour after its mixing because of the secular variation of viscosity.

2. By pulverizing cement very finely, injecting efficiency can be improved, and better injecting and filling efficiencies capable of injection into a fine crack are achieved. Injecting efficiency is also affected greatly by the width of injecting port and viscosity.

3. The viscosity of epoxy resin can be changed variously by a dilute, however, it is necessary to take note of the relationship between viscosity and dilute, because strength decreases suddenly from a 30 % of dilute added quantity.

4. The injecting arrival time and finishing time can be estimated from viscosity and density of injecting materials, and crack depth and the like.

REFERENCES

1. IISAKA, T, SUGIYAMA, A, UMEHARA, H. Studies on Penetration Properties of Repair Materials for Cracks in Concrete, ACI International Conference, SP-128, Vol. 1, pp.727-740, Hong Kong, 1991

2. IISAKA, T, SUMI, T, KIKUKAWA, H. Injecting Properties of Repairing Materials for Crack, JCI,(in Japanese), No.46, pp.988-993, 1992

3. SUMI, T, IISAKA, T, SUGIYAMA, A. Injection Properties of Inorganic Materials, JCA, (in Japanese), No.47, pp.814-819, 1993

4. IISAKA, T, SUMI, T, UMEHARA, H, UEHARA, T AND YOSHIDA, H. Injection Properties of Crack Repair Material, DeCSAT95, Naning, China, pp.945-950, 1995

5. IISAKA, T, SUMI, T, UMEHARA, H. Fundamental Study of Injecting Properties for Inorganic Repairing Materials, JSCE,(in Japanese), No.599,V-40, pp.49-57, 1998

IMPROVING THE PROPERTIES OF CEMENT GROUT USING ORGANIC FIBRE AND SUPERPLASTICIZER

W-H Huang

National Central University

C-C Tseng

Formosa Plastics Inc

Taiwan

ABSTRACT. The physical properties of cement-based grouts containing organic fiber and superplasticizer for isolation of hazardous and low-level radioactive wastes are presented. In order to produce a workable, crack-resisting, impermeable, and durable grout, the grout mixes studied contain cement, fly ash, water, polypropylene fiber, and superplasticizer. Laboratory studies included viscosity, bleeding, setting time, compressive and flexural strengths, pore size distribution, hydraulic conductivity, and durability of grout mixes. The findings indicated that grouts containing organic fiber were more crack-resisting and less vulnerable to volume change, but showed higher viscosity and hydraulic conductivity. With the incorporation of suerplasticizer in grout mixes, the adverse effects introduced by organic fiber were corrected with additional improvements in viscosity, flexural strength, hydraulic conductivity, and durability against wet-dry cycling and sulfate attack.

Keywords: Cement grout, Engineering properties, Polypropylene fiber, Superplasticizer, Waste containment

Dr Wei-Hsing Huang is associate professor at the Department of Civil Engineering, National Central University in the Republic of China on Taiwan. His specializes in sealing materials for isolation of radioactive waste disposal site, including cement-based grout, buffer and backfilling materials.

Mr Ching-Chuan Tseng is design engineer at the construction department of Formosa Plastics Inc., Taiwan. He was formerly a graduate student at the Department of Civil Engineering, National Central University. His masters thesis on cement-fly ash grout contributes to part of this paper.

INTRODUCTION

Cement-based grouts have been used by geotechnical engineers for years to control the movement of ground water and to reduce the permeability of soils or porous geologic media. In recent years, the use of cement-based grouts has been extended to environmental areas in constructing underground curtains and barriers for isolation of hazardous and low-level radioactive wastes. Previous studies on cement-based grouts have postulated the use of fly ash in the grout mixes [1]. However, cement grouts also have been criticized for being susceptible to cracking. To mitigate this problem of cracking and deterioration, incorporation of fiber into cement grouts is considered as a potential solution. The addition of fiber to cement paste has been reported to improve the flexural and tensile strengths of cement paste and concrete, but there is limited information available on cement grouts. In this study, polypropylene (PP) fiber was selected because it is less susceptible to chemical attack and is expected to minimize brittleness of grout barrier thus reducing the potential of cracking in an underground environment. Superplasticizer was used to improve the workability of grout mixes, especially those with low water-solid-ratios.

In a waste disposal site, the primary mechanism for the likely introduction of hazardous and/or radioactive elements into the environment is through physical and chemical deterioration of the isolation barrier. The objective of this study was to determine the effect of the polypropylene fiber and superplasticizer on the physical properties of cement grouts, including viscosity, setting time, bleeding, compressive and flexural strength, hydraulic conductivity, and pore size distribution. In addition, the durability against wet-dry cycling of cement-fly ash grouts was characterized in the laboratory. This paper describes the results of these tests.

EXPERIMENTAL WORK

A standard grout mix prepared using 70 percent (by weight) type I Portland cement and 30 percent fly ash was developed as the control mix in the laboratory program. The fly ash is a Class F fly ash obtained from Taichung power plant in central Taiwan. The PP fiber used is a 3 Denier fibrillated fiber having a length of approximately 10 mm. One percent polypropylene fiber was added to the cement-fly ash mix. The superplasticizer used was a sodium salt of sulfonated naphthalene formaldehyde condensate conforming to ASTM C 494, type F high-range water-reducing agent. Grout mixtures were prepared with water/solid (water to cement+fly ash) ratios ranging from 0.39 to 0.57. The amount of superplasticizer used was 1% of the solid content. Table 1 summarizes the formulas used in the study with their notation used in this paper.

Mixing of all grouts were accomplished using a 3-blade paddle mixer as suggested in ASTM C938. Exact amount of water was first poured into the mixer and the mixer started. Pre-mixed cement-fly ash was added to the mixer in 2 min and kept mixing for 3 min. Delayed addition of superplasticizer is reported to decrease adsorption on C_3A after mixing, resulting in greater availability of superplasticizer to increase dispersion [2]. Therefore, superplasticizer was added to the mixes after cement, at about 4 min from the beginning of mixing.

Finally, PP fiber was added at 1 min before the completion of mixing. The entire mixing procedure took about 5 and a half min.

Hardened grout specimens for strength and permeability tests were cured in moisture room until designated age for testing. Durability specimens were cured for 28 days and then subjected to either wet-dry cycling. They were then tested for changes in hydraulic conductivity and compressive and flexural strengths.

Table 1 Grout formulas and their notation

WATER/CEMENT+FLY ASH	PP FIBER	SUPERPLASTICIZER	NOTATION
0.39	-	-	C-3
0.48	-	-	C-4
0.57	-	-	C-5
0.39	1%	-	C-f3
0.48	1%	-	C-f4
0.57	1%	-	C-f5
0.39	-	1%	SP-3
0.48	-	1%	SP-4
0.57	-	1%	SP-5
0.39	1%	1%	SP-f3
0.48	1%	1%	SP-f4
0.57	1%	1%	SP-f5

RESULTS AND DISCUSSION

Viscosity

The apparent viscosity of grout mixes was measured with a six-speed co-axial cylinder viscometer. Figure 1 shows the shear stress-shear rate relationship of the control grout mix (water/solid = 0.39) with a range of superplasticizer contents. The use of superplasticizer acts to impart a strong negative charge on cement particles, which reduces the surface tension of the surrounding solution considerably and hence improves the viscosity of grouts significantly. As expected, the apparent viscosity decreases as the amount of superplasticizer increases. Greater effect is observed for superplasticizer content of less than 1%. Also, the grouts approach Newtonian behavior as the superplasticizer content exceeded 0.75%. As a result, the superplasticizer content was determined to be 1% throughout the study.

The flow properties of grouts were determined using an ASTM C939 flow cone. In this test, 1.725 litre of grout flows from the discharge tube of the cone and the time of efflux is recorded. Figure 2 summarizes the results. The efflux time of grouts generally increases with increase in the amount of water in the mix. Grouts with low viscosity find applications in penetrating fine fissures or long distances.

If this is desired, however, extra care should be exercised in increasing the amount of water because this may increase the bleed and adversely affect the properties of hardened grout. The addition of PP fiber increases the efflux time, especially at high water/solid ratios.

Time of Set

The initial and final setting times of grout mixes were measured by the Vicat apparatus (ASTM C 953), and the results are summarized in Figure 3. It was found that the addition of PP fiber shows a decreasing effect on the setting times of grout mixes, especially the initial set. On the other hand, increases in setting times due to the use of 1% superplasticizer are observed as expected. In general, the initial set occurs at 5 to 8 hrs, and the final set is less than 12 hrs. For most field applications, the setting times for grout mixes studied are within acceptable range.

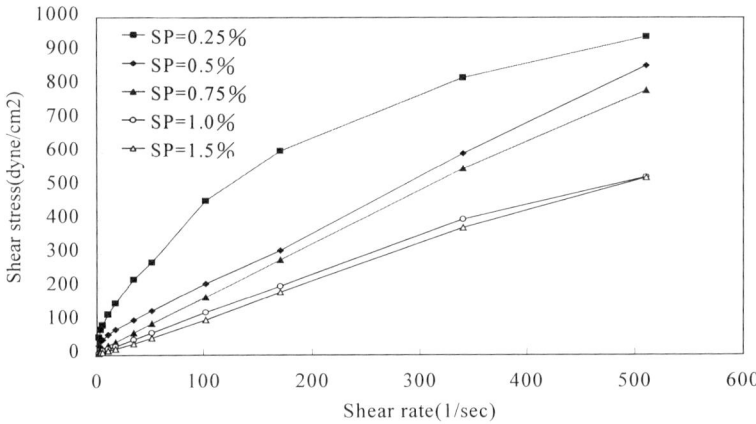

Figure 1 Effect of superplasticizer content on the shear stress-shear rate relationship for cement-fly ash grout (water/solid = 0.39)

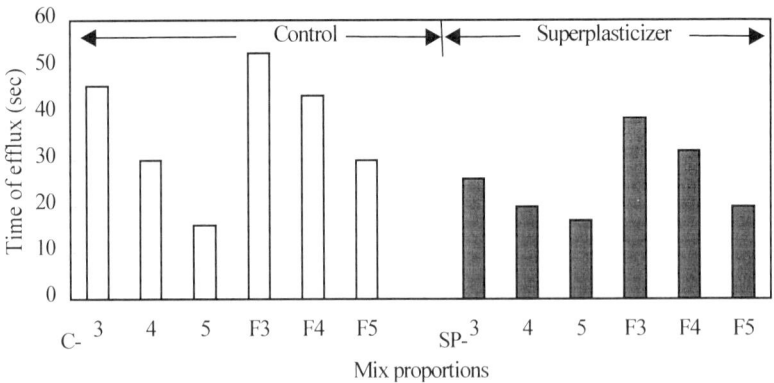

Figure 2 Flow cone time of grout mixes

Bleed

Bleed is the appearance of water on the surface before the grout has set. It is a form of segregation resulting from the inability of the solid particles to hold all mixing water in a dispersed state as the solids settle [3]. Figure 4 shows the final bleeding as percentages of the initial volume of the grout (ASTM C 940). The water/solid ratio has a great effect on bleed of the grout. The data indicate that the addition of PP fiber to the grout increases the final bleed. This can be attributed to the fact that PP fiber does not dissolve in the suspension, and thus the amount of solid (cement + fly ash) particles is relatively reduced in a unit volume of unset grout. The increases in final bleed resulting from the use of superplasticizer are quite noticeable, due to its deflocculating action imparted by superplasticizer in the suspension.

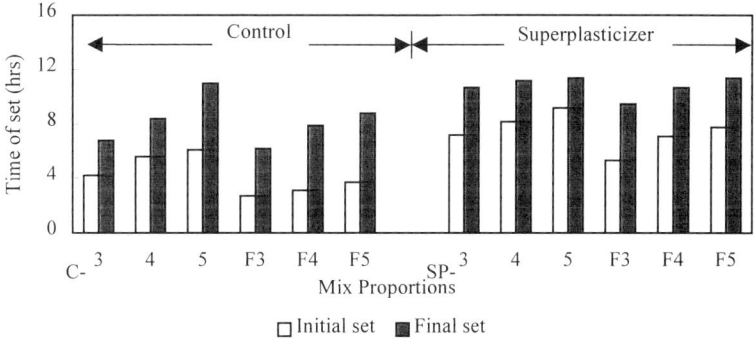

Figure 3 Setting time of grout mixes

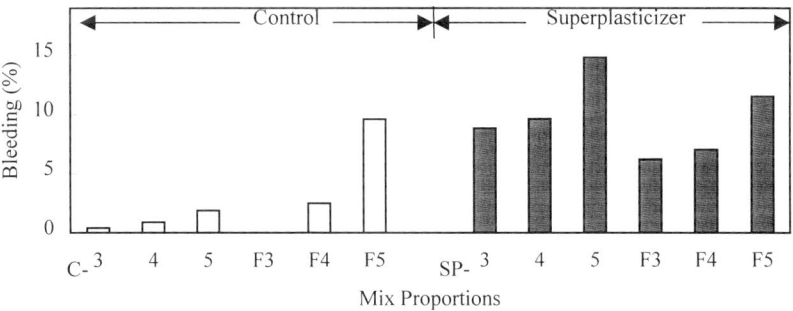

Figure 4 Final bleed for various grouts

Flexural and Compressive Strengths

The compressive strength of the grout provides a good indicator of the quality of the hardened grout. The flexural strength is an important property because underground containment barriers are frequently subjected to high lateral earth pressure. In this study, the flexural and compressive strength tests were conducted on 5x5x16 cm prisms and 5-cm cubes, respectively, at 7, 28, 56, and 118 days. The results are presented in Figure 5 and 6.

In the control group, slight reductions in strength can be observed with the use of PP fiber. The incorporation of superplasticizer increases the flexural and compressive strengths of the grouts, especially for mixes with low water/solid ratio. This indicates that a better homogeneity of the cement grout structure is achieved in the well-dispersed system produced with the addition of superplasticizer. And this effect is more pronounced for fiber-laden mixes with low water/solid ratio.

Pore Structure

The pores in cement grout form a continuum and the pore structure has been used in predicting hydraulic conductivity and durability of cement pastes [4,5]. In this study, pore size distributions of hardened grout were measured using mercury intrusion porosimeter (MIP). The pores were divided into 3 categories according to their size. Figure 7 shows the volume of pores in the grout in each size category determined at 28 days.

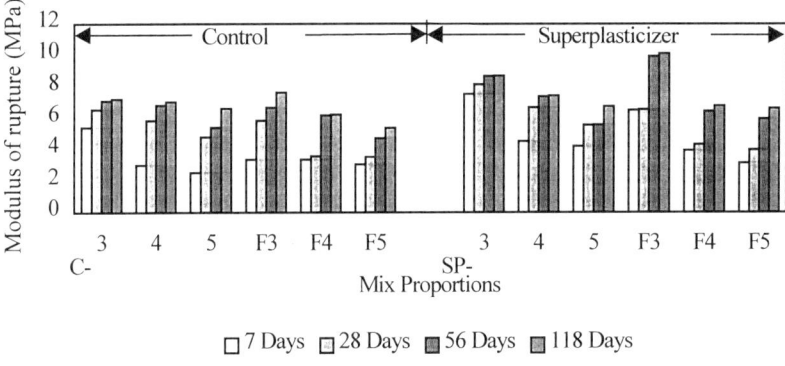

Figure 5 Flexural strength development of grout mixes

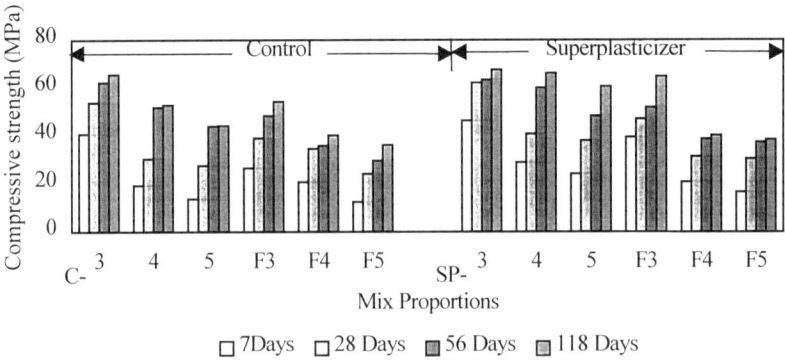

Figure 6 Compressive strength development of grout mixes

In the control group, no significant difference in the total pore volume exists between grouts containing PP fiber and those not. However, the amount of large- and medium-sized pores increases with the addition of PP fiber. This is illustrated in Figure 7 by comparing the proportions of pores in each size category. On the other hand, The use of superplasticizer markedly reduces the total pore volume and pore size. The decreases in the amount and size of pores result in a refined pore structure, which in turn is beneficial to the engineering and durability properties of the grout.

Hydraulic Conductivity

The major function of subsurface grout barriers is to prohibit the migration of flow in or out of the waste. Most deleterious reactions, including sulfate attack, corrosion, alkali-aggregate reactions, and freezing and thawing, involve the ingress of water or aggressive solutions. Therefore, impermeable grouts not only provide hydraulic isolation but also enhance durability. For provision of watertight and durable grout barriers to aggressive underground environments, the most important property is hydraulic conductivity.

Hydraulic conductivity tests were performed on 28-day grout specimens using apparatus similar to that adopted by Soongswang *et al*. [6]. The measured hydraulic conductivity for all grout mixes is presented in Figure 8. The addition of PP fiber shows an increasing effect on the hydraulic conductivity of grouts, especially in the control group. This indicates that the openings at fiber/matrix interface are providing convenient flow paths for water permeation. On the other hand, superplasticizer has more of an impact on the hydraulic conductivity. Due to the reduction in porosity and refinement of pores, the hydraulic conductivity of grouts containing superplasticizer is approximately two orders of magnitude lower than that of control grouts. In general, grouts containing superplasticizer have hydraulic conductivity lower than 10^{-8} cm/s.

Based on the results derived from hardened grout, the addition of PP fiber alone to the mix seems to have adverse effects on the flexural strength, pore structure, and hydraulic conductivity of grout mixes. However, if superplasticizer was added at the same time, these adverse effects were corrected and the expected advantages of superplasticizer on fresh grout were still observed. This can be attributed to the dispersing effect imparted by superplasticizer causing an improved homogeneity in the cement grout system. Therefore, in case improved brittleness is desired for subsurface grout barriers, the incorporation of PP fiber should always be accompanied by admixing appropriate amount of superplasticizer in the grout formula.

Resistance to Wetting-Drying Cycles

Following moist curing of 28 days, grout specimens were exposed to wetting-drying cycles. This was accomplished by alternately submerging grout specimens in water for 24 hours and drying in an oven maintained at 40 C for 24 hours. Grouts were resaturated and tested for their hydraulic conductivity and strengths after 15 cycles of wetting and drying, and the results are compared with those obtained from specimens stored in the curing room with the same age.

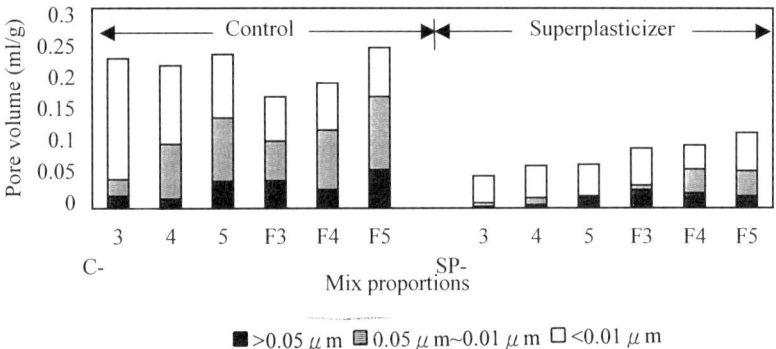

Figure 7 Pore volume distribution of grouts at 28 days

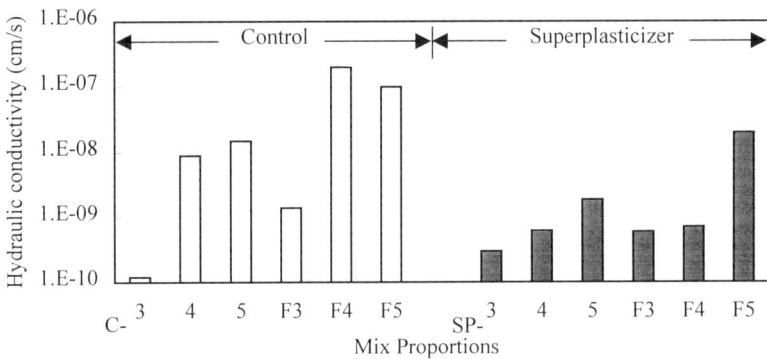

Figure 8 Hydraulic conductivity of grouts at 28 days

Results of flexural strength tests on grout specimens after various wet-dry cycles are summarized in Figure 9. Generally, grouts do not show any significant loss in compressive strength after wet-dry cycling. However, only grout mixes containing PP fiber exhibit no reduction in flexural strength after wet-dry cycling. Figure 10 shows the change in hydraulic conductivity of grouts after exposure to 15 wet-dry cycles, expressed as a percentage of that of the moist-cured specimens. Again, it can be seen that grouts containing PP fiber exhibit no increase in hydraulic conductivity, indicating that PP fiber is effective in resisting the development of cracks resulting from drying shrinkage.

Resistance to Sulfate Attack

After 28-day moist curing, grouts were submerged in 4.2% magnesium sulfate solution and then tested for strength and hydraulic conductivity. The measured strengths at 120-day submersion were compared to those obtained from moist cured grouts. The changes in strength and hydraulic conductivity are expressed as a percentage of those derived from the moist-cured grout in Figure 11 and 12, respectively. Figure 11 shows that no significant differences exist in the change of strength between the control and superplasticized grouts. Grouts with PP fiber show less reduction in compressive strength after sulfate attack.

Figure 12 shows that the change in hydraulic conductivity of control mixes after exposure to sulfate solution was negligible. On the other hand, grouts admixed with superplasticizer alone show some increases, while the hydraulic conductivity of mixes containing both PP fiber and superplasticizer remained unaffected.

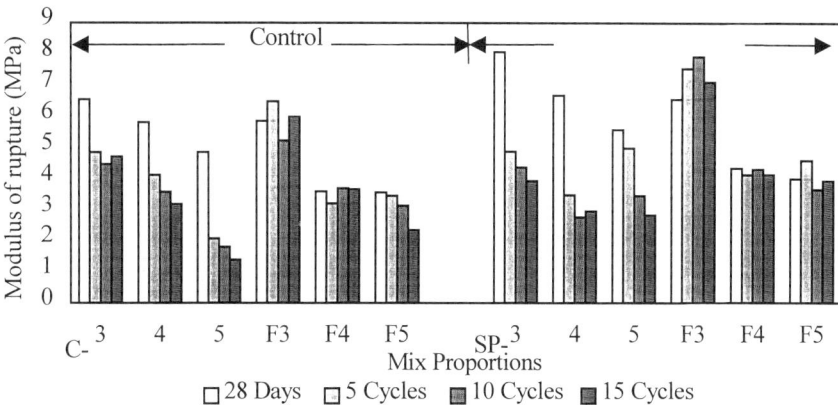

Figure 9 Flexural strength of grouts experienced various cycles of wetting and drying

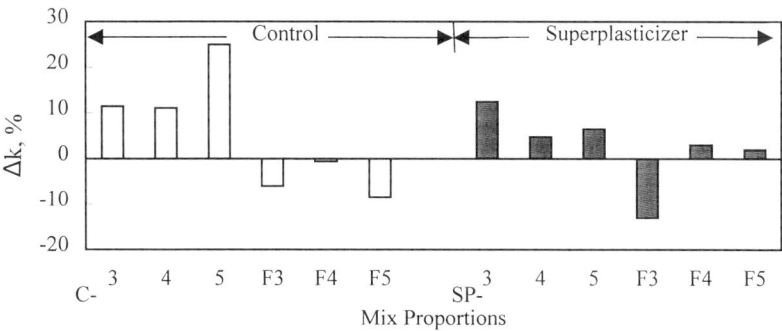

Figure 10 Change in coefficient of permeability of grouts experienced 15 wet-dry cycles

SUMMARY AND CONCLUSIONS

The grouts developed in this study have application in underground situations such as sealing hydraulically active fracture zones and faults, rock-concrete interfaces, sealing containers of radioactive and hazardous wastes.

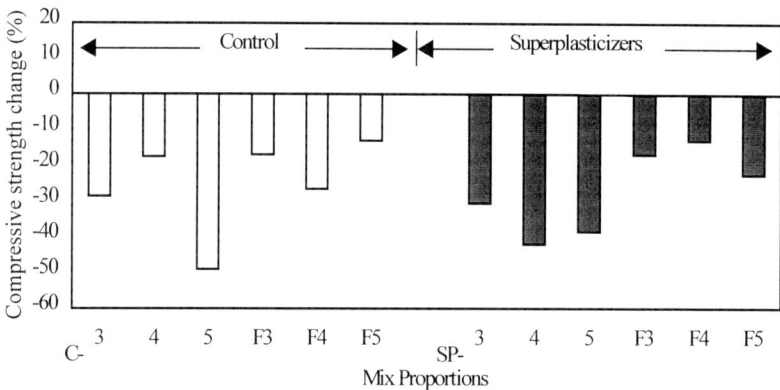

Figure 11 Change in comp. strength after exposure to sulfate attack for 120 days

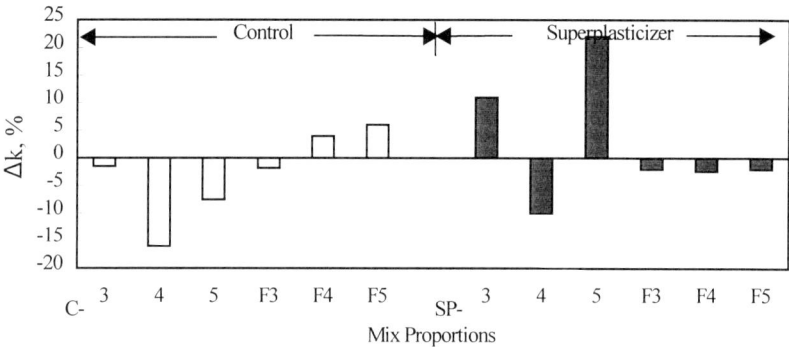

Figure 12 Change in hydraulic conductivity after exposure to sulfate attack for 120 days

Substitution of 30% cement by class F fly ash produces economical grouts with reasonable physical properties. Addition of polypropylene fiber alone may have adverse effects on properties of fresh and hardened grout, but significantly improves the grout's resistance to cracking and changes in volume resulting from wet-dry cycles and sulfate attack. Grouts containing appropriate amount of superplasticizer show enhanced flowability and viscosity, higher strengths, and reduced hydraulic conductivity. Also, the adverse effects resulting from fiber addition are generally eliminated by the use of superplasticizer. To assure enhanced physical properties and durability, it is recommended that the use of PP fiber always be accompanied by incorporation of superplasticizer.

ACKNOWLEDGMENT

This study was supported by the National Science Council under project number NSC 83-0410-E008-012 and NSC84-2211-E-008-033.

REFERENCES

1. HUANG, W H., "Properties of cement-fly ash grouts admixed with bentonite, silica fume, or organic fiber," Cement and Concrete Research, Vol. 27, 1997, pp. 395-406.

2. ALLAN, M L, AND KUKACKA, L E. "Comparison between slag- and silica fume-modified grouts," ACI Materials Journal, Vol. 93, No. 6, 1996, pp. 559-568.

3. MEHTA, P K. Concrete: Structure, Properties, and Materials, Prentice-Hall Inc., New Jersey, U.S.A., 1986.

4. PERRATON, D, AITCIN, P C, AND VEZINA, D. "Permeabilities of silica fume concrete," in Permeability of Concrete, D. Whiting and A. Walitt, editors, SP-108, American Concrete Institute, Detroit, 1988, pp. 63-84.

5. VONDRAN, G, AND WEBSTER, T. "The relationship of polypropylene fiber reinforced concrete to permeability," in Permeability of Concrete, D. Whiting and A. Walitt, editors, SP-108, American Concrete Institute, Detroit, 1988, pp. 85-97.

6. SOONGSWANG, P, TIA, M, BLOOMQUIST, C, AND SESSIONS, L M, "An efficient test set-up for determining the water-permeability of concrete", Transportation Research Record 1204, 1988, pp. 77-82.

CEMENT GROUTING TECHNOLOGY FOR CONSOLIDATING SOILS

M S Akman
Istanbul Technical University
M Mutlu
Sika-Deteks Construction Chemicals Company
Turkey

ABSTRACT. Grouting is employed for rendering impervious the dam foundations and embankments. Polymeric solution grouts are used in the case of fine fissures, and cement grouts are the most common materials for rocks having wide cracks. Injection pressures and grout takes are determined before grouting. To meet the requirements dictated by site conditions the grout slurries should be fluid, stable and injectable. Superplasticizers were added to the cement grouts of a required fluidity to reduce the high W/C and B/C ratios required for a prescribed fluidity and thus to increase strength and durability. Bentonite is added to prevent bleeding in sandy superplasticized slurries. Cement grouting techniques, the properties of grouts used and the tests carried out on grouts during the construction of Berke dam are given.

Keywords: Cement, Bentonite, Grouting, Rock, Rheology, Slurry, Superplasticizer.

Dr Ing M Süheyl Akman is Emeritus Professor of Civil Engineering at the Istanbul Technical University. His research interest covers various aspects of concrete technology, especially the use of mineral and chemical admixtures and durability of concrete.

Mehmet Mutlu, Civil Eng MSc, is in charge of the Technical Department of Sika-Deteks Construction Chemicals Co., Turkey. Formerly chief materials engineer in Atatürk and Berke dam constructions. His works concentrate on fresh concrete technology and concrete admixtures.

INTRODUCTION

Improvement of soils or cracked rocks by grouting is very common in civil engineering practice, especially in dam construction where the cost of grouting may be as high as 80 % of the total cost of construction depending on the quality of the formation to be grouted.

The grouting materials are chosen according to the pore size of the soil or the widths of fissures or cracks in rocks: in very finely fissured rocks or in clayey or silty soils polymeric or chemical grouts should be used. Polymeric solutions may also be preferred for grouting formations with wider cracks. The polymeric grouting materials, which are basically polyurethane, polyacrylamides, aminoplasts or phenoplasts mixed with some organo-mineral resins, are commercially available [1, 2]. In spite of the progress in the technology and quality of the polymeric injection materials, cement grouts containing additions of plasticizers (WR) or superplasticizers (SP) preserve their value especially in the improvement of formations with larger cracks.

Some researchers hold that the use of bentonite is no longer necessary when there is a superplasticizer in the mix. However, the use of bentonite with superplasticizer in sandy cement grouts yielded better results during the grouting works at Berke dam in Turkey.

In this paper, properties of the grouting materials used in the curtain and contact grouting methods are investigated experimentally, and the results obtained in the preparatory tests and in-situ tests carried out during construction of the Berke dam are evaluated.

CEMENT GROUTING IN DAM CONSTRUCTION - AIMS AND MATERIALS

The primary aim of cement grouting in dam construction is to fill the pores of soil or fissures of rocks and render the medium impervious to seepage. This goal should be accomplished economically without causing any deterioration in soil or rock. The grouting work involves many issues. The properties and compositions of grouting materials, injection process and equipment, preparatory laboratory and in-situ tests and the quality control and assurance are just some to cite. They can be managed by a team of geologists, geotechnicians and materials engineers.

Principles in Grouting Processes

Two principal parameters in a grouting process are the volume of grout injected per unit length of the borehole and the allowable injection pressure. The geometry and the pattern of the boreholes, characteristics of geological formations and economic considerations limit the total volume of grout. It is impossible to completely fill large cavities, karstified rocks or very large cracks by grouting since the excessive consumption of grout will be prohibitively uneconomical. The maximum injection pressures are also to be limited so as not to damage the formation being grouted.

Another question in the grouting process is how to determine the pressure P and the slurry volume V/m during the injection operation.

The GIN (Grouting Intensity Number) method proposed by Lombardi consists of fixing some constant value for GIN defined as the product of V/m and P [3]. The constant value of GIN depends on the properties of the formation to be grouted, whose validity is to be verified by in-situ tests.

A graphical representation of the GIN method is shown in Figure 1. The hyperbola obtained is bound by the maximum values of P and V/m. The curve can be used for adjusting the pressure and the flow rate. In practice, however, it is seldom possible to follow this curve during the grouting operation, and therefore the hyperbolic curve is generally transformed into stepwise broken line segments as shown in Figure 1.

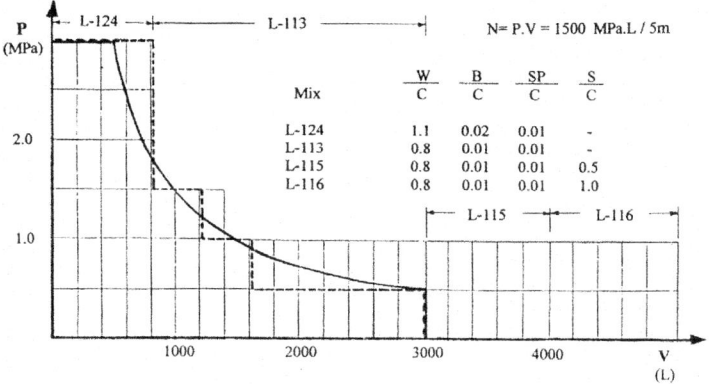

Figure 1 Curtain grouting GIN curve (Berke Dam)

A sudden increase in grout take indicates that the injection front encountered a large crack or a porous zone. The first remedy is to thicken the grout by adding fine sand. Generally the grouting operation begins with thin slurry. Nevertheless, it may be advisable to start with a thick slurry with lower water/cement and higher solids/water ratios to improve the strength and durability of hardened grout.

If a stratum of certain thickness at a certain depth is to be grouted, the borehole can be suitably blocked by using packers. The caulking materials used are lead, wool, saw dust, etc. Regular method for grouting is upstage grouting, i.e. grouting from bottom to top. In the case of extensively fissured rocks, downstage grouting, i.e. grouting from top to bottom, is applied.

Ingredients and Compositions of Cement Grout Slurry

The ingredients of a cement grout are cement, water, bentonite mixture plus one or more of sodium silicate, plasticiser or superplasticiser, water retaining admixtures and fine sand.

The principal component cement is generally of Portland type. Blast-furnace slag cement or sulphate resisting cement are preferred if aggressive ground water is present.

Fineness is the most critical property required of the cement to be used in grouting. The Blaine fineness should be greater than 400 m^2/kg, which is especially important for the stability and injectability of the grout. Recently very fine cements, namely microcements, have been produced with maximum particle sizes not exceeding 20 µm, as compared with 80 to 90 µm of the normal cements. Numerous investigations have been carried out on the use of microcements in cement grouting [1, 2, 4].

Bentonite, a special type of smectite group clays, has been the second solid component of cement grouts for many years. The activities of sodium bentonites are significantly higher than those of other clays [5]. They have high water retentivity, viscosity and thixotropy, which makes them indispensable in cement grouting. Bentonites should be premixed with water using a high speed mixer and left to hydrate for at least 24 hours. Unnecessarily high bentonite contents of up to 5 % by weight of cement had been used in earlier years in some projects. Bentonite may act as a lubricant in the cement grout, but due to its high water sorption the W/C ratio may also increase up to 3, yielding a grout of undesirably inferior strength and durability. Plasticizers or superplasticizers have assumed the dispersive and fluidifying functions of the bentonites and the W/C ratios have been decreased down to 0.6, and the bentonite contents down to 0.5 % or even to 0.2 %, though bentonite still preserving its role in preventing undue bleeding especially in the case of sandy cement grouts.

The third group of ingredients in a cement grout is chemical concrete admixtures as mentioned above. Plasticizers or superplasticizers of naphthalene or melamine sulphonate formaldehyde condensates, carboxylated superplasticizers and ligno-sulphonate plasticizers are commercially available. However, concrete admixture producers add other chemical ingredients to their superplasticizers to impart some additional properties such as stability, thixotropy, set accelerating or retarding and expansive character that may be desirable for injection purposes [2]. Another well-known chemical admixture is sodium silicate used for impermeabilization of soils and to increase the injectability of cement grouts.

Rheology and Stability of Cement Grout

The grout should have high fluidity (defined as the inverse of viscosity) for it to be able to flow at a sufficiently high rate in boreholes, pores of the formations, and especially in fissures of rocks. Stability of a suspension is its capability to retain water and keep solid particles suspended during flow and when standing still. These properties can be simply monitored by measuring bleeding capacity or decantation. The decantation is expressed as the ratio of the water volume collected on the top of grout specimen left to stand still for three hours in a test tube to the total initial volume [6]. The water retaining capacity of a grout can be determined by filter press test in which 400 cc of cement grout slurry placed on filter paper in a filter press cell is filtered under 0.7 MPa pressure. The test is ended when the filtration is over. The time t (second) and the volume V (cc) of filtrated water are measured. The quantity called filter press coefficient (K_{pf}) is taken as a measure of the water retentivity of the slurry and computed by

$$K_{pf} = \frac{V}{400} \cdot \frac{1}{\sqrt{t/60}}, \quad \frac{1}{\sqrt{min}}$$

Two important properties of cement grout, the fluidity and stability are seemingly conflicting properties. The stability against settlement of a suspension is directly proportional to viscosity and the fluidity is inversely proportional. The mechanical behaviour of cement suspensions can be approximated by Bingham body model possessing two parameters: yield value (τ_o) and plastic viscosity (η_{pl}) as shown in Figure 2.

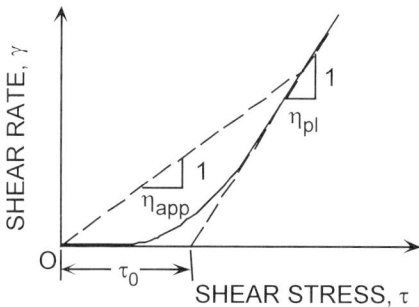

Figure 2 Bingham Body with yield value, τ_o, plastic and apparent viscosities, η_{pl}, and η_{ap}

Stability of the suspension is usually high when η_{pl} is high, but fluidity is high when η_{pl} and τ_o are low. An increase in water content decreases the values of η_{pl}, and η_{ap}, leading to a consequent reduction in stability. In that case, it is necessary to take supplementary measures for stabilising. The first measure to be taken is to decrease the rate of bleeding by using very fine cement and adding bentonite. Higher bentonite contents result in an increase in the value of τ_o and in a decrease in the fluidity, as measured by the Marsh Funnel. In other words, the water requirement for a prescribed Marsh viscosity increases due to the increased fineness of solids. It is strictly necessary to have water saturated grout that does not significantly dilate or flocculate during pumping and conveying in pipes, boreholes, and also in the pores of the formation being grouted. If the pressure needed to convey the suspension in pipes exceeds the segregation pressure of the grout, the internal friction among the solids suddenly increases, mainly due to dilation, the continuous liquid phase pressurised by the pump is no longer able to move the solids and the pipe is blocked [7, 8]. The undesirable effects of the high W/C ratio on the durability and strength of hardened grout should not be overlooked either. The W/C ratios should be 0.4 to 1.0 [8], which can easily be accomplished by using suitable superplasticizers.

The Marsh viscosity of a cement grout measured as the discharge time from the Funnel is required to be 35 to 50 seconds, though this test may not always be adequate for grouts containing superplasticizers.

There are different types of rheometers for determining the rheological parameters η_{pl} and τ_o. Generally the rotational types are used in grouting practice. Instead of obtaining the full diagram of the relation between shear rate and shear stress, only the plastic and apparent viscosities, η_{pl}, η_{ap}, are determined, but, unfortunately, τ_o is not routinely measured probably due to interference with thixotropy.

Plastic viscosity can be considered a material constant but apparent viscosity is not since it depends also on the shear rate. The apparent viscosity measurements on the same cement grout carried out using different rheometers yield different values [9]. However, the use of same instruments, namely the Marsh Funnel and the Baroid Rheometer, provides a standard for comparative evaluation quite satisfactory for practical purposes, though modifications in limits specified may be needed, along with specifications for the test conditions such as temperature, the duration of shearing and the nominal shear rate in the viscosity measurements and vigour of agitation and duration of stand still before the jel strength or yield stress is measured in a viscometer.

The yield stress or gel strength and the rate of gel strength development controls the maximum distance the grout can penetrate as measured from the borehole, limits the extent of the grouted zone, prevents high grout consumption. It also limits the splitting force, and thus dictates the minimum pressure required [10]. The plastic viscosity is used in computing the flow rate under constant pressure and the time necessary for total take. All these computations are based on the hypothesis that the visco-plastic cement grout plug moves in boreholes and the pores of granular formations under constant pressure gradient conforming to Poiseuille rule.

Injectability of Grout Slurry - Groutability

The critical factor affecting the ability of a suspension type grout to penetrate into the voids of a formation to be grouted is the particle size. Mitchell has proposed $D_{15\,(soil)} / D_{85\,(grout)} > 25$ as the condition for the groutability of sandy, silty or clayey soils. For fissured rocks, the requirement is $D_{(fissure)} / D_{max\,(grout)} > 3$ has been specified in many contracts [1,11]. According to these recommendations, the injectability of cement grouts containing only Portland cement and water may also cause problems in grouting fissures narrower than 0.3 mm. Some researchers propose adding sodium silicate to improve the groutability [1]. The penetrating ability of the cement grout is reduced if the permeability of the soil is less than 10^{-4} m/s. The use of microcements in cement grouts has resulted in a reduction in the injectable crack width or pore size; at present 0.2 mm for cement grouting seems to be reasonable. In the solution type grouts, only the viscosity of solution affects the injectability since they practically contain no particulate materials.

CEMENT GROUTING WORKS IN BERKE DAM CONSTRUCTION

Tests and investigations carried out on cement grouts at the Berke dam are presented and evaluated below.

General Information and Description of Berke Dam

The Berke dam is located in south Anatolia on Ceyhan river. It consists of a powerhouse, two cofferdams, two spillway tunnels, a headrace tunnel and a concrete dam body. It is double curvature thin arch dam having 201 m height.
The reservoir capacity is 427×10^6 m^3. The dam is designed to produce hydroelectric energy of 1.67×10^9 kWh per year. The total cost is estimated to be 670×10^6 USD.

The site area and immediate vicinity are entirely situated in the Mesozoic formations mainly composed of limestones underlain by Palaeozoic formations composed of quartzite and phyllite intercalations on the north-eastern side. On the western side the limestone rocks are covered by ophiolite nappes that ensure the reservoir watertightness. The impervious grout curtain is an essential part of the dam construction due to the nature of the bedrock. The depths of the grout holes range from 60 to 150 m. The majority of curtain grouting is accomplished by upstage grouting method. A view of Berke Dam during construction is shown in Figure 3.

Figure 3 Berke Dam Construction

Grouting Materials

Cement grouts of Berke dam consist of cement, bentonite, sand, water, and superplasticizer. The cements used conform to Turkish Standard TS-19 designated PÇ-32.5, equivalent to ASTM Type II Portland cement, with a maximum particle size of 80 μm and Blaine fineness of 400 m^2/kg.

The natural sodium bentonite used had a minimum liquid limit of 350 % and a plasticity index above 400 %. The sand was fairly uniform and had a maximum particle size of 1 mm. The superplasticizer was sulphonated naphthalene formaldehyde condensate type.

The following specifications were set for the grouts:
Viscosity of the grout   (Marsh Funnel)   : 35-50 sec.
Apparent viscosity, η_{ap} ,   (Baroid Rheometer)   : 20-30 cP
Decantation after 3 hours   (1000 ml measure)   : max. 4 %
Coefficient of Pressure Filtration , K_{pf}, (Baroid Filter Press) : max. 0.25 $1/\sqrt{min}$

The grout mixes were routinely tested and improved in the laboratory before each grouting operation. Laboratory tests were carried out on more than 30 trial mixes for mix design purposes. The results obtained on five series of mixes are given in Table 1. The mixes were also tested in-situ trials designed for fixing the limits of Grouting Intensity Number (GIN).

The mix series 3b, consisting of cement, bentonite, superplasticizer and water, gave the most suitable result.

Hence, the mix series 3b with minor modifications was used for grouting throughout the grouting work. During the grouting works, samples of mix series 3b taken from the site were tested in laboratory. A statistical evaluation of the test results is given in Table 2. The modifications caused no significant variations in the properties of fresh mixes, however, the compressive strengths of the mixes with smaller W/C and B/C ratios were higher.

Table 1 Laboratory design test results for grout mixtures

Series n = number of tests	Ingredients	Filter Press Coefficient, K_{pf} minute$^{-0.5}$	Compressive Strength MPa	Viscosity Marsh s	Viscosity Plastic η_{pl} cP	Viscosity Apparent η_{ap} cP	Bleeding Capacity (Decantation) vol. %	Remarks
1 n=8	Water Cement	-	10 - 50 (+)	28 - 50 (+, -)	-	-	0 - 50 (+, -)	Un-suitable
2 n=4	Water Cement Bentonite	0.20 - 0.40 (-)	<10 (-)	35 - 40 (+)	17 - 28 (+)	-	5 - 15 (-)	Un-suitable
3a n=9	Water Cement Bentonite Admix. (WR)	0.20 - 0.35 (-)	10 - 30 (+)	35 - 45 (+)	15 - 27 (+)	20 - 41 (+)	2 - 10 (-, +)	Un-suitable
3b n=12	Water Cement Bentonite Admix. (SP)	0.13 - 0.27 (+)	10 - 30 (+)	35 - 42 (+)	10 - 25 (+)	20 - 40 (+)	1 - 8 (+)	Suitable
4 n=6	Water Cement - Admix. (SP)	0.25 - 0.35 (-)	15 - 20 (-)	35 - 39 (+)	20 - 30 (+)	20 - 30 (+)	0 - 5 (-)	Un-suitable

Evaluation: (-): negative, (+): positive

Table 2 Statistical evaluation of samples (without sand) taken from site

Mix Proportions:	$\frac{W}{C} : \frac{B}{C} : \frac{SP}{C}$	0.9 : 0.01 : 0.012		0.8 : 0.005 : 0.012	
Property	Unit	$\bar{x}$	σ	$\bar{x}$	σ
Density,	Mg/m^3	1.54	0.03	1.57	0.01
Viscosity, Marsh,	s	40.1	4.4	37.5	1.5
Vicosity, Plastic,	cP	13.8	3.5	14.2	3.1
Viscosity, Apparent,	cP	27.8	7.1	27.0	6.3
Decantation,	vol. %	1.4	0.7	1.2	0.5
Filter Press Coeff., K_{pf}	$1/\sqrt{min}$	0.17	0.04	0.15	0.02
Compressive Strength,	MPa	11.9	2.4	16.2	2.4
Number of tests, n		20		45	

$\bar{x}$ = mean, σ = standard deviation

CONCLUDING REMARKS

The main constituents of cement grouts are cement, bentonite and water. To obtain sufficiently stable but also fluid grouts the bentonite content had to be kept high and consequently the water contents and the water/cement ratios became too high to achieve satisfactory strength or durability.

By the addition of superplasticizers into these cement grouts, the water/cement ratio could be reduced significantly without causing any reduction in the required fluidity, durability and strength of the hardened grout. However, the stability of the superplasticized cement grout may be slightly lower. It is necessary, therefore, to add a water retaining admixture or a small quantity of bentonite to compensate for this loss of stability. This addition is especially beneficial in the case of cement grouts containing sand. For example in the Berke dam construction the optimum results for fluidity, stability and strength were obtained with the following composition:

C = 750 kg/m^3,   W = 600 kg/m^3,   S = 375 kg/m^3,   B/C = 0.5 %,   SP/C = 1.2 %

Cement grouts made with bentonite and ordinary Portland cement having a Blaine fineness of 400 m^2/kg can be injected into rock fissures wider than 0.3 mm. This limit can be decreased to 0.2 mm when microcements are used. At present only polymeric and chemical grouts can be injected into cracks narrower than 0.2 mm.

The stability of grouts is monitored by bleeding test and the water retentivity by filter press test that give reliable results having good correlation. Contrary to this, the test results obtained on the flow properties do not quite correlate.

In practice, however, the Marsh Funnel is used for monitoring the fluidity, and coaxial cylinder viscometer Baroid Rheometer is used for determining the plastic and apparent viscosities, though the yield values are not routinely measured probably due to the low repeatability caused by the heavy interference of thixotropy and the dominant hydraulic activity or setting tendency of the cement grout. On the other hand, plastic and apparent viscosities measured also vary with the test apparatus and the method used. Future researches are needed to establish recommendations for measuring the relevant rheological parameters and to specify the critical limits.

ACKNOWLEDGEMENT

The authors express their gratitude to Mr. Altug S, special consultant of the Berke dam project, for supplying valuable documents.

REFERENCES

1. KAROL, R, H. Grout Permeability, Issues in Dam Construction, Proc. of the Session sponsored by the Geotechnical Engineering ASCE, Ed. by Baker W H, 1985, pp 27–32.

2. SIKA SYSTEM DOCUMENT. Technically and Ecologically Advanced Systems for Injection, Ed. by Sika A G, Switzerland, 1998.

3. LOMBARDI, G, AND DEERE D. Grouting Design and Control Using the GIN Principle, Water Power and Dam Construction, June 1993, pp 15–22.

4. MELBYE, T, A. Injection of Rock with Microcements, Grouting in rock and Concrete, Proc. of the Intern. Conf. on Grouting in Rock and Concrete, Ed. Widmann R A A Balkema Publ., Rotterdam, 1993 pp 65-74.

5. AKMAN, M, S, AND GÜNER, A. Bentonite in Cement Grouts and Concrete, Concrete for Environment Enhancement and Protection, Ed. R K Dhir, T D Dyer, E & FN SPON, 1996, pp 617-626.

5. KRIZEK, R, J, SCHWARZ, L, G, AND PEPPER, S, F. Bleed and Rheology of Cement Grouts, Ibid. Ref. 4, pp 55-64.

6. EDE, A, N. The Resistance of Concrete Pumped Through Pipelines, Mag. of Conc. Res. Vol. 9, No. 27, 1957, pp 129-140.

8. LOMBARDI, G, AND DEERE, D, V. Grout Slurries - Thick or Thin? Ibid Ref. 1, pp 156-164.

7. HAKANSSON, U, HASSLER, L, AND STILLE, H. Rheological Properties of Cement Based Grouts - Measuring Techniques and Factors of Influence, Ibid. Ref. 4, pp 491-500.

9. LOMBARDI, G. The Role of the Cohesion in Cement Grouting of Rock, Q 58, R 13, 15th ICOLD Congress, Lausanne, 1985.

10. ÇUKUROVA ELEKTRIK, A, Ş. Berke Dam and Hydroelectric Power Plant, Contract No 1 phase II C, Civil Engineering Works, Vol 2, Adana, Turkey, 1994.

MULTI-BLEND CEMENTITIOUS INJECTION GROUTS FOR REPAIR AND STRENGTHENING OF MASONRY STRUCTURES

E E Toumbakari   D Van Gemert

Catholic University of Leuven

Belgium

T P Tassios

Nathional Technical University of Athens

Greece

ABSTRACT. The repair and strengthening of historical masonry requires the use of materials able to improve the mechanical properties of the structure and physico-chemically compatible to the existing fabric. The correct estimation of the required mechanical properties and the proportioning and combination of inorganic materials accordingly can produce binders that address both requirements. This study deals with the development of high penetrability injection grouts, passing through voids with a diameter smaller than 0.3 mm. It is shown that, with the use of a limited cement content and the combination of lime, natural pozzolans and, for some of the compositions, silica fume, compressive strengths of 15-17 MPa and tensile strengths of 3 MPa at the age of 90 days with W/S as low as 0.85 can be achieved. The effect of sulfate attack (originating from the presence of gypsum) on those systems is also discussed.

Keywords: Composite hydraulic binders, Ordinary portland cement (OPC), Hydrated lime, Natural pozzolans, Masonry, Injection grouts, Ultrasonic mixing procedure, Sulfate resistance, Gypsum.

Ir Eleni-Eva Toumbakari is a Civil Engineer from the National Technical University of Athens. She is currently Research Engineer and Ph.D. candidate at the Department of Civil Engineering, Catholic University of Leuven. Her research concerns the development and the study of the microstructure, hydration and durability of composite binders for the repair and strengthening of masonry structures.

Professor Dionys Van Gemert is professor of building materials science and renovation of constructions at the Department of Civil Engineering of K.U.Leuven. He is head of the Reyntjens Laboratory for Materials Testing. His research concerns repair and strengthening of constructions, deterioration and protection of building materials, concrete polymer composites.

Professor Theodosios P Tassios at the National Technical University of Athens, is working in the fields of concrete and masonry materials and engineering.

INTRODUCTION

The repair and strengthening of masonry buildings, especially in earthquake prone areas, require two types of interventions: structural interventions for the improvement of the response of the structure as a system, and interventions for the protection of the existing materials from environmental aggression and weathering. Masonry buildings repair also require the improvement of their mechanical properties, to withstand external actions, such as existing or new loads and/or imposed deformations. Grouting is one of the most important techniques addressing those issues and is especially suitable to three-leaf walls. Grouting is achieved by using cement or cement-polymer materials, which have the advantage of developing high strengths in a short period of time. They present however physico-chemical and mechanical properties which differ from those of the existing fabric. It is therefore unsurprising that damages and durability failure has occurred, resulting in an increasing suspicion of the conservation professionals towards hydraulic binders in general.

The improvement of the mechanical properties of masonry structures make the use of hydraulic binders necessary, since the use of air hardening binders, mainly hydrated lime, will not permit the development of strength in a period of time meaningful from a structural point of view. On the other hand, the use of materials with a high cement content may be unnecessary for the levels of strengths that are required by a masonry structure. The correct estimation of the required mechanical properties and the proportioning and combination of inorganic materials accordingly can produce very suitable binders, both from a mechanical and a durability point of view. Of major importance here are the requirements relative to the rate of strength development, since it is impossible to achieve acceptable levels of strength with slowly hydrating binders within 28 days. In our research, the time for the development of the required strengths has been set to 90 days, which permits both the effective use of materials that are compatible to existing masonry and the development of strengths sufficient from an engineering viewpoint [1].

The development of injection grouts satisfying the aforementioned requirements presents an additional difficulty, since in the slowly increasing strength of materials is added the necessity of using high water contents, to achieve a high penetrability. The high water contents are not only detrimental for the developing strength levels but may also result in high bleeding and sedimentation rates. In this paper, it is first examined how the use of an ultra-sonic mixing procedure permits to limit the water content required to achieve a defined penetrability. Results relative to the fluidity and bleeding properties of the grouts are not reported here. It should, however, be mentioned that bleeding has been limited to maximum 4% after 120 minutes rest, whereas fluidity was practically constant for 120 minutes [2, 3]. The grouts remained stable through the same time period [4]. Second, the flow- (yield strength, plastic viscosity) and mechanical properties of the developed composite grouts are presented. Finally, the first results relative to the effect of sulfate attack originating from the presence of gypsum, which may be found in old masonries [5], are also discussed.

MATERIALS

Ordinary portland cement CEM I 42.5 LA HSR with low C_3A content, hydrated lime with a BET specific surface equal to 13.32 m^2/g, the natural pozzolan Rheinisch Trass and CSF in slurry were used.

The chemical analyses of the materials and the mineralogical analysis of the cement clinker are presented in Table 1. The choice of a sulfate resistant cement was imposed by the need to study sulfate resistance of the lime-pozzolan mix without interference of the cement. The Rheinisch Trass was always sieved at the 80 microns sieve before use. The mix proportions of the grouts are presented in Table 2. The choice of the lime to pozzolanic materials ratio is based on the estimation that, outside this range, either the lime or the pozzolanic content would be too high to allow for the optimisation of the pozzolanic reaction. In both cases, an important part of the material would remain unreacted and would therefore not contribute to strength through a hydration mechanism. To increase fluidity, a sulfonated naphthalene formaldehyde-based superplasticizer has been used.

Table 1 Chemical composition of the materials and mineralogical composition of the clinker

WEIGHT (%)	HYDRATED LIME	RHEIN. TRASS [< 80 µm]	CSF	CEM I 42.5 LA HSR
LOI	25.3	6.13	1.91	0.92
CaO	73.2	2.16		63.22
CaO free				0.66
SiO_2	0.40	67.56	96.02	21.37
Al_2O_3	0.15	16.02		3.61
Fe_2O_3	0.10	4.56		4.13
MgO	0.35	0.03		2.25
SO_3	0.02	0.35		2.80
P_2O_5				0.18
Na_2O		1.78		
K_2O		1.93		0.52
C_3S				54.15
C_2S				20.04
C_3A				2.59
C_4AF				12.57

Table 2 Mix proportions of the studied grouts

GROUT	COMPOSITION [%-wt]				LIME : (Trass+CSF)
	Lime	Rh.Trass	CSF	Cement	
13b-0	17.5	52.5	0.0	30.0	1:3
14b-0	14.0	56.0	0.0	30.0	1:4
15b-0	11.7	58.3	0.0	30.0	1:5
13b-5	17.5	47.5	5.0	30.0	1:3
14b-5	14.0	51.0	5.0	30.0	1:4
15b-5	11.7	53.3	5.0	30.0	1:5
13b-10	17.5	42.5	10.0	30.0	1:3
14b-10	14.0	46.0	10.0	30.0	1:4
15b-10	11.7	48.3	10.0	30.0	1:5

EXPERIMENTAL DETAILS

Mixing Procedures

In order to study the rheological properties of the grouts, two different mixing procedures have been applied and compared. The first involves mechanical mixing at 2400 revolutions per minute (high-turbulence mixing). All the materials were first mixed dry and then water and superplasticizer were added [6]. The second combines an ultrasonic dispersion at 28 kHz and a mechanical stirring at 300 rpm (ultrasonic mixing). In this procedure, the materials were introduced to the water in a sequential way: the fines were mixed first and then cement was added [1]. Further, all specimens used for the testing of mechanical and durability properties have been prepared using the ultrasonic mixing procedure.

Rheological Testing

The penetrability performance has been tested by means of the sand-column test (AFNOR P 18-891) [7] with minimal and maximal diameters of the sand used to fill the column 1 and 2 mm respectively. Voids with a diameter of 0.15-0.3 mm can thus be simulated [8] . Moreover, the grouts were injected under a 0.8-1.0 bar pressure through this column. Viscosity was measured with a Contraves Rheomat 108 E/R coaxial viscosimeter. Every 15 minutes, a small grout quantity was poured to the viscosimeter recipient and viscosity was measured. Before pouring, the mix (which was resting in a recipient) was gently remixed by hand for 15 seconds, so as to be homogeneous.

Mechanical Testing

The grout specimens were cured at 95% R.H. and 20°C. Due to their fluidity and slow strength increase, the specimens were allowed to stay in the 40x40x160mm moulds for 7 days before demoulding.

Durability Testing

In the absence of standards relative to the testing of materials destinated to the repair of existing masonry, the ASTM C 452-89 standard test method for potential expansion of portland cement mortars exposed to sulfate has been basically used and has been adapted as follows: it has been decided not to work on the pure grouts but on a mixture of sand and grout (thus, on a mortar), in order to better simulate the situation of a grout inside a wall. The preparation of the specimens was done as follows [9] : first, the grouts were prepared with the ultrasonic mixing as described above. Then, they were introduced in a Hobart mixer together with sand and gypsum, and were mixed for 3 minutes. The gypsum content was always 15%-weight of the solid part of the grout and was put as a sand replacement, so as to always maintain a binder (grout) to sand+gypsum ratio of 1:3 by weight. The dimensions of the specimens were 25x25x285 mm. Curing of the mortars was carried out in 95% R.H. and 20°C, and demoulding took place 2 days after preparation.

RESULTS AND DISCUSSION

Effect of the Mixing Procedure

Penetrability

The characteristic required for the comparison of the grouts is equal penetrability. If the compositions were not able to pass successfully through the sand column, the W/S and SP-content were modified accordingly, until a successful combination was found. The results are given in Table 3. The main differentiating factor was the CSF-content. As a matter of fact, grouts containing lime, natural pozzolan and cement are injectable with the same water and superplasticizer content for both mixing procedures. A difference in texture was nevertheless noticeable, since small flocculates were still existing after high-turbulence mixing. When silica fume was added to the grout, the high-turbulence mixing procedure was no longer able to make the grout injectable, unless a substantial change on both the water and the superplasticizer contents was made. On the other hand, when the ultrasonic mixing procedure was applied, the grout was still perfectly injectable without any adaptation of the water and the superplasticizer contents. In order to understand the reasons for this difference, a study of the influence of the mixing procedure on the viscosity of grouts prepared with the two different mixing procedures has been carried out [10].

Table 3 Penetrability values for W/S and SP

GROUT	HT MIXING		US MIXING	
	W/S	SP [%-wt]	W/S	SP [%-wt]
13b-0	0.85	1.2	0.85	1.2
14b-0	0.85	1.2	0.85	1.2
15b-0	0.85	1.2	0.85	1.2
13b-5	1.1	5.0	0.85	1.2
14b-5	1.0	5.0	0.85	1.2
15b-5	1.0	5.0	0.85	1.2
13b-10	1.1	5.0	0.85	1.2
14b-10	1.1	5.0	0.85	1.2
15b-10	1.1	5.0	0.85	1.2

Viscosity

The results related to the shear stress of grouts 13b-0 and 13b-10 are presented in Figures 2-3 at six moments: immediately after mixing (0 minutes), 15, 30, 60, 90 and 120 minutes after preparation. The shear stress has been calculated by means of the apparent viscosity results given by the viscosimeter. Further, the yield stress of each composition, estimated by extrapolation of the descending branch of the shear rate-shear stress curve near the start, is presented in Table 4. The r-square regression coefficients are all higher than 0.98. All the rheograms showed a time increasing yield stress and a thixotropic behaviour.

The behaviour of the grouts, both with and without CSF, can be considered to be Bingham type, at least during the period studied in this research.

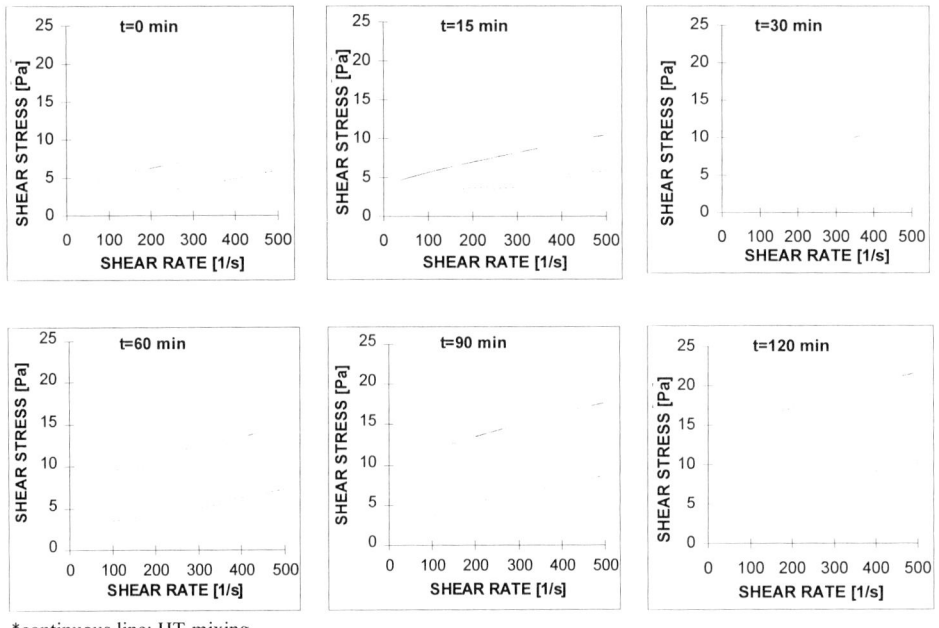

*continuous line: HT-mixing
dotted line: US-mixing

Figure 1 Effect of the mixing procedure on the flow curves of Grout 13b-0

Table 4 Effect of the mixing procedure on the yield stress [in Pa] of grouts

	GROUT 13b-0		GROUT 13b-10	
TIME[min]	HT	US	HT	US
0	3.74	0.84	9.20	6.30
15	4.41	0.99	10.05	6.95
30	5.46	1.14	10.85	7.73
60	8.00	1.58	12.52	9.08
90	10.23	2.27	14.78	10.56
120	13.41	3.01	17.00	12.70

Despite the fact that grout 13b-0 is injectable by both mixing procedures, the rheograms reveal that the rheological characteristics are very different: the shear stress was reduced by approximately 40-60% thanks to the ultrasonic mixing procedure. In addition, the calculated yield stress exhibited by the same grouts was more than 4 times lower thanks to the ultrasonic mixing. Grouts containing CSF have undergone a drastic modification of their rheological parameters, if compared to those without CSF. Thus, the yield stress is, as expected, increased and a pronounced development of a continuous structure is observed after the first 60 minutes, as is suggested by the area of the hysteresis loops of the flow curves (Figure 2). Hysteresis is much less pronounced in the absence of CSF (Figure 1). The effect of the ultrasonic mixing procedure was however more limited in this case. The decrease of the shear stress was between 20-30% and the decrease of the calculated yield stress approximately 25-30%. The action of the ultrasons was nevertheless sufficient to make the grout 13b-10 injectable. On the other hand, grout 13b-10 prepared with the high turbulence mixing procedure is not injectable any more. On the basis of these results, all final grout compositions were prepared with the use of the ultrasonic mixing procedure.

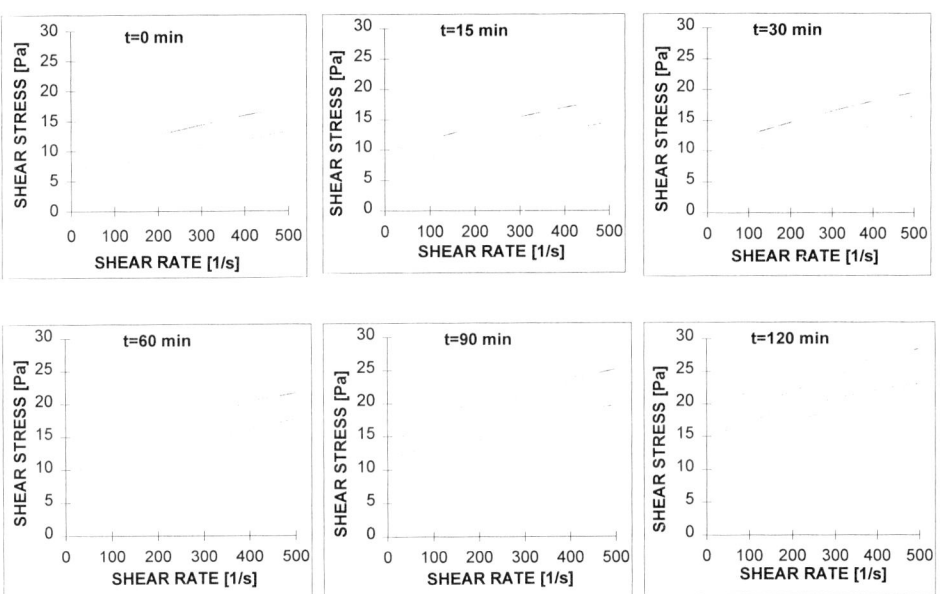

*continuous line: HT-mixing
 dotted line: US-mixing

Figure 2 Effect of the mixing procedure on the flow curves of Grout 13b-10

Effect of CSF and Lime-Pozzolan Ratio

Flow properties

The results relative to the flow behaviour of the grouts appear in Table 5. Yield stress and plastic viscosity increase with time. The rate of increase is more pronounced after 60 minutes, for grouts 14b-10 and 15b-10, and after 90 minutes, for grout 13b-10. It does not seem to change substantially for grouts without CSF. The very low yield stress of the latter, if compared to the yield stress of grouts with silica fume (which are also injectable), suggests that a further reduction in their water content is possible thanks to the ultrasonic mixing procedure. It may also be observed that the yield stress of grouts 14b-0 and 14b-10 is lower than the others' during the first thirty minutes after preparation. It is possible that the grain size distribution of this composition is optimal for flow. This is important, if one observes that the differences in the flow properties between the compositions containing CSF ranges between 18-20%, for the yield stress, and 8-14%, for plastic viscosity.

Table 5 Yield stress and plastic viscosity

Time	13b0	14b0	15b0	13b10	14b10	15b10
YIELD STRESS [Pa]						
0	0.84	0.61	0.82	6.3	5.47	6.77
15	0.99	0.82	1.02	6.95	6.13	7.68
30	1.14	0.82	0.8	7.73	6.76	8.6
60	1.58	1.78	1.47	9.08	8.25	10.53
90	2.27	2.43	2.47	10.56	10.54	12.81
120	3.01	3.03	3.14	12.7	12.96	15.32
PLASTIC VISCOSITY [mPa.s]						
0	10.2	10.7	9.6	14.2	13.4	14.6
15	10.1	10.2	10.1	15.5	13.7	14.9
30	10.5	11.2	10.7	16.1	14.1	15.3
60	11.4	11.5	10.7	17.4	15	17.5
90	12.8	12.7	12.6	18.8	17.2	19.7
120	14.2	13.9	14.5	21.2	19.6	22.3

Mechanical Properties

The results of the mechanical tests (Table 6) confirm the potentiality of multi-blend hydraulic binders. In 90 days, the compositions without CSF achieved a compressive strength superior to the stresses acting on usual masonry structures. The addition of CSF almost doubled the compressive strength. It is worth noting that the rate of strength increase of the first compositions remains constant with time, while it seems to slow down for grouts with 5% CSF and even stop for those with 10% CSF. In 180 days, grout 13b-0 has the same compressive strength as grouts 13b-5 and 14b-5, which contain 5%-wt CSF. A reduction in the compressive strength of grout 15b-10 has however been observed.

Flexural strength also increased with time. The lime to pozzolanic materials ratio does not seem to substantially affect the strength level at 90 days; indeed the measured differences are of the order of 10%. In all cases, grouts designed with a 1:3 ratio exhibited the highest values for flexural strength. The reduction in the flexural strength of the grouts without CSF measured after 180 days ranged from 10%, for grouts 13b-0 and 14b-0, to 18%, for grout 15b-0. A decrease of 3-4% has also been observed for grouts 13b-10 and 15b-5 and 10% for grout 14b-5. The mechanism of this decrease is not clear. The application of phenolphthaleine at the cross section of the tested specimens revealed that an exterior layer with a depth varying from 2-5 mm (increasing with age) was carbonated while the core was not. Thus, the pozzolanic reaction can be considered practically stopped at the surface layer, while it is still continuing at the interior. It can be assumed then, that chemical shrinkage is still taking place at the interior of the specimen. Moreover, the total chemical shrinkage of lime-fly ash mixtures is higher to portland cement [11]. If this is also valid for the lime-natural pozzolan reaction, then the stress gradients induced by the different behaviour between the inner (still hydrating) and outer (practically inert) part of our specimens could lead to an internal microcracking that might explain the observed drop. The fact that grouts containing CSF are generally less affected, could be due to the quicker development of a stronger skeleton (thanks to the presence of CSF), capable to withstand until a certain degree the developing stresses. Additional tests are now under execution, in order to check this hypothesis. In any case, more tests are required before any further interpretation is attempted.

Table 6 Mechanical properties of the composite grouts

Days	COMPRESSIVE STRENGTH [MPa]								
	13b-0	14b-0	15b-0	13b-5	13b-10	14b-5	14b-10	15b-5	15b-10
14	1.8	1.8	1.8	n.m.*	n.m.	n.m.	n.m.	n.m.	n.m.
28	3.2	3.3	3.5	4.3	6.5	4.4	5.4	4.1	5.6
60	6.9	4.5	5.5	8.2	11.9	6.8	8.7	7.3	8.2
90	7.8	6.5	7.3	12.9	15.9	11.5	16	12	17.2
180	13.1	10.9	11.2	13.6	16	13	16.2	15.8	16.1
Days	FLEXURAL STRENGTH [MPa]								
	13b-0	14b-0	15b-0	13b-5	13b-10	14b-5	14b-10	15b-5	15b-10
14	0.8	0.7	0.7	n.m.*	n.m.	n.m.	n.m.	n.m.	n.m.
28	1.5	1.3	1.2	1.2	1.7	1.2	1.5	1.2	1.2
60	2	1.5	1.9	1.6	2	1.3	2	1.2	1.8
90	2.5	2.2	2.3	3	3.4	3	2.6	2.8	2.7
180	2.3	2	1.9	3.2	3.3	2.7	2.9	2.7	3

* n.m. : not measured

Durability Properties Against Sulfate Attack Caused by Gypsum

The results of the expansion of the mortar bars containing gypsum as partial sand replacement and of the evolution of the corresponding dynamic modulus of elasticity (measured by means of ultrasons velocity) are given in Figures 3 and 4. The results indicate that the mortars made with non CSF grouts expanded continuously, reaching levels of 8-9 mm/m. This expansion seems to stabilise only after 120 days and is accompanied by a decrease of the dynamic modulus of elasticity only after that date. It is however possible that expansion might have stopped after 120 days, not because the sulfate attack has ceased but because the development of microcracks (as evidenced by the drop of the Edyn) has permitted to the reaction products (mainly ettringite) to accomodate themselves inside the micro-cracked matrix.

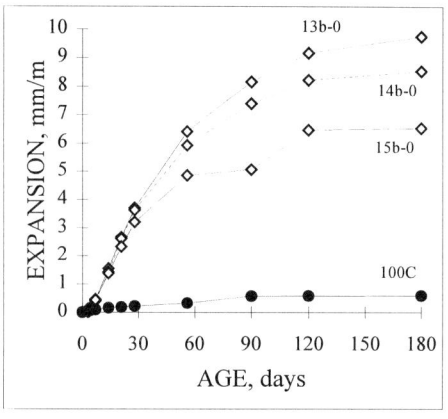 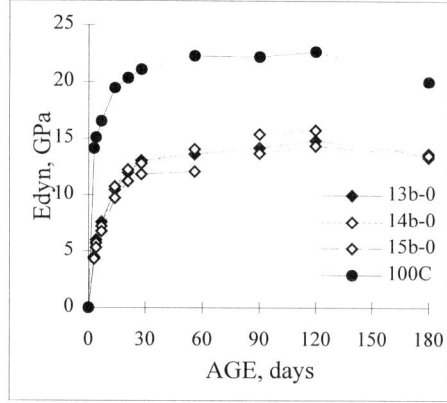

Figure 3 Expansion (left) and dynamic modulus of elasticity (right) of mortars without CSF

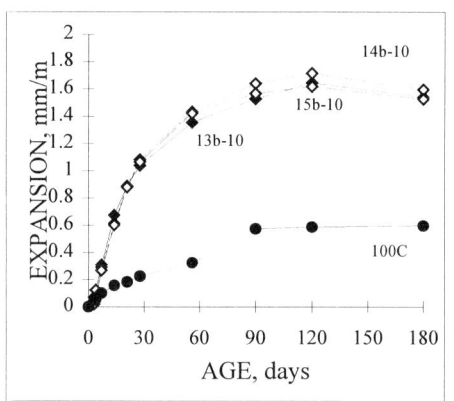 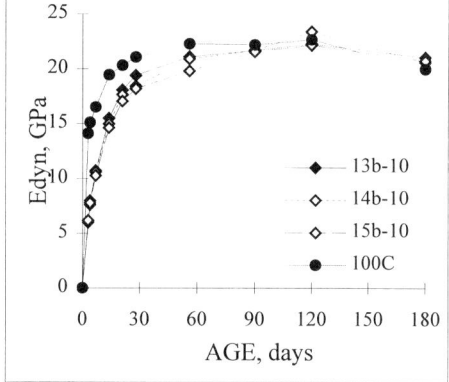

Figure 4 Expansion (left) and dynamic modulus of elasticity (right) of mortars with CSF

The drop in the dynamic Modulus of Elasticity has also been noticed on the sulfate-resistant cement based mortars, despite the very limited expansion of the latter (0.6 mm/m in 180 days). In spite of the very pronounced expansion measured, no cracks have appeared on the surfaces of the specimens before 60 days (for mortar 15b-0) and before 150 days (for mortar 14b-0). The decrease of the expansion of the first is attributed to this early cracking, which must have certainly permitted the accommodation of the reaction products inside the cracks. It is therefore of interest to stress out that expansions as high as 8 mm/m did not result neither in visible external cracking of the specimens (except for mortar 15b-0, as already mentionned) nor in decrease of the dynamic modulus of elasticity 120 days after sulfate attack. These somehow surprising results have already been encountered in the literature [12]. In this research it has been found that expansions up to 7 mm/m caused a small or even no decrease in the resonance frequency of the specimens (40x40x160 mm).

Mortars made with CSF-containing grouts exhibited a similar behaviour, reaching a maximal expansion of 1.7 mm/m after 180 days. After 120 days, a small shrinkage has been measured. The modulus of elasticity reached the value of the modulus of elasticity of the cement mortar after 90 days. A drop in the E-dyn of all mortars has also been measured after 120 days. Up to now, no visible cracks were observed. In any case, the expansion of these mortars, despite a gypsum content as high as 15%-wt, was very limited.

CONCLUSIONS

The research has highlighted a number of issues, which are of interest for an engineering approach to the problem of repair and strengthening of existing, historical, three-leaf masonry structures with the use of composite hydraulic binders.

1. The mixing procedure is of major importance in order to develop a high penetrability grout with a limited water content. In general terms, it seems that the use of ultrasons is a very efficient technique for the homogeneous dispersion of silica fume.

2. An optimal "lime to pozzolanic" materials ratio for all required performances of the grouts does not exist. The optimal flow characteristics were obtained with a ratio of 1:4, while higher strengths with a ratio of 1:3 were observed. The 1:5 ratio must be rejected, since it seems to introduce too high quantity of pozzolanic materials in the mix.

3. The results relative to the presence of silica fume are in agreement with available literature: On the one hand, silica fume enhances strength development and permits a better resistance to attack of sulfates originating from gypsum. On the other hand, the presence of CSF reduces penetrability and increases yield stress. The condition to overcome this difficulty without using a very high water content is the use of an ultrasonic dispersion technique.

4. During the study of multi-blend grouts, some problems have also emerged, namely the drop of flexural strength observed (for some of the grouts) between the 90th and 180th day of hydration, as well as the complex behaviour of the binders in the presence of gypsum. The available data in the bibliography are rather limited; more research is needed for the interpretation of these results.

The results obtained with the combination of lime, pozzolanic materials and OPC highlight the potentiality of multi-blend binders, which seem suitable for many applications.

Their advantages (sufficient compressive and flexural strength, provided a 90 days criterion is adopted, stability in suspension and reduced bleeding, good flow properties, low heat of hydration, physico-chemical compatibility with existing masonry), together with the use of low cost and locally available materials, call for a more extensive research of their properties. It is nevertheless necessary that the study of mixed type binders is followed for long periods of hydration and is combined with a study of the hydration mechanism and related microstructure.

ACKNOWLEDGMENTS

This research program has been made possible with the collaboration of Raf Augustijns, Astrid Van Lerberghe and Nele Louwagie for the penetrability tests, Noël Geets for the mechanical tests and Siska Beghin and Ive Thijs for the durability tests. Their contribution is gratefully acknowledged.

REFERENCES

1. TOUMBAKARI, E.-E, VAN GEMERT, D. Lime-pozzolana-cement injection grouts for the repair and strengthening of masonry structures, Proc. IVth Intern.Conf. on the Conservation of Monuments in the Mediterranean Basin, Rhodos, 1997, Vol.3, pp. 385-394.

2. AUGUSTIJNS, R. Development of a lime-pozzolan-cement grout with an ultrasonic mixing procedure for the repair and strengthening of historical masonry, Dipl.Thesis, De Nayer Inst.-K.U.Leuven, 1997, pp. 94 (in Dutch).

3. LOUWAGIE, N, VAN LERBERGHE, A. Research on the injectability of lime-pozzolan-cement grouts, Dipl.Thesis, K.I.H.O-K.U.Leuven, 1997, pp.186 (in Dutch).

4. TOUMBAKARI, E.-E. Measurement of the stability of suspensions with the use of laser granulometry, K.U.Leuven, unpubl. Report, March 1998.

5. COLLEPARDI, M. Degradation and restoration of masonry walls of historical buildings, Materials & Structures, vol. 23, 1990, pp. 81-102.

6. CHANDRA, S, VAN RICKSTAL, F, VAN GEMERT, D. Evaluation of cement grouts for consolidation injection of ancient masonry, Proc.Nordic Concrete Research Meeting, Goeteborg 1993, pp. 353-355.

7. PAILLERE, A M, RIZOULIERES, Y. Réparation des structures en béton par injection de polymères - Essais d'injectabilité à la colonne de sable, Bull.Liaison Labo. Ponts et Chaussées, 96, 1978, pp. 17-23.

8. DANTU, P. Etude mécanique d'un milieu pulvérulent formé de sphères égales de compacité maxima, Proc. 5th Int. Conf. Soil Mechanics, Paris, Dunod 1956.

9. BEGHIN, S, THIJS, I. Durability of injection grouts. Sulfate resistance and efflorescence tendency of lime-pozzolan-cement grouts for the restoration of historical masonry, Dipl.Thesis, De Nayer Inst.-K.U.Leuven, 1998, pp.195 (in Dutch).

10. TOUMBAKARI, E-E, VAN GEMERT, D, TASSIOS T P, TENOUTASSE, N. Effect of mixing procedure on injectability of cementitious grouts, Cement and Concrete Research (accepted for publication).

11. JUSTNES, H, ARDOULLIE, B, HENDRIX, E, SELLEVOLD, E J, VAN GEMERT, D. The Chemical Shrinkage of Pozzolanic Reaction Products, SP179-11, Int.Conf.on Fly Ash, Silica Fume, Slag and Natural Pozzolans in Concrete, 6th CANMET/ACI/JCI, Bangkok 1998, vol.1, pp. 191-205.

12. MEHR, S. Development of sulfate resistant mortars on the basis of cement-slag-pozzolan, Ph.D. Thesis, Technical Univ.Aachen, 1986, pp. 175 (in German).

THEME TWO:
SPRAYED CONCRETE

Keynote Paper

SPRAYED CONCRETE FOR ARTISTIC INTERIOR SHELLS

P Teichert

Laich SA

Switzerland

ABSTRACT. The 1,000 seat main auditorium of the Goetheanum in Dornach, Switzerland, centre of the worldwide Anthroposophical Society, was entirely rebuilt in 1997/98. The project included the installation of interior elements that reduce the auditorium's volume, decorate it artistically, and improve its acoustics. Essentially they comprise seven large columns along each of the side walls, whose capitals support an architrave 30 metres long and seven metres high. The surfaces of the column bases, the capitals and the architraves are sculptured. These elements are made of a load-bearing steel structure and sprayed concrete of special composition that was treated with hatchets after setting. The sprayed concrete had to meet very demanding requirements. The working conditions and the structural shapes, many of which are very complex, called for exceptional skill on the part of the nozzlemen.

Keywords: Interior elements, Thin shells, Dry-mix sprayed concrete, Structural frame, Formwork, Jobsite-produced dry mix, Glassfibre mesh, Properties, Surfacesculpting with hatchets.

Pietro Teichert is managing director of Laich SA in Avegno, Switzerland. He received his Civil Engineering Diploma in 1961 from Winterthur Polytechnic and has specialized in sprayed concrete work ever since.

THE GOETHEANUM, SWITZERLAND

The Geotheanum in Dornach, Switzerland, is the centre of the worldwide General Anthroposophical Society with its Free University of Spiritual Science (Figure 1). The building also serves as a conference and congress venue and houses several auditoriums for theatrical productions and concerts. It was erected in place of the original Goetheanum structure, which had been built of wood in 1913-1920 and burned down in 1922, in the years 1925-1928 - this time of reinforced concrete. Construction of the monumental, plastically formed structure in concrete was one of the outstanding pioneering achievements in architecture in this century. After the Goetheanum had been opened in 1928 essentially as an unfinished shell, the interiors were completed successively over the next seven decades. The plastically formed interior of the main auditorium, which reflects the highly original style of the building's exterior, is an artistic and structural masterpiece.

The Goetheanum's main auditorium (seating 1,000), originally used for 30 years in the rough, was finished for the first time in 1957. But grave shortcomings (asbestos ceiling, poor acoustics, unsatisfactory design) eventually necessitated its total refurbishment. Extremely high demands were placed on the artistic design, because the auditorium was supposed to convey an internally felt impression of the essence of Anthroposophy. This was attainable only by integrating sculpture, painting and architecture to achieve a holistic work of art in harmony with the artistic events taking place on the stage.

Christian Hitsch, head of the Goetheanum's Art Section and bearer of artistic responsibility for this project, developed a design for the auditorium's interior over a period of seven years (Figure 2). The plastically formed columns, capitals and architraves prompted a search for suitable materials and application methods. Trials with wood, based on the carved surfaces of the original Geotheanum, led nowhere. After a long series of preliminary tests, the decision was made to use reinforced drymix sprayed concrete and to sculpt the resulting layer.

Figure 1 The Goetheanum at Dornach

Figure 2 Model of the interior elements (scale 1:20, halved lengthwise)

New approaches had to be sought to meet the requirements staked out by the project itself, the architects and the artists. The limited load-bearing properties of the Goetheanum's foundations compelled the use of a very light concrete (no heavier than 1,400 kg/m^3) that was suitable for forming thin-walled (about 8 cm thick) shells with complex surfaces. The material had to meet the pre-established aesthetic demands and stand up to treatment with hatchets.

Figure 3 Fabrication of the round steel skeletons of welded reinforcing rods and wire netting

Figure 4 Skeleton section

Figure 5 Fastening the skeleton to the structural steel

The reinforcement had to reproduce the modelled surfaces perfectly while providing the necessary substrate for the sprayed concrete. The interior elements are supported by a framework of steel sections standing on the auditorium floor and anchored to the existing exterior walls. Fastened to this framework are basket-like relief structures of reinforcing rods and welded steel wire mesh called "skeletons of round steel". The skeletons for the most complicated parts of the walls, the 30 metre long and seven metre high architraves, were fabricated on the floor of a metalworking shop (Figure 3), cut into segments (Figure 4), and then reassembled in the auditorium (Figure 5). The remaining skeleton parts were assembled on the spot. To provide a substrate for the sprayed concrete, "lost formwork" of expanded metal was fastened behind the skeletons in such a way that the reinforcing bars and wires would be embedded as thoroughly as possible in the concrete (Figure 6).

Figure 6 Skeleton with expanded metal fastened behind as "lost formwork"

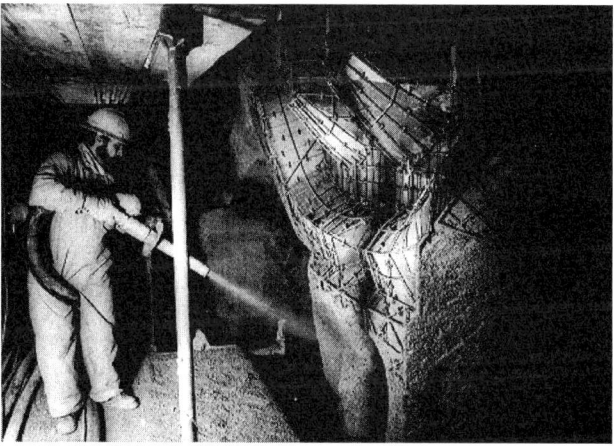

Figure 7 Spraying the concrete

The following dry-mix composition satisfied all of the requirements established for the mix and the hardened sprayed concrete in the Goetheanum:

- pumice aggregate, particle size 0-8 mm expanded clay
- "Leca" 0-3 mm
- white marble sand
- 9 parts white cement
- 1 part hydraulic lime
- red mineral colorant (iron oxide)

Because this mix was not available commercially, it had to be mixed on the jobsite. To achieve uniform colouring, the binder was premixed with the pigment at the factory, packed and delivered to the site. The conditions at the site made it necessary to set up the mixing equipment and spraying gun outdoors next to the building. The mix had to be conveyable over a horizontal distance of 120 metres and upward about 30 metres to reach the nozzle.

As requested by the finishing artisans, the sprayed concrete shell was applied in a layer about ten centimetres thick (Figure 7), or somewhat thicker at ticklish locations to give the sculptors more leeway. For efficient distribution of shrinkage strains, it was reinforced with a glassfibre mesh (Figure 8). The blue mesh, which was covered with about four centimetres of concrete, served as a useful signal later on for the hatchet wielders.

Figure 8 Application of the glass-fibre mesh

The tests carried out for quality control verified the following sprayed concrete properties after 28 days (n = number of tests):

- Dry density (n = 3)   1, 183 kg/m³
- Total porosity (n = 2)   47.2 percent by volume
- Compressive strength (n = 6)   23.8 N/mm²
- Tensile strength, axial (n = 9) 1.0 N/mm²
- E-module (n = 3)   9, 133 N/mm²

This monumental, plastic architecture made unique and extreme demands on the sculptors treating the surface with their hatchets (Figures 9 and 10). The sprayed concrete surface totalling about 2,200 m2 was worked on by an average of 35 sculptors from all over the world for a period of about 6 months.

Figure 9 Sculpting of the sprayed concrete with hatchets

Even though the composition of the dry mix was excellent, the spraying of the concrete for the interior elements in the Goetheanum's main auditorium called for long experience and great skill, care and circumspection on the part of the nozzleman. He also had to cooperate very closely with the sculptors to apply the concrete layer precisely as they wished. Thanks to the fine cooperative spirit of the entire crew, the resulting sprayed concrete shells met the expectations of the artists and sculptors unqualifiedly (Figures 11 and 12).

Figure 10 Surface before (left) and after (right) the hatchet treatment

Figure 11 The totally refurbished main auditorium of the Goetheanum

Artistic Sprayed Concrete 101

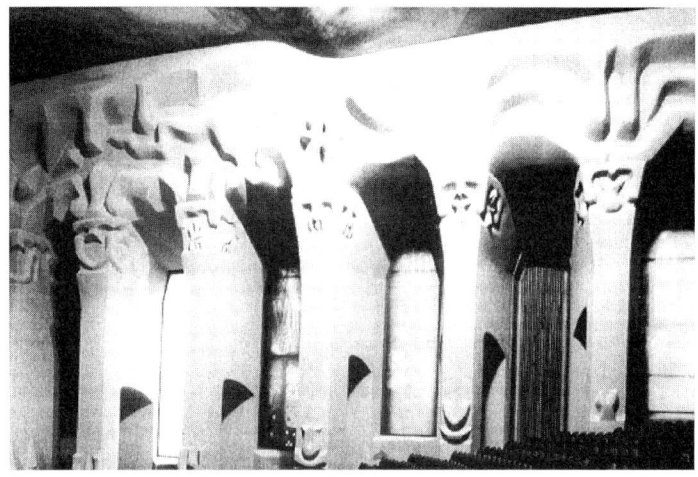

Figure 12 Partial view of an auditoriam wall with columns, capitals and architrave

PICTURE ACKNOWLEDGMENTS

Figures 1, 2, 5, 6, 8, 10 - Laich SA, Avegno
Figures 3 - Peter A. Wolf, Essen
Figures 4, 7, 9, 11, 12 - Jurg Buess, Basel

REFERENCES

1. Der grosse Saal im Goetheanum, Berichte und Bilder der Modelle zum bevorstehenden Ausbau, Stil 4/1994, Sonderheft Michaeli 1994, Buchhandlung am Goetheanum, CH-4143 Dornach 1

2. Zum neuen Ausbau des grosser Saales im Zweiten Goetheanum, Berichte zum Saalausbau, Stil 2/1997, Buchhandlung am Goetheanum, CH-4143 Dornach 1

3. HASLER, H. Spritzbeton als kunstlerisch gestaltete Innenschale. Schweizer Ingenieur und Architekt 10/1997.

4. HASLER, H. Die Neugestaltung des grosser Saales im Goetheanum. Schweizer Baujournal 2/1998.

5. HASLER, H, UND BUESS, J. Der grosse Saal im Goetheanum 1996 - 1998. Herausgegeben von der administration des Goetheanum-Baues, April 1998.

6. HASLER, H, TEICHERT, P, AND MORGAN, D.R. Artistic Shotcrete for a Historic Auditorium. Concrete International, V. 21, No. 3, March 1999

7. HERMANN, K. Gestalten mit Spritzbeton. Cement- Bulletin, Marz 1999, TFB, CH-5103 Wildegg

HYGROTHERMAL STRESS INDUCED PROBLEMS IN LARGE SCALE SPRAYED CONCRETE STRUCTURES

O Hrstka

R Černý

Czech Technical University of Prague

P Rovnaníkovà

Technical University of Brno

Czech Republic

ABSTRACT. A theoretical model of heat, moisture and momentum transport in hydrating concrete is formulated using the methods of continuum physics. Numerical solution of the model is performed, and a computer code in C++ is written. In the practical application of the computational model, temperature, moisture and stress fields in a gas-tight concrete seal of an underground natural gas reservoir built by sprayed concrete technology are calculated. In order to achieve a sufficient accuracy of computed data, the time development of hydration heat of the concrete mixture as the most important parameter of the model is measured using an adiabatic method. The computational results show that in certain time interval, an adiabatic region with a typical dimension of 1 m appears in the central part of the structure. It cannot be affected from outside the structure, and its temperature depends on the hydration heat production only. Contrary to the results obtained with accounting for thermal stress only, hygrothermal stress calculations give an evidence of tensile stress appearance near to the boundary of the structure which can be considered as very dangerous.

Keywords: Hydration heat, Sprayed concrete, Hygrothermal stress

Ing O Hrstka is an Assistant Professor at the Computing and Information Center, Faculty of Civil Engineering, Czech Technical University of Prague, specialist on C++ programming in Linux environment. He is focused on the development of software for numerical simulations of transport phenomena.

Professor R Černý is a Professor at the Department of Structural Mechanics, FCE CTU Prague. He works in the field of mathematical modelling of heat and mass transfer in building materials, he is also active in the development of measuring methods for determination of thermal and hygric properties of building materials.

Dr P Rovnaníkovà is an Associate Professor of Chemistry, Head of the Institute of Chemistry, Faculty of Civil Engineering, Technical University of Brno. Her main research interests are in the field of measuring and analyzing the hydration processes of cement in concrete

INTRODUCTION

Monolithic large-scale concrete structures (i.e. big concrete blocks, in most cases) always suffer problems caused by the hydration heat production because of their limited possibility of heat removal into the surroundings. Temperature increase induces thermal stress, which can lead to the crack appearance, lower final quaJity of concrete, and consequently Wso to worse material properties and worse structure performance. If sprayed concrete technology is applied, which includes application of rapid cementing agents mostly, these effects may be even more serious, because concrete gains material properties, which are close to their "final" values, faster. Practically the only usual way how to deal with the negative consequences of the hydration heat production is slower progress of building these structures. The question, what is already "slow enough", has to be answered case by case in a relation to the properties of concrete mixture applied. We may say, that the most effective method to learn, how long should be the breaks between concreting cycles, is the computer simulation. In this way, different possibilities of geometric parameters of building structures and applied technology can be easily and cheaply modelled and temperature and hygrothermal stress fields can be obtained. Obviously, each computer model needs to get input values which are sufficiently accurate, and accuracy of computer simulation results cannot be better than the accuracy of input values. In this case, material properties of concrete and rock mass, their time development, and as the most important, the time development of hydration heat, are needed.

In this paper, computational modelling of temperature, moisture and stress fields in a case study, which is a gas-tight concrete seal of an underground natural gas reservoir built using the sprayed concrete technology (geometrically a cylinder 3 m long, with a radius of 3.15 m), is performed. The model is formulated using the methods of continuum physics, and it accounts for the heat, moisture and momentum transport in the hydrating concrete. The numerical solution of the model is done by the finite element method, then the computer implementation is performed and a computer code in C++ is written. As for the temperature fields, we had some feedback in the form of measured data on the mentioned concrete seal structure, which was built in Czech Republic in the same time. Therefore, these results can be considered as reasonably precise and reliable. On the other hand, our calculations of hygrothermal stress do not achieve such a good accuracy because we did not have sufficiently precise values of material properties; this modelling should be taken as a kind of demonstration.

MEASURING OF HYDRATION HEAT

Time development of hydration heat was measured using an adiabatic calorimeter. In this way, we can observe temperature development during the hydration process in a concrete mixture. A polystyrene reaction vessel that is heated electrically forms the measuring unit. A sensor located in the vessel cover, while a sensor placed on a metallic holder detects the temperature in the calorimeter walls detects temperature changes in the concrete mixture in the vessel. A regulating unit keeps zero temperature gradient between the mixture and the air inside the calorimeter. This means that thermal loss from the reaction vessel is approximately equal to zero. In the beginning of the experiment, the measuring system is set to the initial temperature of the components of the mixture. Then, water of the same temperature as the mixture is added quickly, and the mixture is put into the reaction vessel which is closed by a thermally insulating cover. Using the adiabatic system, conditions that appear in central parts of a big concrete block can be simulated. This method of measuring also provides a possibility to observe influence of different additives on the development of hydration processes. The measured integral curve of hydration heat production in the real concrete mixture from our case study is presented in Figure 1.

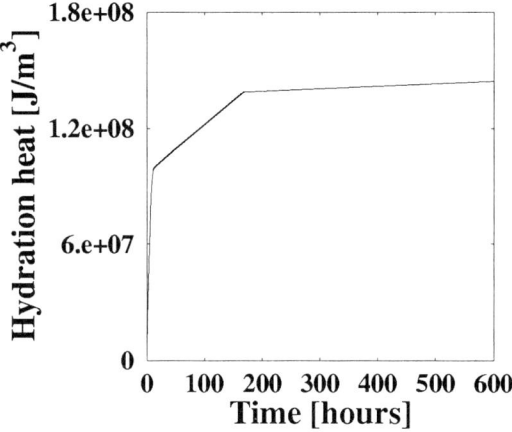

Figure 1 Time development of the hydration heat of the concrete mixture

MATHEMATICAL MODELING

Temperature fields

The heat conduction problem under consideration posseses the following main features:

1. Internal heat source

2. Heat removal to the surrounding environment

3. Special geometry of the object.

Heat conduction equation in a 3D-case has the following form:

$$\rho c \frac{\partial T}{\partial t} = \frac{\partial}{\partial x}\left(\lambda \frac{\partial T}{\partial x}\right) + \frac{\partial}{\partial y}\left(\lambda \frac{\partial T}{\partial y}\right) + \frac{\partial}{\partial z}\left(\lambda \frac{\partial T}{\partial z}\right) + I(x, y, z, t), \qquad (1)$$

where ρ is the density [kg/m^{-3}], c the specific heat [J/kg^{-1} . K^{-1}], λ the thermal conductivity [W/m^{-1} · K^{-1}], T the temperature [K], t the time [s], x,y,z the space coordinates [m]. The term $I(x, y, z, t)$ represents the internal heat source, which can depend on both space co-ordinates and time, in general. Boundary condition can be characterised as follows:

- Newton conditions on the contact of the concrete structure with the air
- ideal thermal contact on the concrete-rock interface
- thermal insulation conditions (zero temperature gradient) on the external part of the boundary (rock massive).

In the practical computations, we used Eq. (1) in the following matrix form (see [1-3] for details):

$$[HP]\{T\}_1 = [PS]\{T\}_0 + \{A\}.$$

Moisture fields

Due to the special conditions appearing in our case study, we could assume the following:

- The only factor affecting the moisture loss in the concrete block is the hydration process.
- The local change of the moisture content due to its loss induced by the hydration process is much faster than the moisture conduction.
- There is no moisture transfer from concrete to the rock massive, because cracks in the rock massive are already full of water being under relatively high pressure of about 1-2 MPa.
- There is no moisture transport from the concrete block to the air in the tunnel the air relative humidity is about 90%.

Therefore, the moisture transfer equation can be simplified to the following form:

$$\frac{\partial w}{\partial t} = J(x,y,z,t), \qquad (2)$$

where w is the moisture content, $J(x, y, z, t)$ is the moisture sink term fitted on the basis of the differential function of the hydration heat production $I(x, y, z, t)$ in the way, that the initial value of the moisture content is determined from the known water/cement ratio of the concrete mixture, and the final value is equal to the maximum hygroscopic moisture content.

Hygrothermal stress fields

As a consequence of the hydration heat production, hygrothermal stress is induced in the concrete structure, as well as in the rock massive which surrounds it. In the mathematical modelling of this process, we employed the simplified quasi-steady version of the momentum equation in the form (see, e.g., [4-7])

$$(\Lambda + \mu)\,\text{grad div}\,\mathbf{u} + \mu\nabla^2\mathbf{u} - (3\Lambda + 2\mu)\,\alpha_T\,\text{grad}\,T - (3\Lambda + 2\mu)\,\alpha_w\,\text{grad}\,w = 0, \qquad (3)$$

where u stands for the displacement vector, α_T is the linear thermal expansion coefficient, α_w is the linear moisture expansion coefficient, material properties Λ and μ are Lame's constants, which are in the following relations to the E a v (Young's modulus and Poisson number):

$$\Lambda = \frac{E\nu}{(1+\nu)(1-2\nu)}, \qquad (4)$$

$$\mu = \frac{E}{2(1+\nu)}. \qquad (5)$$

Matrix form of Equation (3) applied in the computer modelling, is the following:

$$[M_{rr}]\{u_r\}^e + [M_{rz}]\{u_z\}^e + [M_{rT}]\{T\}^e = 0, \qquad (6)$$

$$[M_{zr}]\{u_r\}^e + [M_{zz}]\{u_z\}^e + [M_{zT}]\{T\}^e = 0. \qquad (7)$$

There are no time derivatives in these two equations, translation relates only to momentary values of temperature, not to its time development. However, the temperature varies with time, and therefore this problem can be considered not as simply a steady one, but as a quasi-steady. Consequently, the practical computations need not to be done in every time step, in which we calculate temperature and moisture fields, but only in selected times that are interesting for a designer.

In calculating stress fields from the displacement values determined on the basis of Equation (3), we employed the general Hooke's theorem, which defines the relation between the stress tensor σ and the deformation tensor ε:

$$\sigma_{ik} = \left[\Lambda \sum_m \epsilon_{mm} - \alpha_T(3\Lambda + 2\mu)(T - T_0) - \alpha_w(3\Lambda + 2\mu)(w - w_0)\right]\delta_{ik} + 2\mu\epsilon_{ik}. \qquad (8)$$

COMPUTATIONAL RESULTS

Results of our computations can be summarised as follows:

- Temperature grows to its maximum value at approximately 6 days after the end of the building process. This is a consequence of the fact that hydration heat production continues long time after finishing the spraying and because of the big dimensions of the structure.
- The maximum temperature values in the adiabatic zone (more than 80°C) appear on the longitudinal axis of the structure near to the middle of its length, but not exactly in the middle.
- If the radius of the cylindrical structure is 3.15 m, the temperature is constant until the distance of about 2.0 m from the longitudinal axis. It leads us to a conclusion, that if radius grows, the maximum temperature will not grow with it any more.
- In the longitudinal direction the results are very similar, only the temperature is not constant in such a big space region. This is a consequence of building the structure layer by layer with relatively long technological breaks of 150 minutes.
- The hygrothermal stresses have similar values and similar shapes in all directions. In this case, all directional dilatation of the material competes with the material resistance both in radial and tangential directions.
- If we take into account only thermal stress (i.e., the length changes due to the thermal expansion only), the maximum stress near to the centre of the structure grows over 20-30 MPa, which is much higher than is the plasticity limit in that stage of concrete material properties. Therefore, it seems, that in this region the material begins to plastify much sooner than stress can grow to these values, and higher stress values must be shifted further from the centre of the structure. The stress might be even merely constant in the whole object.

- If also the influence of moisture induced shrinkage during the hydration process is accounted for, the stress fields become completely different. The tensile stress appears near to the boundary of the structure, where the influence of moisture loss is more significant than that of temperature increase.

This summary can be illustrated by several figures. Figure 2 shows an example of calculated results of the time development of the maximum temperature value. Figures 3,4 show examples of calculated results of the σ_r, without (Figure 3) and with (Figure 4) the influence of the moisture induced shrinkage in the $z = 1.5$ m cross section, i.e. in the middle of the cylinder, perpendicular to its longitudinal axis.

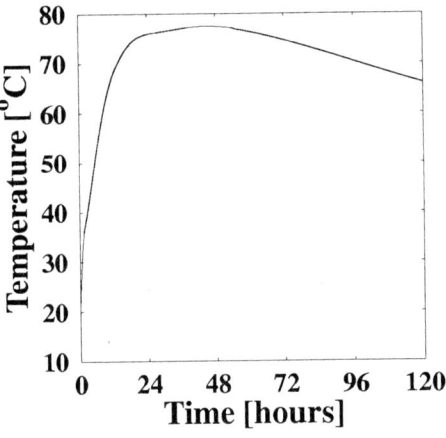

Figure 2 The time development of the maximum temperature value

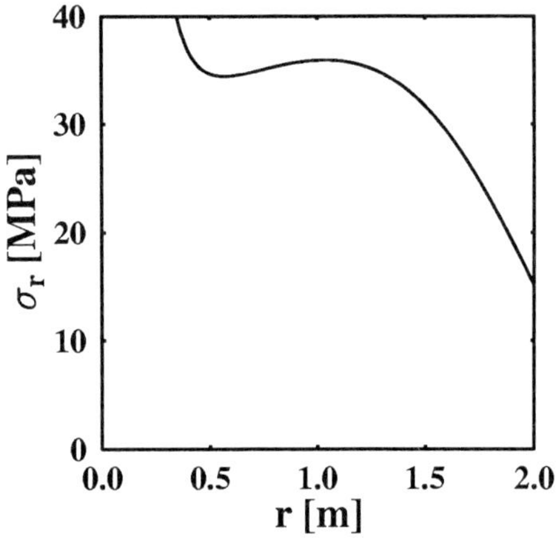

Figure 3 σ_r in the cross section $Z = 1.5$m induced by thermal expansion only

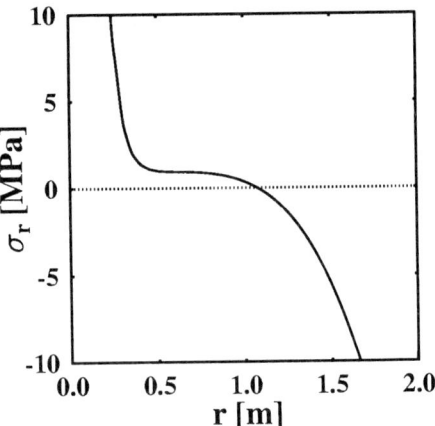

Figure 4 σ_r in the cross section $z = 1.5$ m, induced by both thermal expansion and moisture shrinkage

CONCLUSIONS

The most important result of our computations is the fact that the maximum temperature and hygrothermal stress values and also the time development of these parameters are determined by two main factors: the time development of hydration heat and the dimensions of the concrete block. If the dimensions of the concrete block exceed certain limit (in our case it was about 1 m of the length or the radius), in the central part of the structure an adiabatic region appears. This means, that in certain time interval which is characteristic for the process, there is no heat exchange between this region and the surroundings of the structure. In the adiabatic region, the temperature and also hygrothermal stress fields are not affected by the boundary conditions, and therefore they cannot be influenced from outside the structure for instance by such well-known methods, as is the water spraying. In this region, processes are practically the same as in an adiabatic calorimeter during laboratory experiments, as described in Section 2. Comparison of the stress fields in the two basic calculated cases, namely if the influence of moisture variations during the hydration process is or is not taken into account, shows very significant differences. Contrary to the results obtained with accounting for thermal stress only, hygrothermal stress calculations give an evidence of tensile stress near to the boundary of the structure that can be considered as very dangerous. Therefore, it leads us to a conclusion that taking into account the moisture-induced shrinkage is very significant in the practical calculations.

ACKNOWLEDGEMENTS

This research has been supported by the Grant Agency of the Czech Republic, under grants No. 103/97/0094 and 103/97/K003.

REFERENCES

1. ZIENKIEWICZ, O C, The Finite Element Method in Engineering Science. MeGrawHill, London 1971.
2. ČERNÝ, R, Physics - Transport Phenomena. CTU Press, Prague 1993 (in Czech).
3. ČERNÝ, R, Computational Solution of Transport Phenomena. CTU Press, Prague 1998 (in Czech).
4. VODÁK, F, Continuum Thermomechanics. CTU Press, Prague 1992 (in Czech).
5. WROBEL, M A, Heat and mass flows coupled with stress in a continuous medium. Int. J. Heat Mass Transfer, Vol. 40, 1997, P. 191.
6. NOWACKI W, Thermoelasticity. Ossolineum, Warsaw 1972.
7. SLATTERY, J C, Momentum, Energy and MassTransfer in Continua. R.E. Krieger, Florida 1981.

WET-MIXED SPRAY CONCRETE: IMPROVEMENT VIA GGBS

M F Nuruddin

Institut Teknologi Mara

S A M Yunus

Corroless

A B M Diah

Concrete Society of Malaysia

Malaysia

ABSTRACT. Sprayed concrete for repair works is an excellent placement method for vertical and overhead surfaces. Where formwork is impractical, sprayed concrete repair for large, shallow areas is cost effective.

This paper discusses the performance of shotcrete incorporating ground granulated blast furnace slag (GGBS) as partial cement replacement material compared to normal OPC concrete. Initial surface absorption (ISA), chloride ingress, bonding, and compressive strength tests are employed on samples comprising 0, 30, 40, 50, 60, and 70% replacement of OPC by GGBS. 28-day samples are used in the ISA and bonding strength tests whilst 60-day samples for chloride ingress test. The strength development is monitored for ages 1, 3, 7, 14, 28, and 60 days.

Consequently, mix proportions containing GGBS show improvement in performance with 40% level demonstrating conspicuous positive influence in most of the tests undertaken.

Keywords: Sprayed concrete, Shotcrete, GGBS, Partial cement replacement material, Chloride ingress, Bonding strength, Mix proportion.

Ir Muhd Fadhil Nuruddin is a Senior Lecturer at Institut Teknologi Mara and a registered Professional Engineer with the Board of Engineers Malaysia. His research interests include dignosis of structures/buildings, repair techniques, and durability of concrete. He is also actively involved in the Concrete Society of Malaysia as Council Member in promoting the advancement of concrete technology.

Mr Shaifful Anuar Mohd Yunus is the Assistant General Manager of Corroless Malaysia, a renowned contractor involved with various rehabilitation projects in Malaysia. He is also an active Council Member of the Concrete Society of Malaysia.

Dr Abu Bakar Mohamad Diah is The Secretary of The Concrete Society of Malaysia and has published numerous papers on concrete durability at national and international levels. Amongst his research interests are durable concrete, cement replacement material, and recycled aggregates.

INTRODUCTION

Sprayed concrete, previously often referred to as "gunite", has been a pedigree construction material for sixty years and has been in use for many decades, yet still its capabilities and advantages have not been fully exploited. In general, shotcrete is defined as pneumatically applied concrete or mortar placed directly onto a surface. The shotcrete shall be composed of water, cementitious materials, sand, coarse aggregate, steel fibres (if specified), and admixtures, and shall be placed by either the dry-mix or wet-mix process as specified.

WET-MIXED SPRAYING

Wet process shotcrete by definition involves the pumping of plastic, ready mixed concrete, containing coarse aggregate of maximum size 10mm or greater, to its point of application, where it is finally blown into place by compressed air. Concrete with a slump of 25-50 mm can be sprayed in this way, but the low slump restricts output and pumping distance.

As a consequence, developments over the past decade have concentrated on controlled acceleration of the setting time to enable high slump concrete (100-150mm) to be pumped at a high rate (5-15 m^3/ hour), which is then blown into place by compressed air carrying a metered amount of powder accelerator. As the air blast mixes the powder with the concrete in the nozzle and propels the wet mix into place, the concrete stiffens immediately after impact. The rapid set reduces dust and rebound (5-15%) whether shooting on to vertical surfaces or overhead.

With this process it is possible to have water/cement ratio as low as 0.4, because plasticizers can be used. As the material is being pumped, specifications should be generally as those required for a pumpable mix. However, it is usual to have high cement content between 350 and 450kg/m^3, which gives 28 days cube strengths of between 35 and 55N/mm^2.

Wet sprayed concrete is a structural material and is applied in layers, sometimes built up over several days if necessary. Bonding between layers is excellent if the underlayer is left as sprayed. Trowelling should be carried out with caution, as anything other than a very light application could cause surface cracking and density loss. It is possible, however, to achieve tolerances of +/- 10mm over 3m lengths of plain flat surface. More complicated surfaces require special provision. The wet process has higher outputs:- up to 10m^3/hour and is therefore more suitable for larger volume jobs.

The water content used in wet-mix shotcrete is more precisely controlled. The pressure and volume of material shot in the wet-mix method is more easily regulated and the nozzleman can concentrates more on his job spraying. Using the wet-mix method the admixtures and mixing water are thoroughly mixed with the other ingredients. Traditionally, wet-mixed shotcrete has been used more for high-production shotcreting applications than for repair. But now there are wet-mixed shotcrete machines designed specifically for repair.

Admixtures can be introduced into the premix to improve the workability or modify the setting properties of concrete. A rapid setting admixture is commonly used to facilitate build-up of thick layers of material and this is usually introduced as an additional spray at the discharge nozzle; otherwise early set might take place within the delivery hoses.

Wet-mixed Spraying Repair Techniques

For production spraying, the nozzle is normally positioned at a distance of 600 to 1500 mm from the surface, although larger distances are feasible depending on strength and production requirements. As a rule of thumb the nozzle is held normal to the surface being sprayed, but when shooting through reinforcement, the nozzle should be held close to the work (down to 450 mm) and at a slight angle to permit encasement of the steel [1].

For vertical surfaces it is normal to commence work at the bottom to avoid trapping rebound. In underground work however, depressions in the rock surface in any particular area are generally filled first. It is also noteworthy that no sprayed concrete should be placed on any overhanging surface, other than by remotely controlled nozzle, until the surface has been inspected and declared safe.

Primarily concrete slump and mix design, steel reinforcement and location of surface with respect to the nozzle govern layer thickness. With accelerated mixes, layer thickness of up to 300 mm can be built in one operation but without an accelerator 50 mm is generally the maximum layer thickness that can be safely applied, particularly overhead.

In general, rebound is minimal amounting to an average of 5 to 10% giving a greater yield of sprayed concrete in place and a considerable saving in time for cleaning up and rebound disposal. In addition, the working environment for the nozzle operator is greatly improved with abatement of dust.

Ground Granulated Blast Furnace Slag

GGBS is a waste material which is a by-product of burning of iron. Its oxide contents are essential in producing the pozzolanic reaction which subsequently contribute to the strength and durability of the concrete. It was also found that GGBS increases the resistance to chloride ingress [2].

CHLORIDE DIFFUSION

The main process by which chlorides enter concrete is by diffusion into capillary pores under the action of concentration gradient. These must be water-filled or the chlorides are held up while moisture also diffuses into the pore. Since moisture diffusion is relatively slower, progress of the chloride ion into 'dry' concrete is slower than into saturated concrete.

There is a maximum rate at which chlorides can enter concrete for a given set of environmental and concrete criteria, as defined by steady-state conditions. It is not clear whether such conditions are reached with concrete in-situ but since this represent the worst position then it is reasonable to use the steady-state conditions for diffusion to estimate the relative resistance of concrete.

The physics and mathematics of steady-state diffusion have been well known since the last century and Fick's 1st Law describing the flow of heat energy through metals is still used today to model the movements of chlorides (moisture) through concrete.

Subsequently, based on Fick's 2nd Law, the following equation has been proposed to relate the depth of Cl⁻ penetration, x to time t (after simplification):

$$x = 4\sqrt{Dc * t}$$

where; x = depth of Cl⁻ penetration (m) obtained using the AgNO₃ droplet method
Dc = apparent Cl⁻ diffusion coefficient (m^2/week)
t = exposure period (week) and

Since x is inversely proportional to the ability to resist Cl⁻ diffusion, it becomes a useful parameter to be considered in the comparison of diffusion capacity and thus resistance to Cl⁻ ingress.

EXPERIMENTAL PROGRAMME

The wet-mixed spray process which is being analyzed in this research project consists of thoroughly mixing all ingredients; feeding the mixture into the delivery equipment; delivering the mixture by positive displacement or compressed air to the nozzle; and then jetting the mixture from the nozzle at high velocity onto the surface to receive the shotcrete.

A control mix of grade 25 was designed with a slump requirement of 75 to 100 mm and the water/cement ratio of 0.45. The cement content used was 545 kg/m³. Table 1 shows the oxide composition and physical properties of OPC and GGBS used. This experiment was carried out on concrete samples incorporating GGBS at 0, 30, 40, 50, 60, and 70 % replacement of OPC. Table 2 shows the experimental detail that was carried out.

Table 1 Oxide Compositions and physical properties OPC and GGBS used

PROPERTIES	OPC	GGBS
Physical Properties		
Surface area (m^2/kg)	400	425
Bulk density (kg/m³)	1,400	1,200
Specific gravity	3.2	2.9
Colour	Grey	Off white
Oxide Compositions		
CaO	65	40
SiO_2	20	37
Al_2O_3	6	11
Fe_2O_3	2.5	0.2
MgO	1.5	8
SO_3	2	0.2
K_2O	0.8	0.7
Na_2O	0.5	0.4

Table 2 Experimental Detail On 0, 30, 40, 50, 60, and 70% Inclusions of GGBS

FEATURES	STRENGTH N/mm²	ISAT ml/m²/sec	CHLORIDE TEST m²/sec	BOND TEST N/mm²
Sample Cube	100mm	150mm	100mm	150mm
Test Age, days	1,3,7,14,28,60	28	60	28
Number of tests	108	6	6	6
Curing	Water	Water	Water (1- 7days) Air (7-28 days) Chloride sol (28-60 days)	Water
Standard / Methods used	BS1881 Pt 116:1983	BS1881 Pt 122: 1970	0.1NagNO₃ indicator	BS1881 Pt 201: 1986

RESULTS ON COMPRESSIVE STRENGTH

By and large the GGBS concretes follow the same trend as in OPC concrete. At the early age i.e. before 28 days, all the GGBS concretes give lower strength values compared to the OPC concrete. Refer Figure 1. The reason for this phenomenon is due to the delayed reactivity of GGBS as the OPC has not sufficiently hydrated to activate the GGBS.

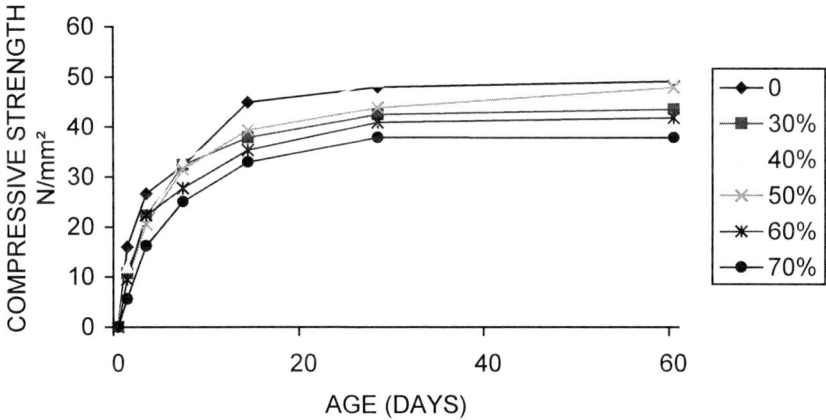

Figure 1 Compressive strength of various GGBS contents against time

It can be concluded that 40% is the optimum replacement level because replacement in excess of 40% showed that the strength did not exceed the OPC control. Nevertheless from the strength development trend it is observed that after 60 days, the strength development of GGBS concrete might surpass their control counterparts. This is especially true for 50% level where at 60 days it fell short of 2.4% from the OPC value.

At 28 days, the 40% replacement concrete surpassed the OPC concrete by 2.6% and at 60 days by 1.5%. Table 3 shows the percentage strength gain of GGBS concrete as compared to OPC control.

Table 3 Compressive strength of GGBS concrete as a % of OPC control at different ages

GGBS (%)	PERCENTAGE STRENGTH WRT OPC AT DIFFERENT AGES					
	1 days	3 days	7 days	14 days	28 days	60 days
30	66.9	83.5	99.7	84.2	88.5	88.6
40	75.0	81.4	99.8	91.1	102.6	101.5
50	59.4	77.3	97.8	87.5	91.4	97.6
60	59.1	83.8	86.0	78.6	85.3	85.1
70	34.7	60.9	77.6	73.3	78.9	77.0

It is also observed that the higher the GGBS replacement level, the lower the rate of strength development especially within the early ages. Figure 2 shows the 28 day cube strength of the various GGBS concretes and it is evident that that 40% GGBS content shows the highest strength i.e. 49.2N/mm^2.

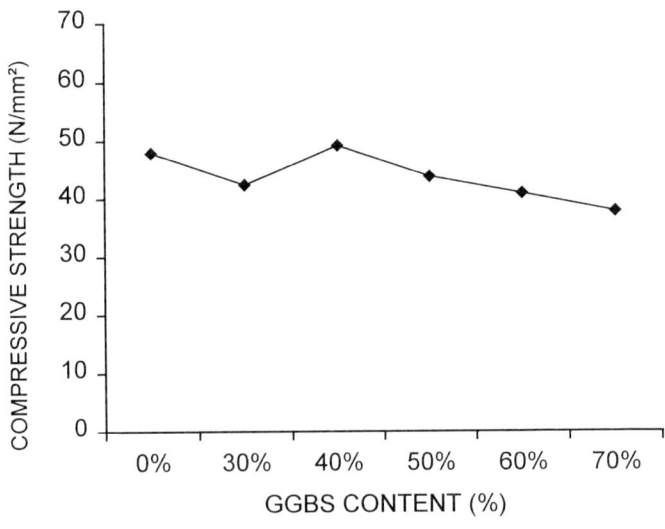

Figure 2 Compressive strength of GGBS concrete at 28 days

RESULTS ON BONDING STRENGTH

Figure 2 shows the bonding strength development of shotcrete made with and without GGBS as a replacement of cement

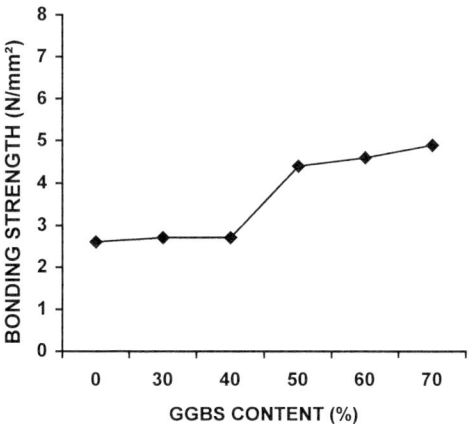

Figure 2 Bonding strength against % replacement of GGBS

The values given herein are the bonding strength of a different percentage of GGBS used in sample cubes. The cube is made in such a manner that the shotcrete is poured into a mould that contains hardened concrete as substrate.

The bonding strength test result shows that there is a marked difference between 30% and 70% level i.e. from 2.7 N/mm^2 to 4.39 N/mm^2. The 40% replacement level of GGBS gives 2.7 N/mm^2 bonding strength, while for 50% to 70% GGBS contents, the graph increases gradually in a parabolic pattern. The figure also shows that the bonding strength development between 0% and 30% GGBS content is slightly lower compared to between 40%, and 70% GGBS contents.

RESULTS ON ABSORPTIVITY

The Initial Surface Absorption Test (ISAT) on 150 mm cubes after 28 days demonstrated that shotcrete specimens with GGBS replacement absorbed water at lower rate than control mix specimen with 40% GGBS content having the lowest ISA value.

Another essential element other than ISA value in characterising shotcrete for its absorption property is the N value. The N value obtained from the slope of the log ISA versus log time graph is shown in Figure 4.

Analysing the relationship between N value and 28 days compressive strength, it illustrates that the N value increases as the compressive strength increases.

Replacement of 40% GGBS to OPC gives the highest value and at the same manner the compressive strength also indicates the highest value.

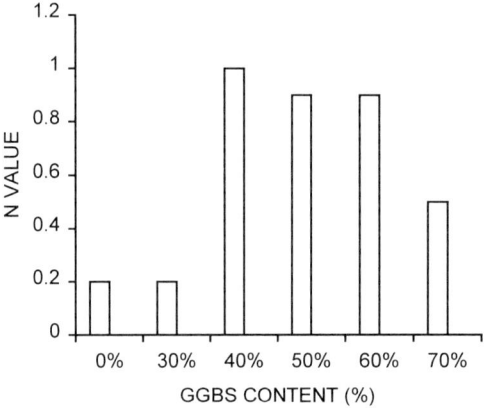

Figure 4 N value versus GGBS concretes

It is evident that the mix proportion of 40% GGBS replacement to OPC gives the optimum value for absorptivity and strength for age of 28 days shotcrete.

RESULTS ON RESISTIVITY AGAINST CHLORIDE INGRESS

Figure 5 shows the depth of chloride penetration of various shotcrete mixes after being soaked in NaCl solution. The graph of coefficient of chloride ingress (D) follows the same trend as in Figure 5 with optimum average coefficient of diffusion at 40% GGBS replacement..

Generally, the coefficient of diffusion is inversely proportional to the compressive strength. Therefore, it can be concluded that the greater the value of compressive strength the lesser the value of coefficient of diffusion. It is also obvious that the chloride attack is lesser in GGBS concrete compared to OPC control.

CONCLUSIONS

The results of compressive test, bonding test, and initial absorption test enable the following conclusions to be made from the study.

Strength Development

1. The compressive strength increases at a slightly slower rate for all GGBS replacement compared to control mix.

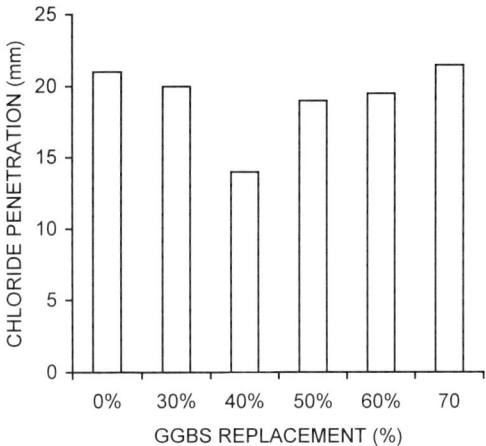

Figure 5 Chloride penetration against GGBS concretes

2. The optimum percentage of GGBS based on the study is 40%. Further increase in the percentage would lead to a slower rate of reactivity of GGBS.

3. In this experiment, 70% of GGBS content shows the lowest compressive strength after 60 days

4. Generally, the higher the GGBS content the lower the strength at early ages but for 40% inclusion, its ultimate strength is higher than OPC shotcrete (control mix).

Bonding strength

1. Bonding strength increases (28-day samples) rapidly from 40% to 50% GGBS contents but at a slightly slower rate for 0% and 30% GGBS content.

2. Mix proportion of between 40% and 70% GGBS replacement to OPC improves the bonding strength tremendously.

Absorptivity

1. The greater the value of N, the greater the compressive strength. It means that the compressive strength increases as the absorptivity decreases.

2. Mix proportion of 40%, 50%, and 60% GGBS replacement to OPC improves impermeability of the shotcrete.

3. Optimum level is at 40%.

Chloride Ingress

1. Replacement of OPC by GGBS can improve the quality of shotcrete in terms of resistivity against chloride especially when expose to marine environment.

2. It can be concluded that the increase in the percentage of the GGBS gives the corresponding reduction in the coefficient of diffusion of chloride.

3. Optimum level is 40% replacement.

REFERENCES

1. NURUDDIN, M F, YUNUS, S A, AHMAD, H, ABDULLAH, M S. Sungai Piah Upper Scheme Low Pressure Tunnel Hydro Electric Project: A Repair Experience. Malaysia – Norway Technological Seminar, Rock Cavern & Tunneling Technology, Kuala Lumpur, Malaysia 2nd & 3rd November 1998.

2. DIAH, A B, NURUDDIN, M F, Improved Durability via Cement Replacement Materials, RBDAM Seminar, 26th April 1997, Kuala Lumpur.

THE INFLUENCE OF ACCELERATOR ON THE LONG-TERM STRENGTH OF STEEL FIBRE REINFORCED SHOTCRETE

G Baker

University of Warwick

United Kingdom

R Hockings   **J Blanck**

Boral Shotcrete

Australia

ABSTRACT. In underground mine support, shotcrete is sprayed with large quantities of activators to achieve very high early strengths, in the order of MPa within 15 minutes. The early successful accelerating agents were caustic and results show that high doses lead to significant long-term loss of strength. More recent compounds claim to give the desired properties through early cement hydration without loss. This article presents results of tests of steel fibre reinforced shotcrete (SFRS) strength over 156 days, for two generic types of accelerator at doses of 3%, 6% and 9%, all compared with a plain control mix. The results were obtained by sprayed a real rock face, coring at the different ages, and testing for compressive strength. Since sprayed concrete is commonly reinforced with steel fibres, SFRS was tested, accepting that strength loss may be inhibited by fibre bridging. This article is aimed at the practitioner who wants to know strength limits for given levels of accelerator, in order to achieve the balance between initial strength, and hence contract speed, and long-term performance.

Keywords: Accelerator, Shotcrete, Steel fibre reinforced shotcrete (SFRS), Long-term performance, In situ strength.

Professor Graham Baker is Professor of Structural Engineering in the School of Engineering at the University of Warwick. His interests cover the mechanics of cementitious materials, the behaviour of reinforced and prestressed concrete structures, including SFRC, and the development of computational models to assess strength and durability of plain and reinforced concrete.

Mr Ralph Hockings is General Manager of Boral Shotcrete, one of Australia's largest shotcreting contractors with major involvement in the mining industry. He has for some time been involved in a nationwide project aimed at providing guidelines for best-practice for underground support, particularly in the use of shotcrete.

Mr Jan Blanck is a Consultant to Boral Shotcrete with many years' experience in the application of wet and dry-mix shotcrete in both civil and mining tunnel facilities throughout the world. His interest cover all aspects of design and application of shotcrete.

INTRODUCTION

Shotcrete is used extensively for underground support in the mining industry, in addition to the many applications in civil engineering such as structural tunnel linings and embankment stabilization. In applications to underground roadway support, significant quantities of admixtures are used, firstly to retard set given that the shotcrete may be batched some distance from the working face, and secondly to accelerate set in order to obtain high early strength gains to provide support in weak or fragmented rock.

It is the latter which concerns this article. Some ground conditions are so poor that support is required almost immediately, and as such contractors often use very high dosages of accelerator to obtain set as quickly as 10-15 minutes after spraying; it may also be that the mining contractor simply wants to blast a new face and so shotcrete support in the wake of the excavation again needs that high early strength. Whatever the reason, the obvious questions to ask are what are the effects of the type and dosage of accelerator on the initial and long-term strength of shotcrete, and what is the performance *in situ*.

Materials suppliers have naturally conducted extensive testing of shotcrete properties using their products - see for example Melbye [1] - but most of the testing has been short-term up to around 28 days. While this is a very good indicator of the material's relative performance, and does satisfy the question regarding the immediate strength in underground construction, the projection of that performance over one year, say, is still an important issue for long-term stability.

Successful accelerating agents act to accelerate the hydration of tricalcium (C_3S) or dicalcium silicate (C_2S) and may simultaneously retard the hydration of alumina bearing phases of cement [2]. However, one of the major concerns [3] is that high doses of caustic accelerator in excess of 5% may actually lead to a significant decrease in strength over time [4], as well as a lower strength relative to a control mix without accelerator, despite the rapid early gains. Figure 1 reproduces these 'classic' test results from Morgan [4] and Dimic [3], from which it would appear that the long-term performance would be seriously compromised by such high doses. Further, Rosskopf et al. [2] tested a wide range of accelerators - calcium chloride, calcium nitrate, sodium nitrate, tri-ethanolamine, sodium and calcium thiosulphate and sodium fluoride - and found that they all caused a loss in compressive strength at 28 days, compared with their achieved 7 day strength.

As a result of this reported evidence, a variety of alternative accelerators have been developed, including a class of rheology agents, such as the modified sodium silicate compound Shotset 250 from Shotcrete Technology or MBT's Meyco T system. These provide an early stiffness without necessarily providing the dramatic early strengths of the caustic activators. Melbye [1] shows that the 1 day strength using Meyco SA140 is as good as their caustic accelerator, Delvocrete, but the early strengths between 15-60 minutes after spraying are not as high. Moreover, preliminary laboratory tests by the authors using Shotset 250 also showed a slow strength gain, but no long-term loss over a 90 day period. Garshol [5] also showed that dosages up to 15% of waterglass (sodium silicate) a reduction in compressive strength over a control mix, but no long-term loss.

On a different tack, Teramura et al. [6] developed an accelerator based on amorphous calcium aluminate, which provides the desired properties through early cement hydration. They report no loss of strength at 28 days for doses of 5%, 7% and 10%.

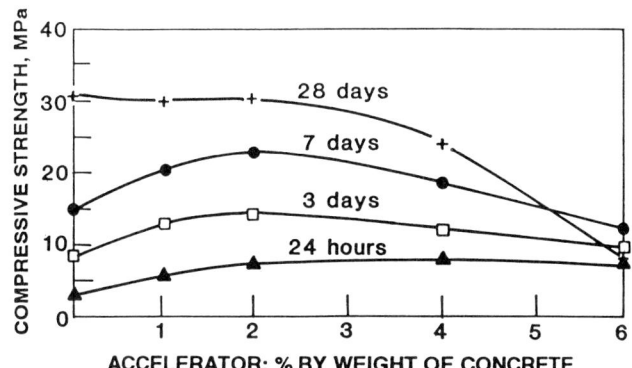

Figure 1 Influence of caustic accelerator on compressive strength of shotcrete
(Reproduced from Dimic [3])

One should remark, however, that other test programmes [7] using caustic accelerators have not shown the same dramatic strength loss. Hence, it may be that there are certain conditions under which caustic accelerators are not absolutely detrimental to strength; it has been noted by Rosskopf et al. [2] that overall performance depends a great deal on the type of compound used. There is of course no question over the safety and environmental issues which have led to the banning of use of these compounds in several countries; Bangzhao [8] in fact commented that most accelerators available are based on carbonates, aluminates and silicates with pH values up to 14.

In this article, we are interested in the long-term performance of shotcrete with different accelerators. In particular, we test the long-term variation in strength of wet-mix shotcrete using both a typical caustic accelerator, and a newer 'acidic' compound. In addition, our shotcrete is reinforced with steel fibres simply because it is common in underground construction: even when accelerator is detrimental to the strength of virgin concrete, perhaps through the formation of porous hydration products, it may be that fibre bridging of voids inhibits growth of damage and hence improves performance.

The other major incentive in this investigation has been to assess the performance of SFRS *in situ*, as opposed to laboratory tests, since field factors may have a significant, or even dominant, influence on long-term performance.

MATERIALS AND TESTING

Concrete Mix

The details of the shotcrete mix design are given in Table 1. This mix is typical for underground support: it achieves compression strengths around 60 MPa in a laboratory mix, and consistently greater than 40 MPa when sprayed underground.

Table 1 Shotcrete mix design

CONCRETE CONSTITUENTS	QUANTITY (per m^3)
Cement (OPC)	450 kg
Aggregate (0-10mm)	1700 kg
10 mm 600 kg	
coarse sand 850 kg	
fine sand 250 kg	
BHP HT Steel Fibres	50 kg
Water Reducing Agent	471 ml/100m^3
Accelerator	3%, 6%, 9%
Water/cement ratio	0.42.

The fibres used were a BHP high tensile fibre: this is a straight, narrow 18mm long fibre with enlarged ends for bond. The volume used is quite modest for this short fibre.

Accelerators

Two typical accelerators were used in the trials: the MBT product Delvocrete Activator S, and the Rhône-Poulenc product Rhoca-jet 50 (RP 750 SC). The Delvocrete accelerator forms part of the MBT chemical cement control system, and is *representative* of the class of caustic accelerators, with a pH of 12; we note that this is less than the early caustic accelerators which often had pH values up to 14. The delvo system is a two component non-chloride chemical system. The first component, delvocrete stabiliser, forms a protective layer around the cement particles keeping the concrete in a plastic state until the activator is added. For shotcrete applications, a special activator 'S' has been developed. The activator breaks up the protective layer, and accelerates the hydration process.

The second accelerator also accelerates hydration, but is typical of a 'new breed' of non-caustic activators, with a pH of just 2.6; another in this class is the MBT accelerator SA160. It is also liquid accelerator, containing amorphous silica suspended in an inorganic aqueous solution. Both accelerators were used at dosage rates of 3%, 6%, and 9% by weight of cement. As a control, one reference mix with no accelerator was also used in the trials.

We emphasize that while we test specific named products, we have no commercial interest: the products adopted represent classes of accelerator compounds.

Shotcrete Application

The site chosen, for convenience, was a rock face in a local open-cut quarry used by Boral for aggregate extraction. The material was a strongly layered quartzite with a slightly weathered face, fragmented by the blasting process, as can be seen in Figure 2.

Figure 2 Hand spraying shotcrete on rock face

In all, 7 panels of shotcrete were sprayed, corresponding to the reference mix and the six sets of accelerated mixes. Because of the limited volume of shotcrete required, spraying was carried out using a hand held 300mm nozzle with shotcrete delivered from a standard pre-mix truck and concrete pump; the nozzle used was a HPS flexible nozzle, developed by Shotcrete Technology. Compressed air hoses were used to pressure spray the shotcrete and for injecting the accelerator through the nozzle assembly.

Accelerator was placed into empty pressure cylinders, with compressed air entering through a one-way valve, forcing the accelerator out along a second line through a flow-meter and valve. The accelerator proportions were controlled through the flow-meter, which was calibrated by hand for the different viscosities of the two products, although it transpired that they were very similar. Figure 3 shows a schematic representation of the equipment set-up.

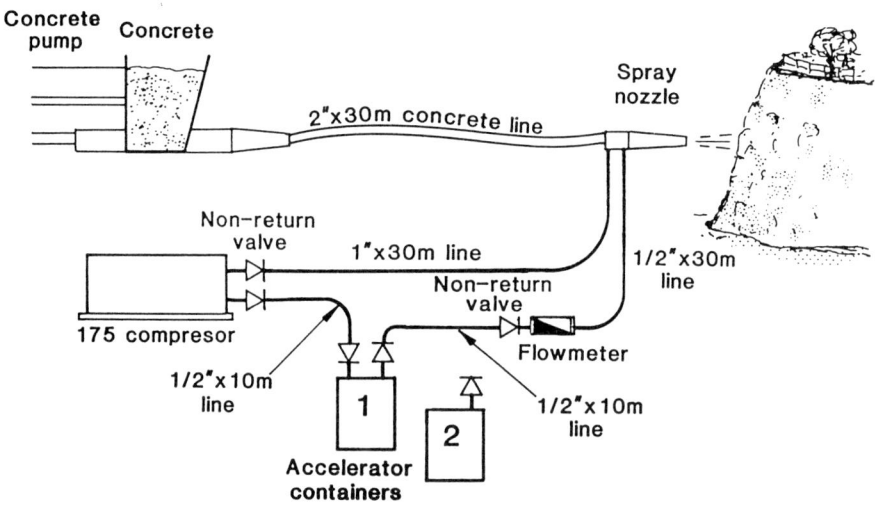

Figure 3 Schematic of equipment set-up

Test Procedure

Cores were taken from the sprayed wall over a period of time at the specific ages: 1, 3, 7, 24, 56, 92 and 156 days. No cores were attempted within the first day because of the likelihood of damage to the cores, particularly from fibres tearing the young concrete in the datum panel. The corer was a Dembicon (450-900 rpm range) using 75 and 80 mm bits.

The cores were taken to the laboratory, sulphur-capped and tested in UCS; the actual diameter, length to diameter ratio, mass and failure load were recorded for each sample. For each age and accelerator dosage, at least two cores and normally three cores were taken and tested. The failure stress was calculated and adjusted for the core length to diameter ratio, in accordance with AS1012 [9], then averaged. In no case did the individual values vary from the average by more than 12%.

RESULTS AND DISCUSSION

Figures 4, 5, 6 show the variation in compressive strength for 3%, 6% and 9% doses of the two accelerated shotcretes in each case plotted against the variation in strength of the reference mix.

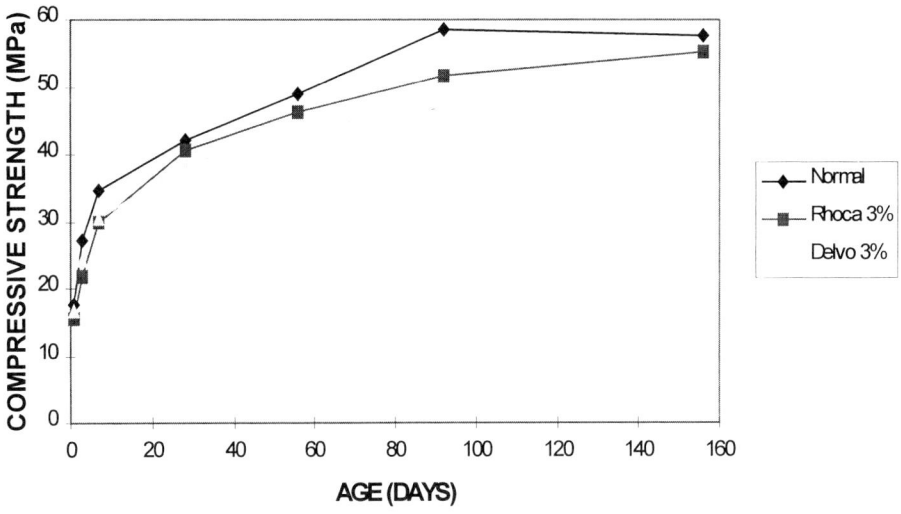

Figure 4 Strength vs age for reference mix and 3% dosages

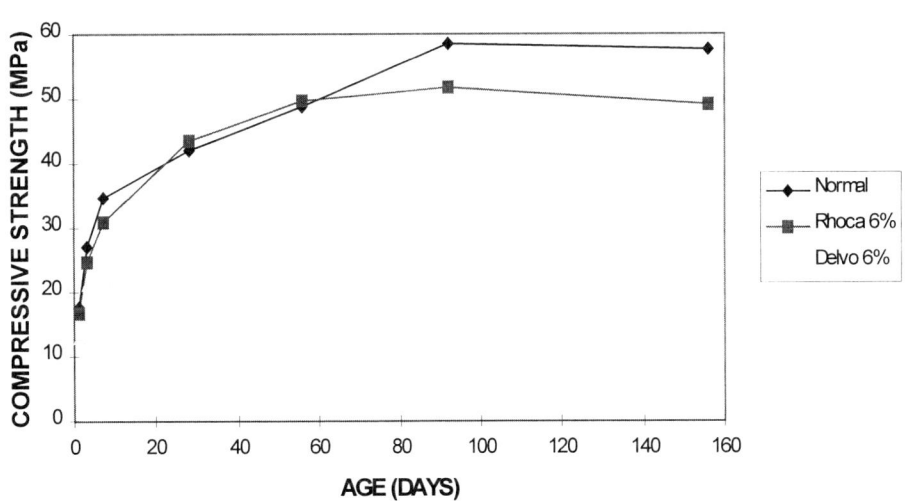

Figure 5 Strength vs age for reference mix and 6% dosages

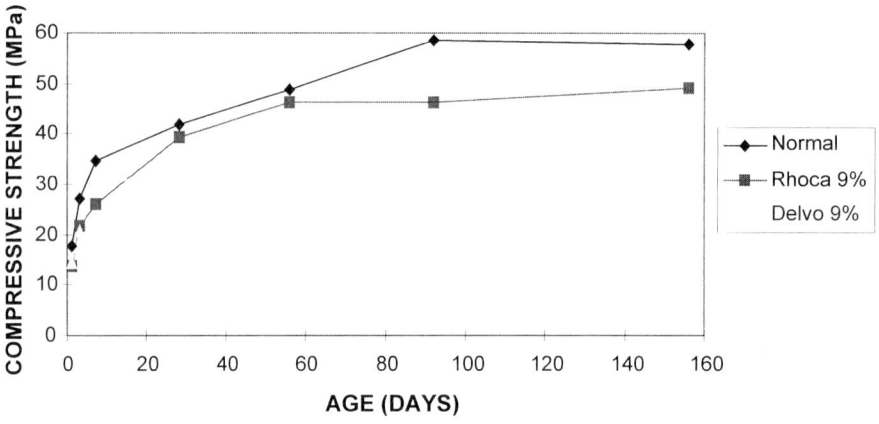

Figure 6 Strength vs age for reference mix and 9% dosages

There are several important features to note. First, the normal mix shows a gradual increase in strength up to the 90 day tests, and then remains essentially constant. The final value achieved was 58 MPa, near to that obtainable in a laboratory test. This suggests that the batch and mix control exercised in the field was very good, and that other results should also be reliable.

From Figure 4 (the 3% dosage) it can be seen that there is little effect when using the Rhoca-jet, apart from some expected scatter; certainly while the accelerated shotcrete consistently showed a lower strength, the difference is marginal. The Delvocrete also performed quite well at this dose, although there is evidence of a decline over 90 days, where the difference is noticeable, but not large.

At 6% dosage, the influence becomes more marked, particularly for the caustic accelerator. We do note, however, that the results do not show a monotonic strength gain, or loss. The 6% Delvocrete gave lower strengths than the 9% Delvocrete between 3 - 92 days, although some aspect of the site location (e.g. rock face angularity) may have been responsible since the strength unexpectedly appeared to decrease from 3 to 7 days. Moreover, the 6% Rhoca-jet dosage at ages less than 92 days showed slightly higher strengths than the 3% dosage of Rhoca-jet.

For the 9% dosage, the Rhoca-jet results were very stable beyond 56 days, and showed only a marginally lower strength than the corresponding 6% value. The caustic accelerator, on the other hand, showed a gradual relative loss in strength from 56 days, falling back to its 7 day value. Also, the final strength at 156 days was only 55% of the reference mix. This is a significant loss, and the fact the results fell consistently from 56 days to 156 days is a concern.

These results are in keeping with previous studies on caustic accelerators, even though the loss was much less than the values given in Dimic [3], perhaps because of the lower pH. Of course the steel fibres (not used in Dimic [3]) may have inhibited void collapse, through bridging, and may have a moderating influence over the loss of strength with time.

Overall the non-caustic accelerator performed very well. The maximum difference between the control and the accelerated shotcrete was 16% for the 9% dosage at 156 days. While significant, there was no long-term reduction. In plotting the results for the 3%, 6% and 9% tests together, it is clear that there is no appreciable difference between the 3%, 6% and 9% dosages. In fact, the difference between the three sets was no greater than those expected from differences due to field application e.g. spraying technique, site location and variations in rock face quality. This suggests that the dosage rate can be chosen to suit the early set requirements, rather than a concern about significant long-term effects.

CONCLUSIONS

Field testing will rarely produce consistent trends from which absolute conclusions can be drawn, but that in itself is an important consideration. Contractors are well aware that variations in ground and environmental conditions, even variations within a site can produce significant differences in properties measured from *in situ* tests. Here, we have noted that the rock face conditions and variations in application can account for as much variation as did a 3% change in accelerator dose, indeed reversing an expected trend with increased dosage in one instance.

However, several trends were evident, and these are:

1. No long-term loss of strength was observed for the Rhoca-jet accelerator, over the 156 day testing period.

2. The long-term strength for the Rhoca-jet accelerator was less than that for the control mix, but there was no appreciable difference between the 3%, 6% and 9% dosages. This indicates that the admixture dosage can be designed to achieve the appropriate early-age properties, without too much concern for long-term performance.

3. The long-term performance of the caustic accelerator (Delvo-crete activator S) was less stable than the Rhoca-jet compound.

4. Generally, the higher dose of the caustic accelerator gave a lower strength at any age, being just 55% of the control mix for the 9% dosage, albeit the effect compared with the control batch was not large for dosages up to 3%.

5. The absence of serious deterioration with the caustic accelerator over time points to a number of important factors: (i) even though fibres do not greatly enhance peak strength, the presence of steel fibres may inhibit *strength loss*, by providing stability across microcracks and pores; (ii) the compound itself may act in a subtly different manner from compounds tested previously; (iii) the testing conditions used for the results reported in Dimic [2] may have had an adverse effect on performance.

ACKNOWLEDGEMENTS

The authors wish to acknowledge the assistance provided by Boral Shotcrete in all aspects of the planning and spraying of shotcrete panels, and the generosity of MBT (Australia) in providing the Delvocrete accelerator. We also wish to thank students Kim Hines and Ken Ng for their effort on this project, and Neil Bellamy (quarry manager) for access to the site.

REFERENCES

1. MELBYE, T A. Sprayed Concrete for Rock Support. MBT: Zürich, Switzerland.

2. ROSSKOPF, P A., LINTON, F J. AND PEPPIER, R B. Effect of various accelerating chemical admixtures on setting and strength development of concrete. Jnl. Testing & Evaluation, 1975, **3**, 322-330.

3. DIMIC, D. Shotcrete admixtures. In Applications of Admixtures in Concrete, A.M.Paillere (ed)., St. Edmundsbury Press: Suffolk, 1995, 108-116.

4. MORGAN, R. Recent developments in shotcrete technology. In The World of Concrete '88, Las Vegas, USA, p54.

5. GARSHOL, K. Development of mechanized wet mix shotcrete application in the Norwegian tunnelling industry. In Shotcrete for Underground Support V, Sharp, J. and Franzen, T. (eds.), ASCE: New York, 1993.

6. TERAMURA, S, MATSUNAGA, Y, HIRANO, H., HANDA, M. AND SAKAI, E. Accelerator for shotcrete based on amorphous calcium aluminate. In Shotcrete for Underground Support VI, D.Wood and R.Morgan (eds.), ASCE: New York, 1993, 9-16.

7. MORCH, A. Wet mix shotcrete used in subsea tunnel environments. In Shotcrete for Underground Support V, Sharp, J. & Franzen, T. (eds.), ASCE: New York, 1993.

8. BANGZHAO, L. The causticity of accelerator for shotcrete and its impairment on strength of shotcrete. In Shotcrete for Underground Support VI, D.Wood & R.Morgan (eds.), ASCE: New York, 1993, 17-24.

9. AS1012. METHODS FOR TESTING CONCRETE, Part 14, Standards Association of Australia, 1991.

NUMERICAL MODEL FOR ACCELERATING ADMIXTURE IN DEM MODELLING OF SHOTCRETE

U C Puri
T Uomoto
University of Tokyo
Japan

ABSTRACT. The use of the accelerating agent is very important for the perfect insurance of concrete on the target wall. In this regard, rapid setting cement or the accelerating agent (coagulant) is quite often used at the point of nozzle in either dry process of shooting or wet process of shooting. For numerical modeling of the accelerating agent, a factor, $\alpha(t)$ is considered for the increment in work done or change in force-displacement law after adding accelerating agent. All DEM parameters are equally influenced by the accelerating agent and considered by the factor. Experimental evidences are reproduced through numerical modeling. It is observed that, although the accelerating agent is necessary in shotcrete, its quality and dose must be properly judged before using in order to avoid its bad effects.

Keywords: Rheology, Shotcrete, DEM parameters, Bingham model, Accelerating agents.

Mr Umesh Chandra Puri, is a Ph.D. student in the Institute of Industrial Science, the University of Tokyo, Tokyo, Japan. His research interests include concrete rheology numerical modeling of fresh concrete and shotcrete.

Professor Taketo Uomoto, is a faculty member in the Institute of Industrial Science, the University of Tokyo, Tokyo, Japan. His major research works include durability of concrete material, new composite material, evaluation of deteriorated concrete structures. He is the chairman of the Asian Model Code Committee.

INTRODUCTION

Shotcrete is a very old technique of applying concrete. Shotcrete is an important component of modern underground construction and tunneling activities. In tunneling procedures employing the New Austrian Tunneling Method (NATM), the tunnel walls are shotcreted immediately after excavation with a concrete setting within few seconds to protect the danger of fall of materials. Despite lots of benefits of shotcrete, main problems in shotcrete are huge rebound loss and change of mix proportion of shot concrete (due to rebound loss) and its durability. Underlining these problems, authors had done the Distinct Element Modeling (DEM) [1] of shotcrete for the numerical prediction of shotcrete behavior. In this research, main focus is given to the modeling of the effect of the accelerating agent in shotcrete. In shotcrete for underground construction, the accelerating agent is highly employed to develop high early strengths that provide structural safety during construction. Addition of accelerating agents elevates the shooting stiffness of fresh shotcrete to enable build-up of thicker sections and prevents the danger of fallout of large masses of fresh material from walls and overheads. For all types of accelerating agents, although the early strength is higher, the strength reduction is the most pronounced at the later stage. And also in the case of large or excessive accelerator dose leads to more frequent spraying faults and more rebounds. The durability of accelerated shotcrete in aggressive exposure environment is also adversely affected [2]. Due to this fact, recently, many attempts are being made to the development of the accelerating agent to enable with less dust formation during shooting, less strength reduction at the later stage, durable shotcrete, alkali-free, etc.(for environmental consideration)[3].

DEM MODELING OF SHOTCRETE

DEM is simply a tool, which models a prototype in a number of rigid balls. DEM is originally proposed for rock and granular flow simulation [4]. The basic principle of DEM is from the equation of motion used in particle dynamics. The equation of motion of an element, i, having the mass, m_i, the moment of inertia, I_i, and linear and rotational displacement vectors u, ϕ, respectively can be written as

$$m_i \ddot{u} + C_i \dot{u} + F_i = 0 \tag{1}$$
$$I_i \ddot{\phi} + D_i \dot{\phi} + M_i = 0 \tag{2}$$

In this equation F_i is the sum of all forces acting on the element, i. M_i is the sum of all moments acting on i. C_i and D_i are damping coefficients. In shotcrete before shooting, concrete is in a fresh state. So, first fresh concrete state must be modeled. In simulation, fresh concrete is assumed to be represented by a collection of distinct elements. The smaller the particle size, the larger CPU time is required and computation is long. So, fine aggregate particle can not be considered as distinct element and must be considered within mortar. Also, the interaction between fine aggregate and cement paste can not be considered. Therefore, each distinct element comprises inner core of gravel (circular) and surrounded by mortar layer; a simple two-phase model is proposed as shown in Figure 1.

The principle of calculation for the inside gravel is as same as for granular modeling. As for mortar, the principle of calculation is also same as granular modeling but there is an additional coefficient used to identify the allowable value of extension of the surrounding mortar layer. This coefficient takes into account cohesion (or adhesion) effects of fresh concrete mix [5]. This coefficient is illustrated in Figure 2.

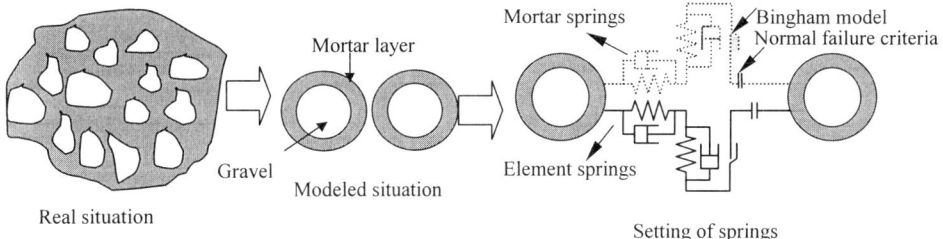

Figure1 DEM modeling of shooting material (concrete)

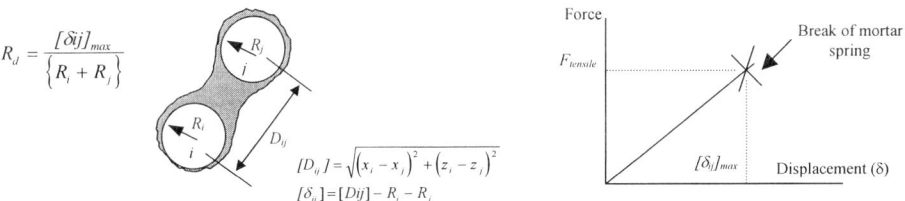

Figure 2 Definition of R_d Figure 3 Force-displacement relation

The force-displacement relation is shown in Figure 3. R_i, R_j are the radii of two contacting elements i, j. If $[\delta_{ij}] > [\delta_{ij}]_{max}$ the mortar spring fails and the elements detach from each other independently. Both the contact forces in the normal and tangential direction vanish automatically as such $[f_{pn}] = [f_{ps}] = 0$.

Rheology Consideration in DEM

It has been widely accepted that a Bingham model conforms well as the rheological model of wet-consistency concrete [6]. Effects of mix composition including admixtures are clearly indicated by these rheological constants [7]. In DEM modeling of fresh concrete, it is also regarded that fresh concrete behaves as Bingham model with two rheological constants namely the yield stress and plastic viscosity as in Figure 4(a).

Two distinct elements under shear show first elastic stage, then yielding and finally reach maximum shear stress as in Figure 4(b). At the contact of two distinct particles, mortar layer comes first at the contact and once its deformation range is overcome, the inside element comes into play. In two-phase modeling, each distinct element consists of inner core of gravel of high stiffness and is surrounded by mortar of low stiffness.

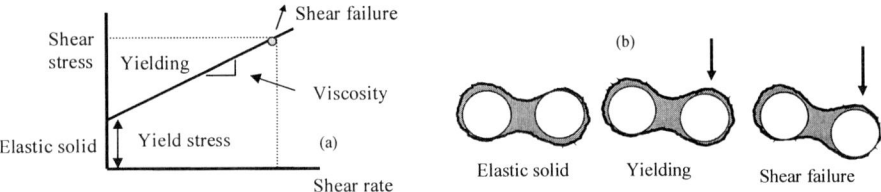

Figure 4 (a) Bingham model, (b) distinct elements under shear

The inside gravel must also be clearly despised. In two-phase modeling, much attention is to be paid to the modeling of the interface between gravel and mortar layer in each distinct element whose DEM parameters (stiffness and dashpot, etc.) are entirely different [8]. It is recommended to go through reference 8 for details.

Consideration of an Accelerating Agent in DEM

For the numerical modeling of the accelerating agent, various models are investigated [8]. The essential features of all models are increasing the work done to its value before the addition of the accelerating agent and changing the force-displacement law. A model is found appropriate that enables the work done $\alpha(t)$ times and displacement $1/\alpha(t)$ times to their values before adding the accelerating agent [8]. This is shown in Figure 5.

$$\frac{1}{2}\alpha(t)F_1 S_1 = \frac{1}{2}F_2 S_2 \tag{3}$$

Imposing the relation, $\alpha(t)S_2 = S_1$. Following equations are obtained.

$$\alpha(t)^2 F_1 = F_2 \tag{4}$$
$$\alpha(t)^3 K_1 = K_2 \tag{5}$$

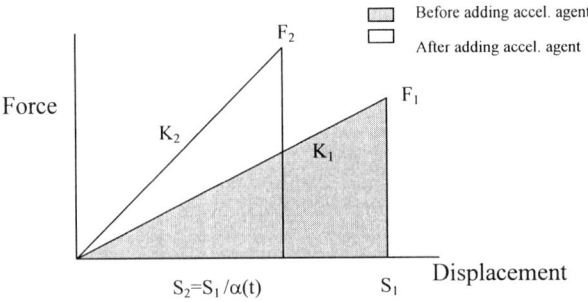

Figure 5 Consideration of the accelerating agent in DEM

So, conditions in Eqs. (4) and (5) are necessary to impose in the calculation. In all shotcrete simulations, the effect of the accelerating agent has been taken using this model. According to this, DEM parameters for fresh concrete are modified as such after the addition of accelerating agent.

$$Q_{stk}(t) = \alpha(t)^2 Q_{STK} \tag{6}$$

$$k_{pn}(t) = \alpha(t)^3 k_{pnINI} \tag{7}$$

$$\eta_{pn}(t) = \alpha(t)^3 \eta_{pnINI} \tag{8}$$

$$k_{ps}(t) = \alpha(t)^3 k_{psINI} \tag{9}$$

$$\eta_{ps}(t) = \alpha(t)^3 \eta_{psINI} \tag{10}$$

Where Q_{STK}, k_{pnINI}, η_{pnINI}, k_{psINI} and η_{psINI} are the initial DEM parameters set for the flow simulation of fresh concrete.

Discussion on the Factor α(t)

α(t) is a factor to consider the increment in work done or change in force-displacement law after the addition of accelerating agent. It is assumed that, all DEM parameters are equally influenced by the accelerating agent and can be considered by the single factor, *α(t)*. One to one effect to each DEM parameter is not investigated in this research. In this modeling, the expression for *α(t)* is taken by simulating the exponential increment of *the stiffening effect of the accelerating agent*, which is assumed similar to the setting of fresh concrete during hydration [9]. The following exponential increment as shown in the Eq. (11) is assumed.

$$\alpha(t) = Q_a - (Q_a - 1)\exp[-Q_b(t-t_o)^2] \tag{11}$$

where, $\alpha(t)$ is rate of increment of DEM parameters, Q_a, Q_b are properties of the exponential curve as shown in Figure 6 (usually, $Q_b = 1.0$), t_o it total time required for elements to reach the nozzle for shooting (from the start of simulation), t total time from the start of simulation to the instant of time when shooting is finished or stopped.

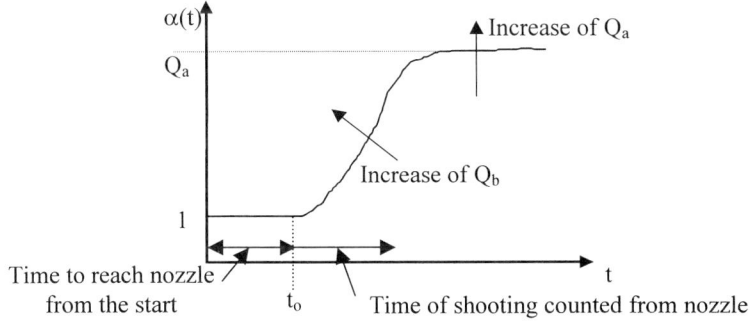

Figure 6 Stiffening upon the addition of the accelerating agent

The programming should be done in such a way that DEM parameters for each element do not change until they reach at the nozzle. When elements are about to exit out the nozzle, the modification of DEM parameters is done using Equations (6) to (10).

SHOTCRETE SIMULATIONS

Now, it is going to simulate shotcrete performance with and without accelerating agent. The value of $\alpha(t)$ is 1 for no accelerating agent. The value of $\alpha(t)$ is higher than 1 for accelerating agent. In all shotcrete simulations, the shooting pressure is 0.44 MPa, FM of gravel is 6.34, distance between nozzle and target wall is 1.4 m. The total simulation time is 10 s.

Shotcrete Without Accelerating Agent

Shotcrete simulations without accelerating agent, at 2s and 4s out of 10s, total simulation time are shown in Figures 7a-7b. Shot concrete attaches on the wall even for no accelerating agent. Rebound loss is small for 2s time of shooting but it increases for 4s time of shooting. So, for no accelerating agent, the shooting thickness should be small to minimize the loss.

Shotcrete with Accelerating Agent

The simulations for $\alpha(t=10s)$ equal to 1.5 and 2 are shown in Figures 8a-8b. On increasing the value of $\alpha(t)$, rebound loss is less and also shooting is possible to a large thickness with minimum rebound loss. But it should be noted that, performance is enhanced only to some level of increase in accelerating dose. Although the accelerating agent reduces redound loss, its high amount results in rise of rebound loss due to bond failure as shown in Figure 9[1].

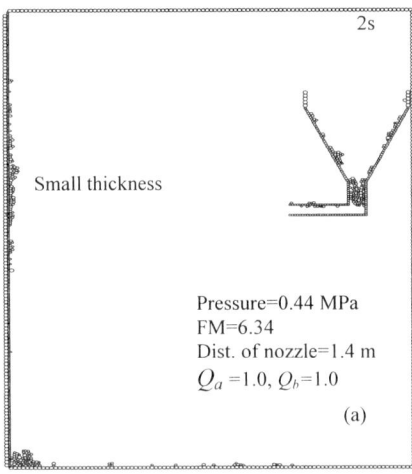
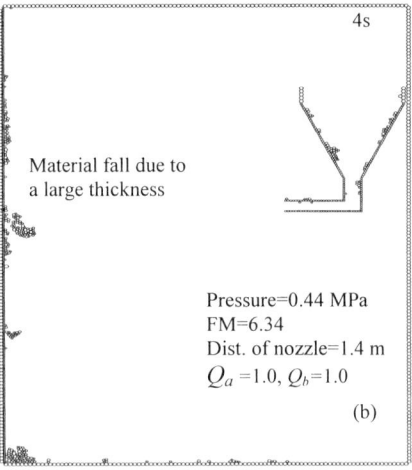

Figure 7 Shotcrete simulation without accelerating agent shown at (a) 2s at (b) 4s out of 10s total simulation time

The value of Q_a resembles to $\alpha(t=10s)$, indicating the effect of stiffness (maximum value). On the other hand Q_b controls the rate of stiffening effect (rate of rise). Q_b causes a similar effect as the power of $(t-t_o)$ does. The effect of Q_a and Q_b is same provided the shooting thickness is same. Quite often, higher Q_b immediately stiffens shot concrete. This is beneficial to reduce the rebound loss to some extent only. Sometimes it may increase rebound loss due to a large thickness build-up at same position as shown in Figure 10 (simulates high amount of the accelerating agent case). So, the dose and quality of accelerating agent must be inspected to avoid its bad effects.

Regarding the value of $\alpha(t)$, the research does not define it quantitatively. It is unity for no accelerating agent. It has been assumed that the value of $\alpha(t=10s)$ is 1.5 for normal dose of accelerating agent (for 10 s simulation). This means that, the work done is increased by 1.5 times to its value before the addition of accelerating agent and displacement is reduced by 1/1.5 times to its value before the addition of accelerating agent.

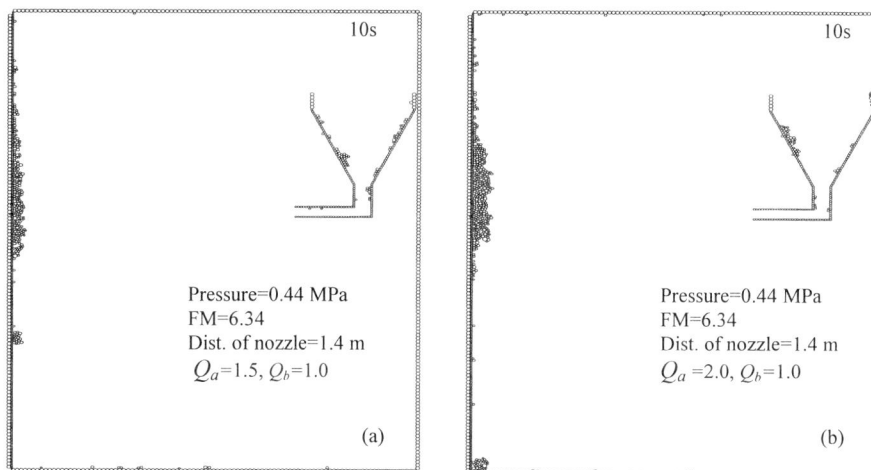

Figure 8 Shotcrete simulation with accelerating agent for (a) $\alpha(t=10s)=1.5$ and (b) $\alpha(t=10s)=2.0$

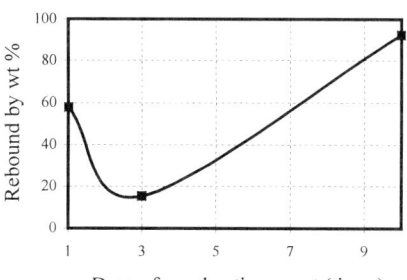

Figure 9 Rebound by wt. % versus dose of accelerating agent [1]

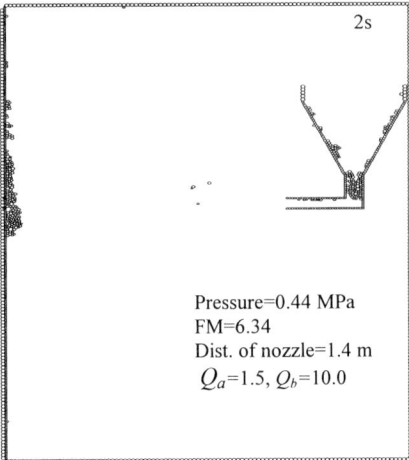
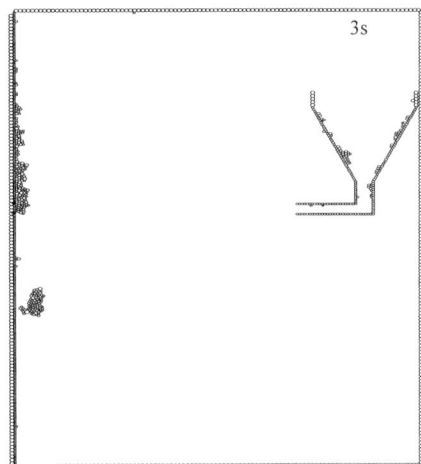

Figure 10 Same conditions as for in Figure 8a but Q_b increased to 10 (shown at 2s and 3s out of 10s, total simulation time)

Due to the limitation of computer CPU time, the calculation has been done only up to maximum 10s. So, within this small time, it is difficult to quantify the effect of the accelerating agent by mechanical tests. Chemical test might be helpful for the verification of this value.

Luiz, P.R. et. al. [10] study on shooting stiffness after the addition of different accelerating agents. Procter test is performed. Procter resistance actually measures the yield value of fresh shotcrete. Higher the Procter resistance, the higher shooting stiffness. Proctor resistance has a reverse relation with water by dry material ratio (by mass). An attempt to explore the connection between setting times (obtained from Gillmore apparatus) and shooting stiffness or compressive strength is unsuccessful [10]. However, this Procter test result does not give any help in simulation technique presented here.

CONCLUSIONS

1. DEM tool proves to have great potential for shotcrete simulation.

2. The effect of Q_a and Q_b is same provided that shooting thickness is same. Higher values of Q_a and Q_b may increase the rebound loss due to large thickness build-up.

3. The numerical analysis reproduces the experimental evidences of shotcrete. That is, the increase of accelerating agent helps to reduce rebound loss to some extent but the high amount of accelerating agent has a reverse effect giving high rebound loss due to the bond failure at the target wall.

REFERENCES

1. PURI, U, C, AND UOMOTO, T. Numerical modeling—a new tool for understanding shotcrete. Accepted for publication in RILEM Materials and Structures.

2. MORGAN, D, R. High early strength blended-cement wet-mix shotcrete. Concrete International, May, 1991, pp 35-39.

3. LUHAS, W, et. al. Innovation in shotcrete technology. Shotcrete for underground support, pp 155-164.

4. CUNDALL, P, A, AND STRACK, O, D, L. A discrete numerical model for granular assemblies. Geotechnique, Vol. 2, No. 1, pp 47-65, 1979.

5. PURI, U, C, AND UOMOTO, T. Numerical simulation of shotcrete rebound and its void inspection by 2D-DEM considering real size distinct element particle grading. Sixth East Asia-Pacific conference on structural engineering and construction, Taipei, 14-16 Jan., 1998, pp 1911-1916.

6. JIRO, M, AND KIKUKAWA, H. Viscosity equation for fresh concrete. ACI Material Journal, Vol. 89, No. 3, 1992, pp 230-237.

7. GJϕRV, O, E. High strength concrete. Advances in concrete technology, energy mines and resources, Otawa, Canada, MSL 92-6(R), 1992, pp 21-77.

8. PURI, U, C. Numerical simulation of shotcrete by distinct element method. Ph.D. thesis submitted to the University of Tokyo, 1999.

9. MAKI, T. Fundamental study on mechanism of shotcrete using 2-D distinct element method. Master's thesis submitted to the University of Tokyo, March, 1997 (in Japanese).

10. LUIZ, P, R, et. al. Interaction between accelerating admixtures and Portland cement for shotcrete: the influence of the admixture's chemical base and the correlation between paste tests and shotcrete performances. ACI Material Journal, Vol. 93, No. 6, 1996, pp 619-628.

WORKABILITY, SHEAR STRENGTH AND BUILD OF WET-PROCESS SPRAYED MORTARS

S A Austin

P J Robins

C I Goodier

Loughborough University

United Kingdom

ABSTRACT. This paper, which reports on part of a three year research project into wet-process sprayed concrete for repair, examines the influence of rheology on the pumping and spraying of sprayed mortars. The workability properties of seven commercially available pre-packaged repair mortars and six laboratory designed fine mortars were examined using the Tattersall Two-point viscometer, the slump test, a build test and a vane shear strength test. The Two-point apparatus was successful with low-workability mortars and the flow resistance and torque viscosity of the mortars was determined. The vane shear strength test provided an instantaneous reading of the shear strength of the mortars and is compared with their slump. The build value, a measure of sprayability, is then compared with these two workability parameters and the flow resistance in order to determine their inter-relationship.

Keywords: Mortars, Rheology, Wet-process, Sprayed concrete, Two-point test, Workability, Slump, Build.

Dr S A Austin and **Dr P J Robins** are Senior Lecturers in the Department of Civil and Building Engineering at Loughborough University.

C I Goodier is a Research Associate in the Department of Civil and Building Engineering at Loughborough University.

INTRODUCTION

This paper presents the results from work that is being undertaken as part of a three year research programme at Loughborough University, funded by the Engineering and Physical Science Research Council and supported by substantial industrial collaboration from Balvac Whitley Moran, Fibre Technology, Fosroc International, Gunform International Ltd and Putzmeister UK Ltd. The main aims of this project are:

1. to gain a fundamental understanding of the influence of the pumping/spraying process, mix constituents and proportions on the fresh and hardened properties of wet-mix sprayed concrete;
2. to improve the wet-mix spraying process, in particular operator environment, maximum conveying distances and stop-start flexibility;
3. to specify, measure and optimise in-situ properties, particularly strength, bond and durability;
4. to disseminate information in appropriate form to practising engineers to promote and accelerate the use of wet-mix sprayed concrete for repair in the UK.

The main emphasis of the research project is on mortars and small aggregate concretes (<10 mm) applied in thin layers (<100 mm) at controlled low/medium output rates (< 5m^3/hr), in some cases with mesh or fibre reinforcement. This paper concentrates on the fine mortars which contain aggregates with a maximum size of 2-3mm and this group has been further divided into pre-packaged proprietary mixes (designated P1 to P7) and designed mixes (D1 to D6). A large range of pre-packaged proprietary mortars have been developed for hand application and there are a number of pre-packaged proprietary mortars being developed specifically for wet spraying. We have been pumping, spraying and testing both the relatively sophisticated pre-packaged materials and the more basic designed mixes in order to characterise their performance and hence identify the constituents and proportions within the mixes that produce sprayable mortars with adequate hardened properties.

RHEOLOGICAL TESTING OF MORTARS

Recent work conducted by Beaupré [1] investigated the rheological properties of sprayed concrete and the relationship between pumpability and sprayability, including the development of predictive models based on yield and flow resistance determined from tests conducted with a rotational viscometer. Sprayability can be defined as a property that incorporates parameters such as adhesion (ability of plastic mix to adhere to the surface), cohesion (influencing the thickness that can be built-up), and rebound. Beauprè termed this shootability and found a linear relationship between build-up thickness and the yield value of the mix after spraying, and concluded that shootability increases with flow resistance, and is thus in conflict with pumpability which has the opposite relationship. This research examines further the relationship between build-up thickness, pumpability and shear resistance.

Most authors (Tattersall and Banfill [2] and Beaupré [1]) use the simple Bingham model to express cement paste flow curves as this has been proved to give reasonably accurate and repeatable results within the boundaries of accuracy of the apparatus. For a Bingham fluid the relationship between the shear stress (τ) and shear rate (γ) is given by:

$$\tau = \tau_o + \mu.\gamma \tag{1}$$

where τ_o is the yield stress, above which there is a linear relationship between τ and γ characterised by the plastic viscosity μ. Mortar can be observed to be a shear thinning liquid in which the viscosity decreases when the shear rate increases. It also possesses a yield value: a minimum shear stress that must be applied before the mortar can begin to flow. If this shear thinning effect is permanent then this behaviour is known as irreversible structural breakdown, whereas if the structure reforms after shearing it is said to be thixotropic. This structural breakdown, together with Equation 1 is shown in Figure 1(a).

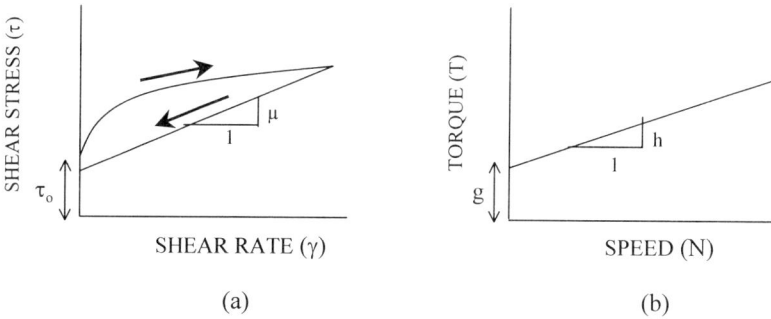

Figure 1 Typical flow curve for mortars (a) Stress-strain, (b) Torque-speed

Tattersall first used a Hobart food mixer to plot flow curves based upon the power needed to drive an impeller in fresh concrete [3]. He later developed a more accurate rheometer with a hydraulic transmission, termed the two-point test apparatus (Mk II). Tattersall found that when the torque (T) was plotted against the speed (N), the relationship was almost linear (Figure 1(b)):

$$T = g + h.N \tag{2}$$

where g is the intercept on the torque axis and h the slope of the line. Beaupré referred to g as the flow resistance, and h as the torque viscosity. This equation is of the same form as the Bingham model (Equation 1) and thus it can be said that g is a measure of yield value, and h of plastic viscosity. In principle it is possible to convert g and h to fundamental units equivalent to τ_o and μ by calibration with standard fluids [4] but most investigations work with the direct parameters (which are of course equipment dependent).

MATERIALS AND MORTAR MIXES

The research has investigated a range of proprietary repair mixes (mainly developed for hand application) and six designed mixes. For the latter, the ordinary Portland cement conformed to BS12:1989 [5] and the silica fume was a proprietary undensified powder. The sands were a crushed Portland stone sieved to a maximum size of 3mm and a building sand graded between 75μmm and 2.36mm.

Some mixes also included an SBR in a 3:1 water:SBR suspension. The proportions of the mixes designed for the project are given in Table 1 and the constituents of the pre-packaged mortars are shown in Table 2.

The mortars were mixed using a 0.043m^3 capacity forced action paddle mixer. The pre-packaged mortars were mixed according to the manufacturers instructions with 3.3 to 4.0 litres of water per 25Kg bag and a mixing time of approximately 4 minutes. The designed mixes were mixed in the same way and in all cases the water was added until the desired consistency for spraying was achieved. i.e. workable enough to be pumped but stiff enough not to slough after being sprayed onto a vertical substrate. The mortar was pumped through a Putzmeister TS3/EVR variable speed worm pump and then down a 25mm diameter rubber hose at an approximate rate of 6 l/min, depending on the mortar. The mortar was then sprayed with an air pressure of approximately 300 kPa.

Table 1 Proportions of designed mixes (by weight)

MIX	CRUSHED STONE	BUILDING SAND	OPC	SILICA FUME	SBR:WATER	LIQUID/CEMEN-TITIOUS RATIO
D1	3	0	1	0.05	1:3	0.65
D2	2	1	1	0.05	1:3	0.55
D3	1	2	1	0.05	1:3	0.48
D4	0	3	1	0.05	1:3	0.44
D5	3	0	1	0.05	0:3	--
D6	4	0	1	0.05	0:3	--

Table 2 Composition of pre-packaged mortars

MIX	POLYMER MODIFIED	FIBRES	SHRINKAGE COMP.	LIGHTWEIGHT FILLERS	MORTAR DESCRIPTION
P1	No	No	Small amount	No	Basic repair mortar
P2	Yes	Yes	Yes	Yes	High build repair mortar
P3	Yes	Yes	Yes	Yes	2-part re-profiling mortar
P4	Yes	Yes	No	Yes	Basic repair mortar
P5	Yes	Yes	No	Yes	Render/repair mortar
P6	Yes	Yes	Yes	Some	Repair mortar
P7	Yes	Yes	Yes	Yes	Lightweight repair mortar

TESTING PROCEDURE

The test methods are described briefly below, two for workability, one for pumpability and one for sprayability and, taken in this order, they enable a rheological audit to be made of a mix as it progresses through the mixing, pumping and spraying process.

Workability

The workability was measured by the slump test [6] and by a modified form of the shear vane test for soils [7]. Two slumps were measured immediately after the mortar had been mixed and if slumps were significantly different (>15mm) then a third was taken and the average of the two closest values calculated.

The shear vane test was used as it is a simple, portable apparatus which could give an indication of mortar workability at various points in the pumping and spraying process. It consists of a torque measuring device at the head of the instrument together with a set of enlarged vanes to provide sufficient shear resistance to register on the torque scale. The maximum torque was then used to calculate a shear strength for the mortar (in kPa.).

Pumpability

The two-point apparatus was the Mk II version developed by Tattersall [3] which has been found to be satisfactory for medium- to high-workability concretes. The mortars tested here had slumps of between 45 and 80mm. It has been suggested that the apparatus might not be sensitive enough for mortars if the torque exerted on the impeller are too low to give a significant increase in pressure, but sufficient change was observed in this work.

During preliminary trials with the apparatus empty it was found that the recorded pressure at a constant speed decreased over time. The apparatus was therefore always warmed up prior to testing for a period of 2 hours at a speed of 0.9 rev/s, after which the change in recorded pressure with time was negligible. The idling pressures were then recorded between the speeds of 0.6 and 2.6 rev/s at increments of 0.2 rev/s. With the bowl rotating at 0.6 rev/s the bowl was gradually filled with approximately 25Kg of mortar to a level 75mm below the top of the bowl. The speed was then increased incrementally and the corresponding pressures recorded. Once 2.6 rev/s had been reached the speed was reduced incrementally in the same way and the corresponding pressures again recorded. The decreasing results that follow the structural breakdown (Figure 1(b)) were used for calculating g and h.

Sprayability

This was assessed both qualitatively (did the material pass through the nozzle) and quantitatively in terms of the amount of material that could be built up on a standard grit-blasted 500x500x50mm concrete substrate.

The mortar was sprayed horizontally onto a vertical 300x300mm target area to obtain as large an amount of material as possible on the substrate whilst keeping within the 'target'.

The mortar would then fail under its own weight either cohesively or adhesively and the total weight of mortar was recorded, together with the failure mode and the maximum depth of build.

TEST RESULTS

Shear Vane

The shear vane provides a basic measure of the shear strength (in kPa) of a mortar and this can be plotted against slump (in mm), as shown in Figure 2. The shear strength has been calculated using the British Standard formulas for the measurement of soil shear strength multiplied by a conversion factor for the increased vane size. This shear strength can, in principle, be related to the yield stress (τ_o) in Equation 1. As expected, the shear strength decreases as the slump increases. It can provide an instantaneous result exactly where the rheological properties of the mortar needs to be measured, i.e. in the hopper of the pump.

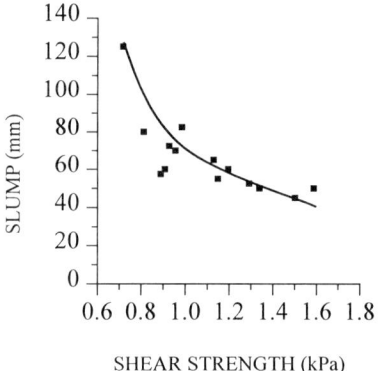

Figure 2 Shear vane Vs slump

Tattersall Two-Point Test

Figure 3(a) shows the results obtained from the two-point test on the mix P1. The figure shows a distinct upcurve and downcurve which was typical for all the mortars tested. However, approximately half way along the downcurve the torque appears to increase as the impeller speed decreases. This is due to the mortar not falling into the impeller sufficiently and therefore not creating a high enough reading above the idling pressures. A regression line drawn through these points, as shown in Figure 3(a) provides misleading values of g and h. The points from the initial part of the downcurve (Figure 3(b)) have therefore been used in this paper for plotting the downcurve, and hence calculating g and h. The values of g and h for the mix D1 at different slumps are shown in Figure 4(a). As would be expected, the mix with the lowest slump (50mm) had the highest yield value and the lowest plastic viscosity. A greater distinction between the values for g and h for the 82.5mm and 120mm slumps would be expected but these results suggest that the apparatus is less sensitive for mortars at higher slumps. Figure 4(b) shows the g and h for the mortar P2 after it has been mixed, pumped or sprayed. The increase in both g and h as the mortar is pumped and then sprayed would be expected as the excess air is forced out of the mortar during the pumping and compacting operations.

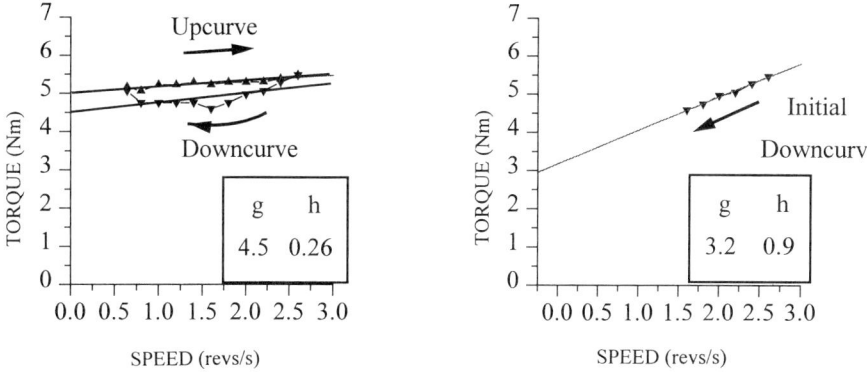

Figure 3 Two-point test, mix P1 (a) Upcurve and downcurve (b) Initial downcurve

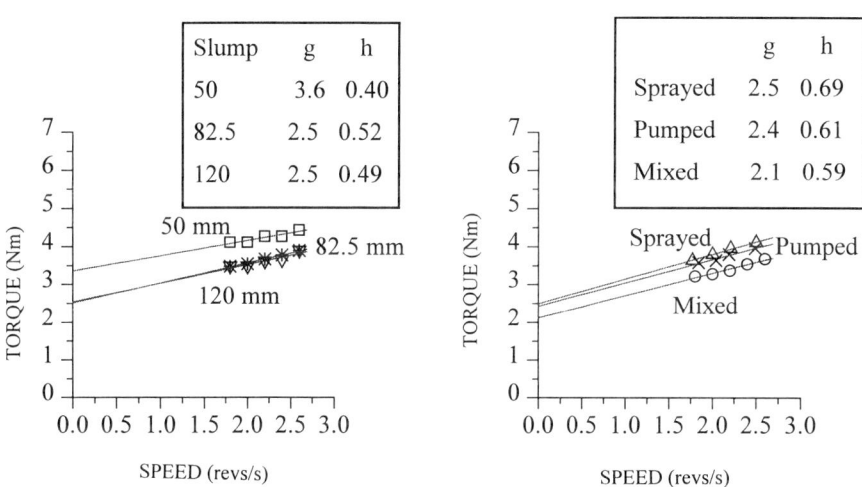

Figure 4 Two-point test (a) Effect of slump on mix D1
(b) Effect of mix P2 being mixed, pumped and sprayed

The values of g and h for the mix D1 at different slumps are shown in Figure 4(a). As would be expected, the mix with the lowest slump (50mm) had the highest yield value and the lowest plastic viscosity. A greater distinction between the values for g and h for the 82.5mm and 120mm slumps would be expected but these results suggest that the apparatus is less sensitive for mortars at higher slumps. Figure 4(b) shows the g and h for the mortar P2 after it has been mixed, pumped or sprayed.

The increase in both g and h as the mortar is pumped and then sprayed would be expected as the excess air is forced out of the mortar during the pumping and compacting operations.

The two-point test results for all the mortars, both the pre-packaged and the designed mixes, are shown in Figures 5. They were all mixed with water prior to testing until the desired consistency for pumping and spraying had been achieved. Of the pre-packaged mortars, the mortar with both the highest g and highest h is mix P1 which had the most 'basic' mix design of all the pre-packaged mortars tested, and contained no polymers, fibres or lightweight fillers. The mix with the next highest value of g, mix P4, was also known to have a relatively basic mix design. These two mixes were also the cheapest commercially of all the pre-packaged mortars tested. The two mixes which were known to be highly polymer-modified (P6 and P3) had the lowest values of g, although their corresponding values of h were very different. The mix P3 is a two-part (powder and liquid) re-profiling mortar which has been formulated to enable it to be applied in thin layers without it separating or being too 'sticky', which could explain why it had the smallest value of g. The designed mixes in Figure 5(b) show a clear trend dependent upon the mix design: the greater the proportion of crushed Portland stone within the mix compared with the building sand then the greater the value of g. The addition of SBR to a mix, in this case mix D5 having no SBR and mix D1 being an identical mix containing a 3:1 water:SBR solution, appears to have little effect on either g or h. This is in contrast with the pre-packaged mortars where the highly polymer-modified mortars possessed a lower value of g.

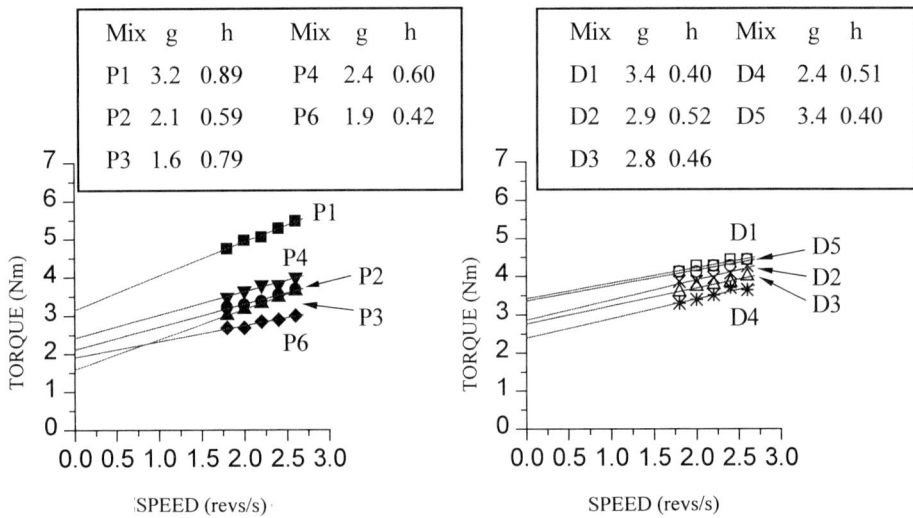

Figure 5 Two-point test (a) Pre-packaged mortars (b) Designed mixes

Build Test

The build values (in mm) obtained for each of the mixes are shown in Table 3. The mass of the mortar sprayed onto the substrate was also measured and this was used, together with the cross-sectional area of the base of the mortar (usually 300mm square) to calculate the maximum shear force produced between the mortar and the substrate. The bending stress was calculated by idealising the mortar on the substrate into the frustum of a square-based pyramid (i.e. a square-based pyramid with the top 'sliced' off parallel with the base). The volume, and therefore the dimensions of this frustum, could be calculated using the mass, the fresh wet density, the area of the base and the height of the frustum (i.e. the build value). This shape was then used to calculate the maximum moment and therefore the maximum bending stress of the mortar.

Table 3 Build test results.

MIX	BUILD (mm)	MASS (kg)	MAX. SHEAR (N/m2)	BENDING STRESS (N/m2)	FAILURE MODE
D1	210	21.4	2571	2120	Adhesive
D2	300	27.3	3279	4922	Cohesive
D3	280	24.2	2907	3872	Adhesive
D5	270	26.8	3219	3521	Cohesive
D6	220	23.2	2787	2357	Adhesive
P1	320	41.5	3816	3476	Cohesive
P2	270	13.0	3147	3002	Cohesive
P3	230	----	2308	2375	
P4	290	26.6	3728	3662	Adhesive
P5	300	49.5	5946	4374	
P6	200	32.2	3868	1566	
P7	350	----	2460	3853	

Figure 6(a) shows the relationship between the build-up thickness and the slump of the mortar before pumping. This agrees with the results presented previously by Beaupré [1] who showed that it is not possible to predict the build-value of a mix simply by measuring the slump immediately before pumping. However, the results seem to indicate an increase in build for an increase of slump. This seems the reverse of what would be expected but at the low workabilities tested here, an increase in slump would produce a slightly wetter, and therefore more cohesive mix, thereby increasing the build.

Beaupré also reported a good relationship between 'g' (the flow resistance, obtained from the Two-point test) and the build value. The relationship between these two parameters in this study are shown in Figure 6(b). The trend is not as strong as that found by Beaupré (who tested 10mm aggregate sprayed concretes with build values from 10 to 350mm) compared to the mortars presented here which have build values between 200 and 300mm. It can be assumed that the line of best fit passes through the origin as a material with zero g (e.g. water) will also have a build-value of zero. It can also be noted that the pre-packaged re-profiling mortar designed to be easily trowelled (P3) had the lowest value of g.

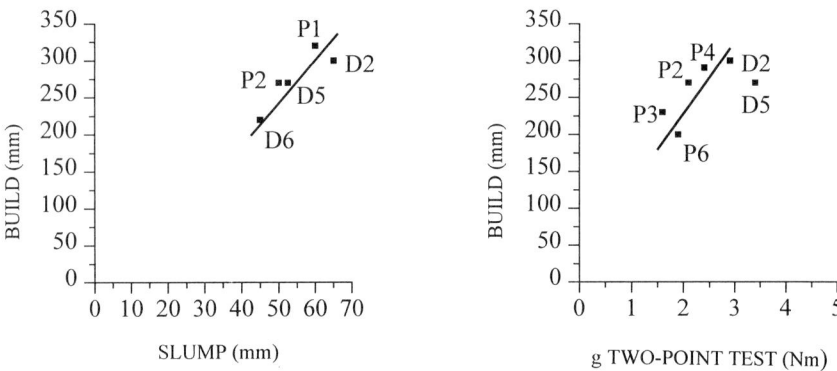

Figure 6 Build value (a) Slump (b) g Two-point test

Figure 7 presents the relationship between the build-value and the vane shear strength immediately before pumping. These results indicate an increase in build for a decrease in vane shear strength. As in Figure 6(a), this seems the opposite relationship to what would be expected but at these low workabilities a decrease in shear strength could produce an increase in the cohesiveness of the mortar, and therefore a corresponding increase in build. As the vane shear strength decreases further (due to an increase in workability) a point is reached where the mortar no longer fails due to the tensile stresses being exceeded but by a shear (i.e. flow) failure. At this point the maximum build is obtained. This point is difficult to establish here due to the workabilities of the mixes being within a narrow range. It can be noted that the mix D5 possesses a higher build value than the mix D1(which is identical to the mix D5 except for the addition of SBR), yet approximately the same shear value. The mix P1 also contained no SBR yet possessed the highest build value of all the mixes tested (except for the lightweight mortar, P7 (not shown)) which suggests that the presence of a polymer could reduce the build-value for a given shear strength.

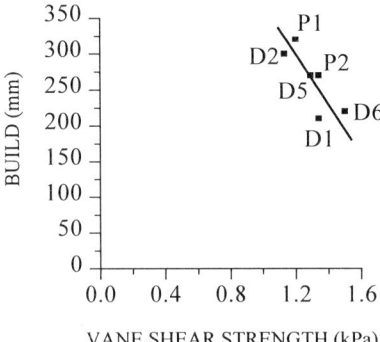

Figure 7 Build value Vs vane shear strength

CONCLUSIONS

A variety of data has been presented and discussed on the rheological performance of wet-sprayed fine mortars. A rheological audit was developed and tests for each stage used to characterise the pumpability and sprayability of each mortar. A shear vane test has been developed which can give an instantaneous measurement of the shear strength of mortar wherever this property needs to be assessed. A good correlation with the slump of a mortar was found and a relationship presented relating the vane shear strength to the build of mortar.

The Two-point test apparatus produced satisfactory results with fine mortars with low workabilities, although care needs to be taken in the conduct of the test and interpretation of the results. The grading of the constituents and the presence of polymers both had a significant effect on the results obtained.

ACKNOWLEDGEMENTS

The authors are grateful for: the financial support of the EPSRC (Grant number GR/K52829); the assistance of the industrial collaborators Balvac Whitley Moran, Fibre Technology, Fosroc International, Gunform International Ltd and Putzmeister UK Ltd; and the supply of additional materials by CMS Pozament, Flexcrete Ltd and Ronacrete Ltd.

REFERENCES

1. BEAUPRÉ, D. Rheology of high performance shotcrete. PhD Thesis, University of British Colombia, 1994.

2. TATTERSALL, G H AND BANFILL, P F G. The rheology of fresh concrete. Pitman, London, 1983.

3. TATTERSALL, G H. Workability and Quality Control of Concrete. E&FN Spon, London, 1991.

4. BANFILL, P F G. Rheological methods for assessing the flow properties of mortar and related materials. Construction and Building Materials, 1994, Vol. 8, Number 1, p 43-50.

5. BRITISH STANDARDS INSTITUTION. Specification for Portland cements. BS12:1989, British Standards Institution, London, 1989.

6. BRITISH STANDARDS INSTITUTION SPECIFICATION, Method for determination of slump, BS1881:Part102:1983, British Standards Institution, London, 1983.

7. BRITISH STANDARDS INSTITUTION. Specification for soils for civil engineering purposes, Part 9. In-situ tests, BS 1377:Part 9:1990, London, 1990.

INNOVATIONS IN THE FIELD OF SHOTCRETE REPAIRS

D Beaupré   N Dumais
S Mercier
Université Laval
P Lacombe
Service d'Expertise en Matériaux
M Jolin
University of British Columbia
Canada

ABSTRACT. This article presents the results obtained during a demonstration project where repair work was conducted using shotcrete. This project included the testing of new types of shotcretes. Two mixtures were shot using the dry process while six other mixtures were shot using the wet process. One of the dry-mix shotcrete mixtures contained a powdered air-entraining agent. All the wet-mix shotcretes were shot using the temporary high air content concept, which is described in this paper. Since the water/cement ratios of those mixtures are relatively low, they are also considered as high performance shotcretes. Four of the six wet-process mixtures included steel or polyolefine fibres. This paper also describes the nozzleman qualification program conducted in this project. Are also presented the various properties of the eight shotcrete mixtures. The overall results indicate that the new techniques tested in this project (the powdered air-entraining agent and the temporary high air content concept) are efficient and well adapted to the repair field, and also lead to durable shotcretes.

Keywords: Shotcrete, Sprayed concrete, Spacing factor, Durability, Scaling, Freeze-thaw resistance, Nozzleman certification, Temporary high air content, Fibre, Shrinkage

Dr D Beaupré Assistant professor at Université Laval and a principal researcher for the inter-university research center on concrete Sherbrooke-Laval and a director of the American Shotcrete Association. His field of interest includes rheology of fresh concrete, pumping and shotcrete technology. Most of the research carried out on the shotcrete area is part of the Industrial chair on shotcrete and concrete repair.

P Lacombe is a Professional engineer working in the field of shotcrete and concrete technology for the Service d'Expertise en Matériaux (S.E.M.). At the time of this study, he was a research engineer for the Industrial chair on shotcrete and concrete research at Laval University

N Dumais is a Graduate student at Université Laval. Her master thesis is on the practical use of the high initial air content concept

S Mercier is a Graduate student at Université Laval. Her masters thesis is on the use of powder air-entraining admixture in dry-mix shotcrete

M Jolin is a Ph.D. student at the University of British Columbia. He received his masters degree from Université Laval in the field of dry-process shotcrete accelerator

INTRODUCTION

This paper summarises the shotcrete repair work done during the refection of the Webster parking (Sherbrooke, P.Q.) in fall 1996. The condition of the old parking was such that major interventions were needed at the level of the beam reinforcement and repair of damaged elements (beams, walls, and columns).

The majority of the observed surface damages appeared as cracks and was caused by the corrosion of the reinforcing steel. Many new techniques and materials were applied to repair the structure. For example, shotcrete was used in many areas of the parking principally to repair and strengthen various beams. The repair technique used consisted in removing the old concrete by hydrodemolition, which proved to offer an excellent bond between the structure and the new repair shotcrete [1]. After, reinforcement is either replaced or added before the concrete is sprayed onto the surface.

The two processes (wet- and dry-mix) were used to apply the shotcrete. Note that the dry-mix process consists in conveying the dry material into the hose with compressed air to the nozzle. The water, controlled by the nozzleman, is added at the nozzle. The wet-mix shotcrete process, on the other hand, consists in pumping ready-to-use concrete and accelerate it at the nozzle to the required shooting velocity by mean of compressed air.

This article is divided in three parts. The first part describes the new technologies used in this project. The second one presents the nozzleman qualification program implemented before the work started. Finally, the third part presents the mix design, the research program and the results obtained from the various tests on the eight shotcrete mixtures.

NEW TECHNOLOGIES

Use of Powdered Air-Entraining Agent

The technique generally used to entrain air in dry-mix shotcrete is to add a liquid air-entraining agent to the shooting water [2]. However, this method has some disadvantages. First, the amount of air-entraining agent in the shooting water is difficult to control on site. Second, since the amount of water added at the nozzle is controlled by the operator, the amount of air-entraining agent in the fresh shotcrete will vary. Also, this method requires an additional step on the construction site (when the liquid air-entraining agent has to be added to the water), which is another potential error situation, and requires additional equipment (pump and tank). Finally, the control of the amount of air in the fresh shotcrete is easily achieved by the nozzleman: he simply has to increase the amount of water in the mix.

Recent studies conducted by the Industrial Chair on Shotcrete and Concrete Repairs of Laval University showed the effectiveness of powdered air-entraining agents in increasing the frost durability of dry-mix shotcrete.

This type of admixture has the great advantage of being bagged with the other oven-dried materials (cement, sand, coarse aggregates and fibres). With a dosage large enough, the powdered air-entraining agent produces an adequate air void system, which generates a dry-mix shotcrete resistant to freeze-thaw cycles [3].

Powdered air-entraining agent can help resolve many of the inconveniences encountered with the traditional liquid admixture: one less step on the construction site and improved quality control since the amount of air-entraining agent is constant throughout the concrete. Field tests were therefore conducted to validate the findings made in the laboratory.

Temporary High Air Content

Until now, the durability of shotcretes produced by the wet-mix process have been considered inadequate to allow its usage in repairs exposed to freeze-thaw cycles where deicer salts are present [4]. Note that the use of wet-mix shotcrete is relatively more complicated than the use of dry-mix shotcrete since there is a compromise to reach between the required criteria of pumpability and shootability [4]. Generally, the more pumpable the concrete is, the more difficult it will be to shoot it correctly: concrete with a large slump value will be easily pumped, but it will be difficult to obtain an adequate build-up thickness. Conversely, concrete with a small slump value, even if it permits to attain a larger build-up thickness, will be difficult to pump.

In order to compensate for the apparent impossibility of the wet-mix shotcrete, set accelerating admixtures are generally added at the nozzle to improve the shootability. However, the use of these admixtures is generally avoided in regions where the concrete is to be exposed to freeze-thaw cycles since they can be harmful to its durability. With this in mind, the concept of temporary high air content represents a valuable option to set accelerators.

The concept of temporary high air content, introduced by Beaupré [4], was used on the Webster parking to produce high performance concrete using the wet-mix process. With this concept, a high air content (from 15% to 20%) is used to increase the fluidity of the concrete, hence improving its pumpability. During the pumping and particularly during the shooting, a large amount of air is lost due to compaction. A lowered fluidity is obtained by the compaction effect, hence increasing the shootability of the shotcrete. For example, concretes with 15% of air content before entering the pump will have an air content of 3% to 5% after shooting. This technique permits to shoot concretes that have a low water/cement ratio, which should increase their durability.

Some wet-mix concretes used at the Webster parking contained steel fibres or polyolefine fibres. Those synthetic fibres are non-corrosives and help increase the mechanical properties of the hardened concrete [5].

NOZZLEMAN QUALIFICATION PROGRAM

Shotcrete technology requires workers that are especially skilled and experienced [6]. To ensure that the workers selected for the repair of the parking would possess the required skills, a nozzleman qualification program has been designed. This qualification program was conducted approximately two weeks before the actual work in order to carry out the required tests. The nozzlemen tested had to fill panels in the vertical and overhead position containing reinforcement as prescribed in the specifications. In order to qualify, a nozzleman must embed the reinforcement adequately, produce an adequate air void system in the shotcrete and obtain a minimum compressive strength. Only the nozzlemen who succeeded to all the requirements were allowed to place shotcrete during the repair work.

RESEARCH PROGRAM

Mixture Design

Overall, two mixtures were shot using the dry process and six mixtures were shot using the wet process in this project. One of the dry-mix shotcrete was shot using a liquid air-entraining agent while the other one already included a powdered air-entraining agent with the pre-bagged materials (Table 1).

The wet-mix shotcretes vary by the presence of fibres and their water/cement ratio (Table 2). Because fibres are recognised for their crack control, they were added in this case to verify if their presence can replace the welded wire mesh actually used in the majority of the thin shotcrete repairs. The design of those mixtures was done according to the tests performed in the laboratory of Laval University.

Table 1 Shotcrete composition (Dry-mix process)

COMPONENT	MIXTURE	
	Powder AEA	Liquid AEA
Cement 10FS (kg/m^3)	450	450
Water (kg/m^3)*	180	180
Sand (kg/m^3)	1500	1500
Stone (10-2.5 mm) (kg/m^3)	265	265
Air entrainer agent	Powder: 0,95 (kg/m^3)	Liquid: 12 ml/l of water

* Estimation based on a water-cement ratio approximately of 0,40

Table 2 Shotcrete composition (wet-mix process) and fresh concrete properties

	MIXTURE					
	W/B = 0,3			W/B = 0,35		
	30SF	30FA	30FP	35SF	35FA	35FP
Cement HSF (kg/m^3)	447	440	460	453	432	452
Sand (kg/m^3)	1126	1140	1155	1110	1105	1145
Coarse aggregate 10 mm (kg/m^3)	440	430	415	455	455	430
Water (kg/m^3)	134	132	138	155	155	162
Fibres (kg/m^3)	nil	45 steel	9 poly.*	nil	45 steel	9 poly.*
SF (kg/m^3) **	nil	nil	nil	10,0	10,0	10,0
WR (l/m^3) **	nil	nil	nil	1,5	1,5	1,5
SP (l/m^3) **	5,3	11,0	11,0	5,0	5,0	5,0
AEA (l/m^3) **	7,0	5,0	3,3	2,5	2,5	2,5
Slump (mm) before pumping	180	150	200	220	170	220
Air on fresh concrete (%) before pumping	16	13	15	17	19	18
Air on hardened concrete (%)	6,5	5,9	5,0	5,3	5,0	6,8

* Polyolefine ** FS:silica fume, WR: Water reducer, SP: superplasticizer, AEA: air entrainer agent

It should be mentioned that the wet shotcrete mixtures were adjusted during the project in order to increase the adhesion of the new shotcrete to the old concrete. Before those changes, the viscosity of the concrete was too high, hence the slump was increased in order to permit pumping. However, this increased slump was incompatible with the overhead shooting position. The water/cement ratio was then increased from 0,30 to 0,35 in order to lower the viscosity and facilitate the pumping. The silica fume content was also increased to improve the adhesion of the fresh shotcrete.

Sampling and Curing

All the mixtures shot during this project were sampled from 500 mm × 500 mm × 150 mm panels. Those panels were cured by covering them with dampened cloths and a plastic sheet to avoid water evaporation. After that, the panels were carried to the laboratory of Laval University for sampling and testing.

For the wet-mix shotcretes, samples were also taken before and after pumping (but before shooting) and tested for their compressive strength and air void characteristics.

Testing

In the case of the wet-mix shotcretes, slump (ASTM C 143) and air content (ASTM C 231) before pumping were measured on the fresh concrete. Air-entraining agent and superplasticizer adjustments were sometimes required. The results of those tests are presented in Table 2. When one compares the value of the air content before pumping and after shooting (measured on hardened shotcrete), the effect of the pumping and compaction due to the shooting onto the surface is obvious: generally, the air content goes from 16% to 6% (in average). This shows clearly that the concept of the temporary high air content can be used on a construction site.

For all mixes, compressive strength tests at 7 and 28 days (ASTM C 42), scaling tests with deicer salts (ASTM C 672), shrinkage tests (ASTM C 157) and air void system characterisation (ASTM C 457) were conducted. Also, flexural and toughness tests (ASTM C 1018) were carried out on the wet-mix shotcretes that contained fibres and on the sample mixture.

RESULT PRESENTATION AND DISCUSSION

Tests on Dry-Mix Shotcretes

All the results obtained from tests on the hardened dry-mix shotcretes are presented in Table 3. Compressive strength at 7 and 28 days, air void characteristics, mass of residues after 50 cycles of scaling tests and shrinkage measurements after 88 days are presented in this table.

Compressive Strength

Compressive strength tests were carried out after 7 and 28 days according to the ASTM C 42 procedure on samples with a length of 150 mm and a diameter of 75 mm. The samples had been cured in a 100% relative humidity environment at 23° C until the time of the test. The compressive strengths obtained for both mixtures are the same and are representative of concretes usually used in repair: 33 MPa or 29 MPa at 7 days and 43 MPa or 41 MPa at 28 days.

Table 3 Tests on hardened shotcrete (dry-mix process)

AEA MIXTURE	COMPRESSIVE STRENGTH (MPa) 7 d	COMPRESSIVE STRENGTH (MPa) 28 d	AIR (%)	SPACING FACTOR (μm)	SPECIFIC SURFACE (mm^{-1})	SCALING 50 CYCLES (kg/m^2)	SHRINKAGE 88 d (μm/m)
Liquid	29	43	5,2	220	25,8	2,0	862
Powder	33	41	4,4	160	38,6	0,6	767

Air Void System Characterisation

The air void characteristics were measured according to the ASTM C 457 procedure on samples of 100 × mm × 100 mm × 20 mm. The powdered air-entraining agent was efficient since this shotcrete obtained a spacing factor of 160 μm (Table 3). The concrete shot with the liquid air-entraining agent also obtained a good spacing factor with 220 μm. In both cases, the shotcretes respect the 300 μm limit taken in the specifications. The values of the specific surfaces for the mixtures with a liquid and a powdered air-entraining agent are respectively 25,8 mm^{-1} and 38,6 mm^{-1}. The higher value obtained for the powdered air-entraining agent shotcrete (38,6 mm^{-1}) indicates that the air void system is composed of smaller bubbles, thus generating a smaller value for the spacing factor.

Deicer Salt Scaling Test

The deicer salt scaling tests were conducted according to the ASTM C 672 procedure. The samples, sawed from the test panels, were cured (100% R.H.; 23° C) until the age of 14 days. After, the samples (300 mm × 225 mm × 50 mm) were dried (50% R.H.; 23° C) for 14 more days. After that, the samples were prepared and covered with salt water (3% of $CaCl_2$) for 7 days after which the deicer salt scaling test was started for 50 cycles. During the test, residues were collected and weighted every other 5 cycles.

The results presented in Table 3 show that the mixture containing the powdered air-entraining agent behaved in a better way when compared to its companion shot with a liquid air-entraining agent: 2,0 kg/m^2 for the liquid and 0,6 kg/m^2 for the powder. This difference is probably due to the difference between the spacing factors (160 μm versus 220 μm) or to a difference between the finishing of the surfaces of the testing panels.

Shrinkage

The shrinkage tests were conducted according to the ASTM C 157 procedure. Two prisms (400 mm × 100 mm × 75 mm) for each mixture were sawed from the test panels shot at the construction site. Afterwards, the samples were stored in water until the age of 7 days, which corresponds to the beginning of the tests. The results presented in Table 3 are the shrinkage values after approximately 3 months of drying in a 50% relative humidity environment. They are of the same magnitude as those generally found for this type of concrete. The mixture containing the powdered air-entraining agent yielded a slightly smaller shrinkage value, which is positive.

Tests on Wet-Mix Shotcretes

All the results obtained from the tests on the hardened wet-mix shotcretes are presented in Table 4. Compressive strength (7 and 35 days), flexural strength (28 days), air void characteristics, mass of residues after 50 cycles of scaling tests and shrinkage measurements after 104 or 120 days are presented in this table.

Compressive Strength

Compressive strength tests were obtained following the same procedure as for the dry-mix shotcretes. The strengths are obtained is relatively high and vary from 36 MPa to 50 MPa at 7 days. Unfortunately, no results are available at 28 days because of an equipment breakdown. The tests scheduled for the 28th days were carried out at the age of 35 days instead and the results vary from 52 MPa to 71 MPa. The low water/cement ratio and the compaction caused by the shooting are responsible for those high compressive strengths.

Table 4 Tests on hardened shotcrete (wet-mix process)

MIX	COMPRESSIVE STRENGTH 7 d 35 d (N/mm²)	FLEXURAL STRENGTH 28 d (N/mm²)	AIR (%)	SPACING FACTOR (µm/m)	SPECIFIC SURFACE (mm^{-1})	SCALING 50 CYCLES (kg/m^2)	SHRINKAGE 120 d (µm/m)
30SF	48 60	8,3	6,5	170	27,8	0,9	616
30FA	50 70	9,0	5,9	245	18,0	2,4	647
30FP	37 70	9,0	5,0	224	24,4	0,3	614
35SF	38 52	9,0	5,3	210	21,6	1,8	598*
35FA	37 71	7,5	5,0	196	25,3	0,5	590 *
35FP	36 58	9,5	6,8	225	18,7	1,2	666 *

*104 days results instead of 120 days

Flexural Strength

Flexural strength tests were carried out according to the ASTM C 1018 procedure on samples of 100 mm × 100 mm × 400 mm. The samples were cured in a 100% relative humidity environment at 23° C until the age of 28 days. The values of flexural strength (Table 4) obtained are very high and vary from 7,5 MPa to 9,5 MPa. Figure 1 shows the behaviour of the concretes after rupture. It can be seen that the concretes containing the polyolefine fibres had a post-peak behaviour similar to the concretes containing the steel fibres. Following the definition given by Morgan et al. [7], those concretes possess a toughness index of 2.

Air Void System Characterisation

The values of spacing factors measured on the hardened wet-mix concretes are generally below the specified 230 µm. The values obtained vary from 170 to 245 µm. Only the 30FA mixture is slightly above the prescribed limit. These results show that it is possible to generate an adequate air void system with this shooting technique.

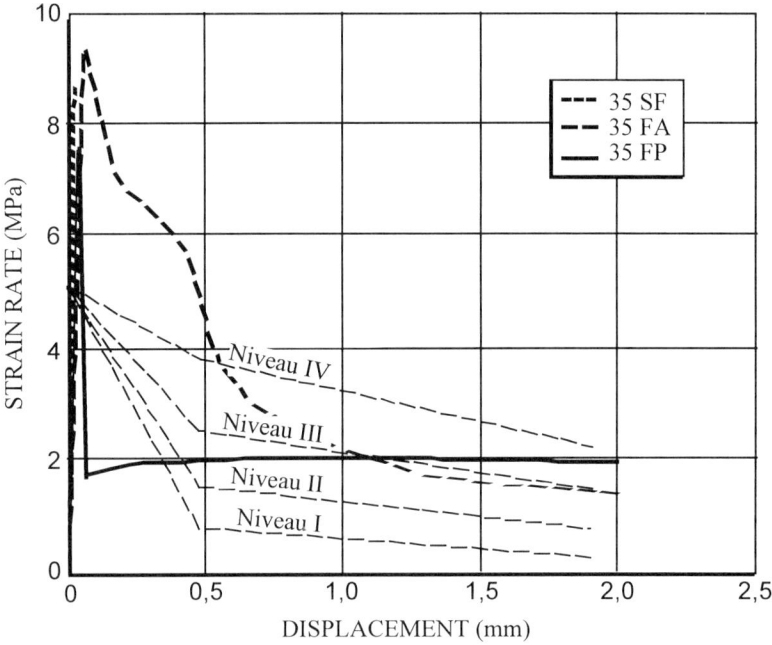

Figure 1 Flexural resistance test results (W/B = 0,35)

Deicer Salt Scaling Test

The deicer salt scaling test results are, on average, above the usual limit of 1 kg/m². The total amount of residues collected for the concretes with a water/cement ratio of 0,30 and 0,35 are relatively constant and equal to 1,2 kg/m². However, those concretes still have a better durability than concretes with a higher water/cement ratio [8], but have a lower durability than concretes with a similar design placed by traditional vibration. The variability of the results is probably imputed to the instability of the air void system. In order to verify this hypothesis, air void size-distribution analysis are presently under way. The results of those analyses will hopefully give an answer to this problem. The quality of the finishing of the tested surface can also play an important role in the variability of the results. Tests under way in the laboratory of Laval University show an important influence of the quality of the finishing of the surface on the salt scaling resistance results.

Shrinkage

The shrinkage tests were conducted in the same manner as for the dry-mix shotcretes. The results presented in Table 4 show that the shrinkage after 3 months of drying at 50% relative humidity is on average 600 µm/m, which is inferior to the value obtained for the dry-mix shotcretes even for a longer period of drying. A more complete analysis will be possible when the shrinkage will be stabilised, which usually takes 6 months.

CONCLUSIONS

This demonstration project verified, in a real construction environment, the use of new types of dry and wet-mix shotcretes. The results show clearly the efficiency of the powdered air entraining agent to produce an adequate air void system and an excellent salt scaling resistance. Moreover, the handling of a powder instead of a liquid simplifies the on-site operations and quality control, even if the obligation to pre-bag the dry materials slightly increases the cost of the shotcrete.

For the wet-mix shotcrete, it has been possible to pump, on site, concretes with a low water/cement ratio which, without the high initial air content, could not have been pumped. Similarly to what was obtained in the laboratory, the pumping and especially the compaction obtained during the shooting brought the air content down to an acceptable value. The test results show that it is possible to produce concrete with a good air void system that withstands well the salt scaling tests.

The addition of the polyolefine fibres allowed post-peak behaviour similar to the one obtained with steel fibre reinforced concrete. Since the work performed at this parking is carefully monitored, it should be possible to evaluate the long term performance of the repair. This evaluation will be particularly aimed at assessing the importance of the welded wire mesh in the control of the early cracking and at the bound between the new and old concretes. The authors hope that the information presented in this paper will help engineers to produce better shotcrete repairs.

ACKNOWLEDGEMENT

The authors wish to thank Concrete Canada, the city of Sherbrooke and the ministère des Affaires municipales du Québec for their financial support. The authors also thank Les consultants SM, Le Groupe Sherko, Le Groupe Lefebvre, Lafarge Concrete and Béton Mobile du Québec for their collaboration.

REFERENCES

1. TALBOT, C, PIGEON, M, BEAUPRÉ, D, MORGAN, D R. [1994] Long Term Bonding of Shotcrete, ACI Materials Journal, vol. 91, november-december, p. 560-566.

2. LAMONTAGNE, A. [1993], La durabilité au gel des bétons projetés par voie sèche: influence du type de ciment ainsi que du type et du dosage de l'agent entraîneur d'air, M. Sc. Thesis, Laval University, 92 p.

3. DUFOUR, J.-F. [1996], Effets des entraîneurs d'air en poudre sur l'entraînement de l'air et la durabilité du béton projeté par voie sèche, M. Sc. Thesis, Laval University, 176 p.

4. BEAUPRÉ, D, TALBOT, C, GENDREAU, M, PIGEON, M, MORGAN, D R. [1994] Deicer Salt Scaling Resistance of Dry- and Wet-Process Shotcrete, ACI Materials Journal, vol. 91, n° 5, september-october, p. 487-494.

5. MORGAN, D R, RICH, L D. [1996] Polyolefine Fibre Reinforced Wet-Mix Shotcrete, ACI/SCA International Conference on Sprayed Concrete, Edinburg, September 10-11 p.127-138.

6. ACI 506R-90, Guide to Shotcrete, American Concrete Institute, Detroit, 41p.

7. MORGAN, R, CHEN, L, BEAUPRÉ, D. [1995] Toughness of Fibre Reinforced Shotcrete, Engineering Foundation Conference: Shotcrete for Underground Support VII, Buchen-Telfs, Austria, June 11-15, p. 66-87.

8. BEAUPRÉ, D. [1994], Rheology of high performance shotcrete, Ph.D. Thesis, University of British Columbia, 250 p.

SHOTCRETE:
INTERNATIONAL PRACTICES AND TRENDS

K F Garshol
MBT International Underground Construction
Switzerland

ABSTRACT. A rapid development within shotcrete technology, equipment, working environment and capacity/economy has taken place and continues at a rapid pace. The most significant steps the last two decades are the change to the wet mix method and to steel fibre reinforcement, using mechanised equipment. The extremely rapid change in Scandinavia, followed by other European countries is demonstrated and the reasons are explained. Quality, capacity, working environment, safety and overall economy are the main issues discussed. Present trends in the development of shotcrete pumps, dosing equipment for accelerator and mechanisation and automation are described. Control of the final quality of shotcrete is a key objective. Concrete technology today offers a number of solutions to everyday practical problems and quality questions. Examples are hydration control admixture, alkali free accelerators, consistency control systems, curing by admixture added in the batching plant and super active water reducing admixtures. Chemical admixtures may compensate even poor aggregates that would otherwise cause pumping problems. Rebound well below 10% and total dust below 5 mg/m^3 of air at shotcrete output above 15 m^3/h are good illustrations of the benefits. State of the art wet mix shotcrete technology supports the tendency to install permanent final linings by shotcrete application, the so-called Single Shell lining approach. The drive for lower cost of tunnelling is one reason for this tendency. The features and benefits of Single Shell solutions are discussed.

Keywords: Shotcrete, Wetmix, Steelfibres, Equipment, Admixtures, Alkalifree, Robot, Dosing, Curing, Accelerator.

Knut F Garshol is Assistant Director of MBT International Underground Construction in Zurich. He is a Geological Engineer from the Norwegian Institute of Technology, Trondheim, with mining and tunnelling experience from contracting and consulting during 28 years. He has been actively involved in all aspects of shotcrete for rock support, particularly the wet mix method with steel fibres and he was a member and chairman of the Norwegian Shotcrete Committee for several years. Garshol is currently taking part in the CEN standardisation work for sprayed concrete, representing Switzerland. He has presented a series of papers on shotcrete, rock support, pressure injection and rock mechanics in national and international conferences. (Contact through: knutfg@online.no or http://www.ugc.mbt.com/).

INTRODUCTION

Over the years since the start of shotcrete application, one fact cannot be discussed: There has been a rapid development within all aspects of the method, from concrete technology and equipment to working environment and capacity. It is also clear that the development continues at a rapid pace.

In 1911 Carl Ethan Akeley, founder of today's Allentown Pneumatic Gun (a MBT company in USA), patented the first cement gun (dry spraying).

In 1957 MEYCO Equipment introduced the first rotor machine (GM 57), which was a revolution due to possibility of continuous shotcrete application.

During the 1960's the wet mix method started to develop and during the 1970's had already in some countries taken over the market (Scandinavia).

During the 1980's the steel fibre reinforcement as an alternative to mesh (or welded wire fabric, WWF) increased rapidly in countries where the wet mix method was already established.

During 1980's and 1990's the global change in the direction of wet mix application has continued, with the steel fibre reinforcement following as a natural next step.

When trying to describe international trends and practices, the problem is that there are significant differences from region to region and the state of affairs are also different in mining compared to civil construction. To describe this variation in detail would go too far and is not really interesting. This paper, therefore, is focusing on what this author feels are important developments. Only examples will be given to support the subjective views presented.

DEVELOPMENT IN SCANDINAVIA

A brief presentation of the shotcrete technology development in Scandinavia and why it happened, will at the same time illustrate why countries like France, Italy and in later years also Switzerland, UK and some other European countries have been and are moving in the same direction.

Between 1971 and 1980, the shotcrete application in Norwegian tunnelling turned from dry mix method only, to 100% wet mix application. A similar, but somewhat slower change took place in Sweden and Finland.

The next dramatic change was from mesh reinforcement to steel fibres. Again, the fastest turnaround happened in Norway. Between 1978 and 1988, the steel fibre consumption increased from about 50 tons per year to about 2800 tons. During these 10 years, the mesh reinforcement was completely replaced by steel fibres and the market share of reinforced shotcrete to unreinforced increased. In 1986 the total volume sprayed in Norway was 56,000 m^3, out of which 70% was fibre reinforced and the rest was unreinforced. A bit later, a similar development took place in the rest of Scandinavia.

The main reasons for these changes are clear enough:

- Rapid increase in shotcrete output per pump hour
- Rebound reduction from typically 30% down to 10% and less
- Substantially reduced transport and set-up times per application
- Reduced number of people per equipment
- Less compressed air and total energy consumed for shotcrete on the wall (< 50%)
- As a direct consequence, the cost reduction was dramatic. In fixed value, the cost of shotcrete on the wall was reduced by 75% from 1972 to 1986

Such a cost reduction may seem unrealistic at first glance. But, if the pump output increases from less than 5 to about 25 m^3/h, the rebound reduces by 66%, robot application and integrated self sufficient equipment units cut the set-up time from average 2 hours to 20 minutes and the necessary staff is reduced by about 50%, it may all seem a bit more plausible.

Furthermore, the steel fibre reinforcement can be added to the concrete batch, basically without any changes to the overall system and the fibre rebound in the wet mix method is about equal to the rebound of concrete material (see Figure1). Figure 1 also shows why steel fibres and the dry mix method is a poor combination; more than 50% of the fibres are normally lost in rebound [1]. Even though the total materials cost per m^3 may double when fibres are added, this is more than compensated for by improved quality, safety, convenience and capacity. Structurally equivalent reinforcement by steel fibres has been back calculated to cost 50-60% of the WWF alternative. When the complete cycle time for a drill and blast operation may be reduced by 25 to 50% (see Figure 2) the potential project savings overrule the differences in materials cost between mesh and fibres by a large margin.

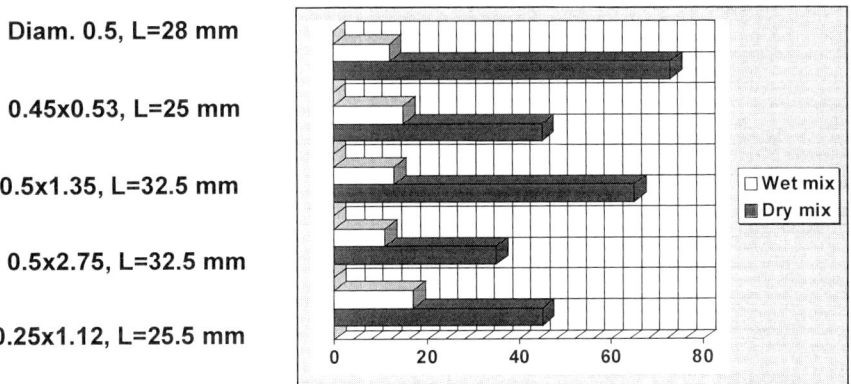

D. Wood et.al.

Figure 1 Fibre rebound, wet and dry method

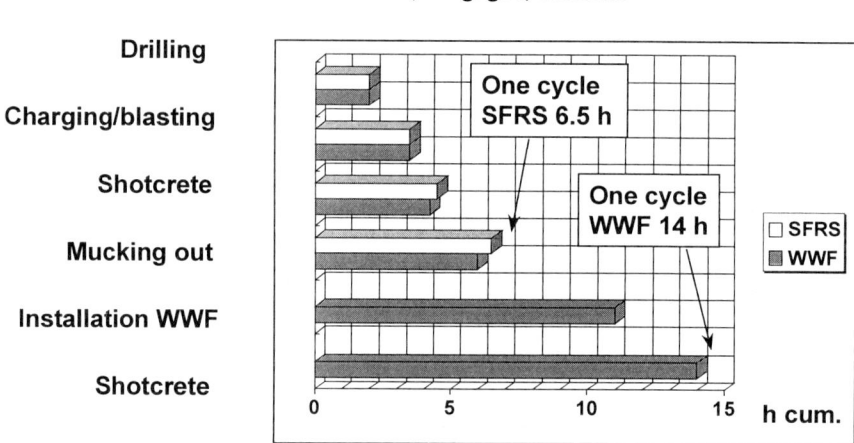

Figure 2 Cycle times using mesh and steel fibres

SOME HIGHLIGHTS FROM OTHER COUNTRIES

The data presented in Table 1 are not accurate, but are still reasonably correct and can be used for the purpose of showing today's situation and the tendencies. Some comments are given as a supplement.

When the construction period started in London Underground for the Jubilee Line and other well known projects in the area, the market was practically 100% dry mix. Within slightly more than a year, this was turned around to wet mix, illustrated by the fact that MBT alone had 17 Suprema wet mix pumps in operation.

It is sometimes claimed that in the Third World countries, only the simple, "low cost" dry mix method can work properly. Reality is different, since there is no relation between the choice of method, local development tendency and such popular regional tags. This is well illustrated by the figures in Table 1 for China, India, Brazil compared to Austria and Germany.

Austria and partly Germany have been traditionally very strong dry mix areas, as was Switzerland some years back. Switzerland turned around to wet mix, while especially Austria has developed new systems within the dry mix method, using extremely quick setting specially developed cement types. This allows spraying of thick layers at pretty high capacity, without the use of any admixtures or accelerators.

This quick cement alternative appears favourable at first (lower material costs per m^3 mixed), but the downsides are also there. The most serious negative aspects being rebound and dust, sensitivity to cement quality, high to very high energy cost and no realistic chance of using steel fibres.

Table 1 Shotcrete methods and volumes per year (approximate)

COUNTRY/REGION	WET, %	DRY, %	m³/YEAR	TENDENCY
Japan	90	10	2.0×10^6	Wet
China	60	40	600,000	Wet
Asia/Pacific	60	40	1.0×10^6	Wet
USA	70	30	500,000	Wet
Brazil	80	20	400,000	Wet
Latin America	60	40	500,000	Wet
United Kingdom	80	20	100,000	Wet
Germany	20	80	600,000	Wet??
Austria	1	99	100,000	?????
Switzerland	90	10	150,000	Wet
France	90	10	200,000	Wet
Spain	90	10	300,000	Wet
Italy	100	0	700,000	Wet
Greece	80	20	200,000	Wet
Scandinavia	100	0	250,000	Wet
India	50	50	200,000	Wet

For a number of local reasons, this is the situation and it may well remain so for some time more. However, this author regards the method as a blind track and even though it cannot be named a tendency, there are still clear signs that a change in direction wet mix is brewing.

Another interesting area may be Australia, which is not mentioned in the above table. In this area the mining industry was ahead of civil construction in the move from dry mix to wet mix and partly to steel fibres. This has lead to a substantial market share for wet mix in both mining and civil construction in Australia. The road tunnelling ongoing in Sydney and Melbourne is all carried out wet mix

GENERAL TRENDS

Equipment

The dedicated piston pumps for wet mix shotcrete application are gaining on other pump alternatives. Today, the capacity of such a pump must be up to about 20 m³/h, but it must also function well down to 25% of this capacity. As far as possible, the concrete flow should be pulsation free.

Shotcrete quality requirements are increasing and to be able to control the quality it is necessary to control dosage of accelerator. An integrated dosing system, linked to the concrete pump capacity, is more and more requested. The more advanced version of this, delivered as current technology from MEYCO Equipment, includes flow measurement and computer controlled dosage. See Figure 3. Under development is a similar system, where also the actual concrete flow will be continuously checked by a flow-metering device.

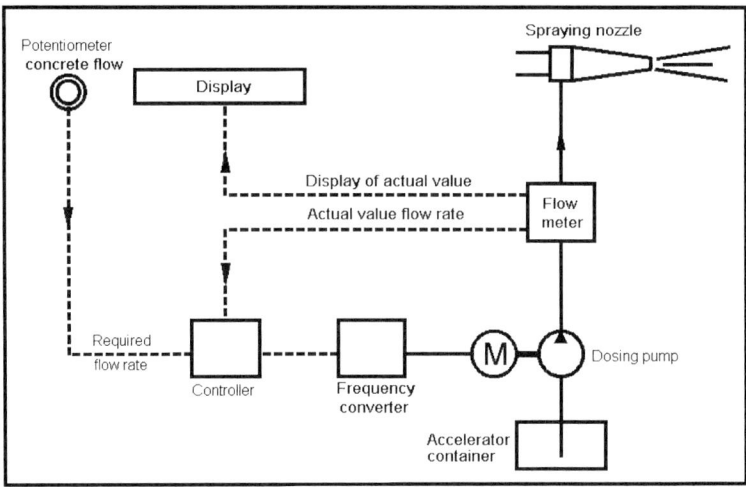

Figure 3 Control system for accelerator dosage

The high capacities are only possible with mechanised nozzle movement, the so-called robots. There are a number of such equipment designs on the market, but not all of them suitable for wet mix. The tendency, furthermore, is to integrate in single units all the necessary functions for shotcrete application. This means on a carrier: Concrete pump with dosage system and tank, robot arm, compressor for spraying air, hydraulic power pack, support legs, working lights, water tank and high pressure cleaning device, remote control of all main functions, cable and cable drum etc. In civil construction an increasing number of such systems are being sold, as illustrated by 7 MEYCO units to Norway alone within a 6 months period.

Similar compact units for mining and for smaller civil construction tunnels are also available. See Figure 4 and 5.

In the future, we have identified major possibilities within computer aided robot operation. We can already run fully automatic shotcrete application to pre-defined thickness and location, as well as semi-automatic spraying, where the operator is controlling the spraying pattern while the computer takes care of nozzle distance and angle.

Environmental Aspects

Traditionally, shotcrete operators were used to excessive dust and health problems, skin burns, risk of loosing the eyes and sometimes high risk of injury from falling rock (dry mix application, manually, using caustic aluminate accelerator and mesh reinforcement on unsupported ground).

Shotcrete: International Practices and Trends 169

Figure 4 Self contained complete spraying unit (robot)

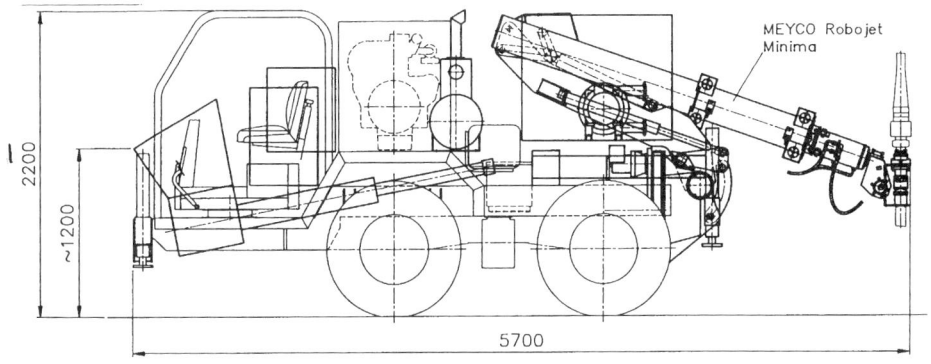

Figure 5 Compact robot unit

It is an international trend, that these negative conditions are not accepted any more (with large local variation). As an example: In France it is now in reality not allowed to use caustic aluminate as accelerator, due to the health risk to personnel. According to information given in the ITA Working Group Shotcrete, in Washington 1996, one important reason for the high wet mix market share in Brazil is the health problems caused by dust.

One step forward is the choice of the wet mix application method. More is gained if also the caustic aluminate is exchanged by the modified silicate based system TCC. Dust and rebound is reduced, it works with all cement types and even though caustic, an accidental spray in the eyes will not cause permanent damage if washed out with water immediately. Skin burns can be made a matter of the past.

The latest development is the liquid alkali free accelerator. The working environment is one reason for the development and use of such products. Within the wet mix application method, results from the North Cape tunnel indicates total dust less than 3.5 mg/m^3 air using MEYCO SA 160 alkali free (with silicate based accelerator, about twice as much). The other reasons are quality and durability of the shotcrete and requirements regarding ground water and the outside environment. The trend to increase the use of alkali free accelerators is now very clear.

CONCRETE TECHNOLOGY

The primary areas of present day shotcrete development can be split in two main groups:

1. Wet mix with the use of admixtures, additives and accelerators tailored to the needs of the project. Reinforcement by steel fibres.

2. Dry mix based systems using quick reacting cements, avoiding all admixtures and accelerators. Reinforcement by mesh.

The first group covers what is today the main volume of shotcrete production and also the expected future of shotcrete. Apart from the wide range of existing technical solutions and modifications, there is every reason to expect a further rapid development primarily within this area. The group 2 issues are therefore not further commented.

Some of the present possibilities may be used to illustrate the situation.

Hydration control admixture may be used as a logistics tool, by stopping the cement hydration for whatever necessary between a couple of hours, up to days. Be aware, this can be done without restricting the freedom of choosing when to spray, even when accelerator and high early strength is required. This technique furthermore enhances the final concrete quality and even the early strength.

Wet mix shotcrete may be applied with crushed aggregate only, poor sieve line with lack of fines, low cement quantity and long pumping lines. Of course, there are limitations, but chain polymer admixtures can stabilise the mix, lubricate the pumping line and compensate most of these problems.

When high quality is required, curing of concrete is of great benefit. In shotcrete applications this is normally never done, due to a number of practical problems. Today, an admixture added in the concrete mixer can take care of this and thus increase bond strength, reduce cracking and improve other quality parameters.

Alkali free liquid accelerators are available and a constant research activity is ongoing in search of further improvements. Already today, early strength in wet mix as high as above 5 N/mm^2 after 30 minutes has been reached. More typical is 1 N/mm^2 after 30 to 60 minutes. Final strength is currently in the range of 35 to 60 N/mm^2, depending on specified strength.

Wet mix spraying with a fluid concrete mix (slump measure of 150 to 200 mm), without accelerator has been considered more or less impossible, unless just 20 mm is applied per layer. With the TCC system, spraying of up to 100 mm, even in the roof, is possible using no accelerator. One important reservation must be noted: The next layer may have to wait for a few hours, depending on type of cement and on the temperature.

SINGLE SHELL PERMANENT SHOTCRETE LINING

The term "single shell sprayed concrete lining" is probably not understood in the same way by everybody. The following definition can be used:

> **Single-shell construction method** - All static and structural requirements to be met by the tunnel lining are fulfilled by a single shell element consisting of one or several layers forming a composite structure [2].

The practical consequence of linings carried out as single shell structures is that all concrete placed, will be part of the final structure. Compared to traditional methods, where the concrete spraying carried out at the tunnel face is considered temporary and structurally not contributing, this is probably the single most important difference.

Compared to the traditional use of a temporary sprayed concrete support (already alone close to structurally satisfactory) and a mass concrete permanent lining, the cost savings potential could be higher than 50% [3]. One reason is concrete volume savings, as illustrated in Figure 6.

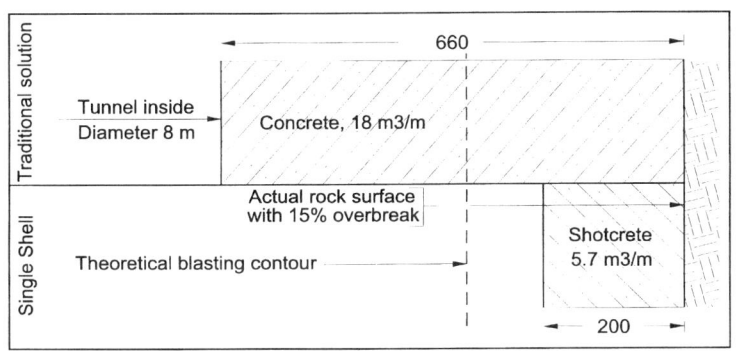

Figure 6 Concrete quantities for a 8 m diameter tunnel, example calculation

In all parts of the world there is a growing demand for underground construction, especially in urban areas. Typically, this demand is higher than the available financial resources. This is the case in modern countries as well as in third world countries. The economic realities are one reason for an internationally growing interest in the single shell lining concept. One example is the research program on the subject proposed to the EU by six companies from the industry, with a three years budget of more than CHF 3 million. The Geofronte organisation in Japan is currently evaluating the single shell concept, as well as the highway road authorities in the UK.

CONCLUSION

Expressed as short as possible, the visible international practices and trends are the following:

- Wet mix shotcrete application is dominating, with a clear trend for further increase
- A general use of mesh reinforcement, but with an increasing volume of steel fibres
- A drive in the direction of mechanisation, automation and higher capacity
- More focus and stricter regulations on safety, working environment and external environment
- Today, mostly temporary support shotcrete, but increasing interest and use of higher quality shotcrete for permanent linings

A presentation of practices and trends with a content like in this paper, will probably generate some resistance from those professionals, being more, or only familiar with the dry mix method. This is understandable. For the sake of balance it must be emphasised: The dry mix method has a number of good features that makes it the right tool in a number of situations. This is likely to remain so. At the same time, the situation of today and the trends described are clear and cannot be claimed to happen just by chance. If a recommendation may be allowed:

Keep the balance, use dry mix when this method is the best choice, but there is a wealth of experience and information available about modern wet mix application and the reasons why it should mostly be preferred. Before deciding which way to go for a given project, make sure to get hold of such information.

REFERENCES

1. WOOD, D. et al. Concrete International, May 1992 and June 1994.

2. AUSTRIAN CONCRETE SOCIETY. Guideline Shotcrete, Final Draft Issue, Vienna 20. February 1997.

3. BARTON, N. Investigation, Design and Support of Major Road Tunnels in jointed Rock using NMT Principles, IX Australian Tunnelling Conference, August 27-29 1996, Sydney, Australia.

THEME THREE:
FOAMED CONCRETE

Keynote Paper

MICRO-PROPERTIES OF FOAMED CONCRETE

E P Kearsley
M Visagie
University of Pretoria
South Africa

ABSTRACT. Foamed concrete is manufactured by adding a foaming agent to a cement based mortar. Research on the material properties of foamed concrete has been conducted on an ongoing basis at the University of Pretoria since 1992. The mechanical properties of foamed concrete are dependent mainly on the micro-properties of the material and therefore microscopic studies on the micro-properties have also been conducted. The air-void size distribution will be discussed in this paper as it is one of the most important micro-properties influencing the strength of foamed concrete. The aim of this paper is to find parameters to explain and quantify relationships between air-void distribution and structural properties. Preliminary results will be discussed.

Keywords: Foamed concrete, Air-void size distribution, Image analysis, Micro-properties, Lightweight concrete, Aerated concrete.

Mrs Madeleine Visagie is lecturer in Civil Engineering at the University of Pretoria in South Africa. She is currently busy with research on the Micro-Properties of foamed concrete for her Masters degree in Civil Engineering. This paper is based on the research project being a prerequisite for completion of a master's degree in civil engineering.

Mrs Elsabé P Kearsley is a Senior Lecturer in Civil Engineering at the University of Pretoria in South Africa. Her main interests are the development of affordable alternative building materials and methods.

INTRODUCTION

Foamed concrete is produced by introducing large voids (0.1mm to 1.0mm diameter in size) into the mortar mass, which consists of cement, water and filler. [1] The introduction of these voids is achieved by adding foam to the mix. A foaming agent consisting of hydrolized proteins is diluted with water and aerated to form the foam.

According to Neville [1] the strength, as well as the durability and volume changes of hardened cement paste, appears to depend not so much on the chemical composition as on the physical structure of the products of hydration of cement and on their relative volumetric proportions. Therefore an in-depth look into the parameters of the air-void structure of foamed concrete is necessary to establish the relationship between micro-properties and material properties.

This investigation forms part of the research project being a prerequisite for completion of a master's degree in civil engineering. The aim of this paper is to find parameters to explain and quantify relationships between air-void size distribution and structural properties. Preliminary results will be discussed.

GENERAL BACKGROUND

Research on the material properties of foamed concrete has been conducted on an ongoing basis at the University of Pretoria from 1992. Tests to determine the relationships between compressive strength versus water / cement ratio, - ash / cement ratio and - sand / cement ratio indicated that there appears to be an optimum mix for specific percentages of foam. The density of foamed concrete is directly related to the percentage of foam that is added to the cement – based slurry. Foamed concrete with densities varying between 700 and 1500 kg/m^3 has been investigated. [2].

Wainwright and Olorunsogo [3][4] stated that the pore structure of a cementitious material, predetermined by its porosity, permeability and pore size distribution, is a very important micro-structural characteristic as it influences such properties of the material as: - strength, fracture toughness and durability. The mechanical properties are dependent mainly on the distribution of the pores within the hardened cement paste. The micro-properties could therefore be a primary factor influencing the material properties of foamed concrete.

Two foamed concrete mixes A and B (see Table 1) with the same wet density and water / cement ratio were cast, but the compressive strength for the mixes differed (by approximately 30%). Microscopical images from samples of the two mixes were prepared (see Figure 1). In photo A, an image of the mix with the higher compressive strength, the spherical air-voids are destinct with an even distribution through the matrix.

Whereas in photo B, the mix with the lower compressive strength, some of the air-voids are arranged together to form large uneven openings and in other places the air-voids' perimeters are irregular. From this microscopical images it appears that some of the most important micro-properties of foamed concrete are the air-void shape, - spacing factor and - size distribution.

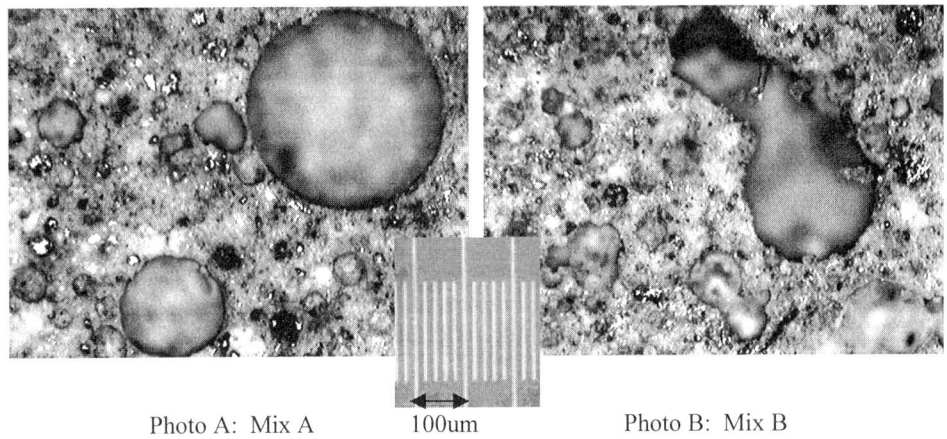

Photo A: Mix A   100um   Photo B: Mix B

Figure 1 Microscopical images

Methods to Determine Air-Void Content

The ASTM C457 [5] describes a "linear traverse"or "modified point-count method" for microscopical determination of air-void content and parameters of the air-void system in hardened concrete. It can be used to predict the durability of normal concrete in freeze-thaw conditions.

The European standard prEN 480 Part II [6], based on linear traverse measurements has been adopted as an efficacy test for air entraining agents. The linear traverse method described in both standards involves identification, counting and, in the European Standard, measurement of air-voids.

Cahill [7] developed a new method where image analysis in conjunction with the linear traverse method are used to determine the air-void parameters of air entrained concretes.

Nasser and Singh [8] developed a new apparatus and method to determine the air-void characteristics of fresh and hardened concrete. The apparatus consists of a video camera, lighting system, computer and image processing software. The method consists of video images taken of the concrete surface to be tested and these are analyzed to determine the size, distribution, spacing and percentage of the air bubbles in the concrete.

For the purpose of this research a similar image analysis method as developed by Nasser and Singh [8] was selected to be used in determining the air-void parameters of foamed concrete. The image processing and analysis system used consists of a high-resolution monochrome camera connected to an optical microscope and computer with the image analysis software "Optimate 5.21"[9][10].

The microscope has a movable X,Y stage and by moving the sample manually in both the X and Y directions, representative images of the sample are collected. The software has a function, which allows the defining of areas on each image. The data of every defined area in the image is stored and analyzed separately [9].

The output data of the defined areas, consisting of the total area in calibration units, total length of perimeter in calibration units, x – and y position of area centroid, circularity, longest axis in calibration units, width perpendicular to longest axis, mean and standard deviation of the grey value of the pixels enclosed by the area, is stored in an ASCII file for use by any data processing computer program [9].

EXPERIMENTAL DETAILS

Sample Preparation

Six different foamed concrete mixes are used in this investigation – see Table 1 for composition, densities and 28-day compressive strengths of mixes. The foaming agent used is " Foamtech", consisting of hydrolysed proteins and it is manufactured in South Africa. Rapid Hardening Portland cement from PPC, Herculus, Pretoria and pulverized-fuel ash from Letabo is used in the mixes.

Table 1 Foamed concrete mixes

MIX NO	WATER / CEMENT RATIO	ASH/CEMENT RATIO	DRY DENSITY (kg/m3)	AVERAGE 28-DAY CUBE STRENGTH (MPa)
A	0.87	2	1471*	21
B	0.87	2	1475*	16
1	0.58	1	1235	24
2	0.58	1	901	6.5
3	0.58	1	811	5
4	0.58	1	735	3
5	0.58	1	646	2
6	0.58	1	559	1.5

* Wet density

Image analysis specimens are taken from the 28-day strength cubes and the dimensions of these samples are 50mm x 50mm x 12-15mm. The intended test surfaces of the samples are wet ground with a 180 and 600 grit to produce a smooth flat surface. It was cleaned with compressed air to remove any residue and prepared in the oven at 50°C to ensure a dry surface for microscopical analysis [9].

Identification of Air-Voids

The focus of the optical microscope is adjusted until the cement paste is in focus and all the air-voids are blurred (see photo A and photo B). This contrast is sufficient to distinguish between air-voids and cement paste and thus every air-void's perimeter can be manually traced on the image [9].

MEASUREMENT OF AIR-VOID PARAMETERS

Air-Void Size Distribution

The essential feature of entrained air-voids, is that they are spherical and therefore circular in section [6],[7]. From the output data of the defined air-voids, the area is used to determine the diameter.

Air-void diameter = $\sqrt{\dfrac{4A}{\pi}}$ where A = area of void

Only voids with diameters greater than 0.01mm were counted. The size distribution of air-voids in foamed concrete follows a log-normal distribution [9]. The size distribution of air-voids for the mixes shown in photo A and B were determined and histograms of these mixes are shown in Figure 2 and Figure 3.

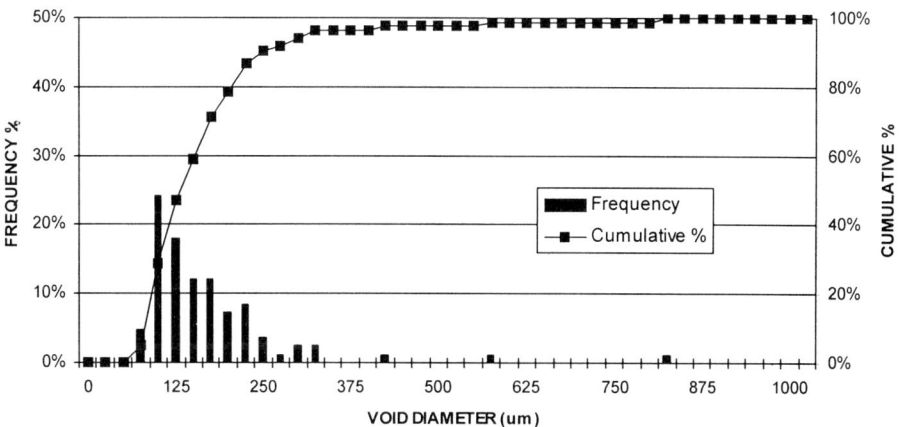

Figure 2 Void distribution of mix A

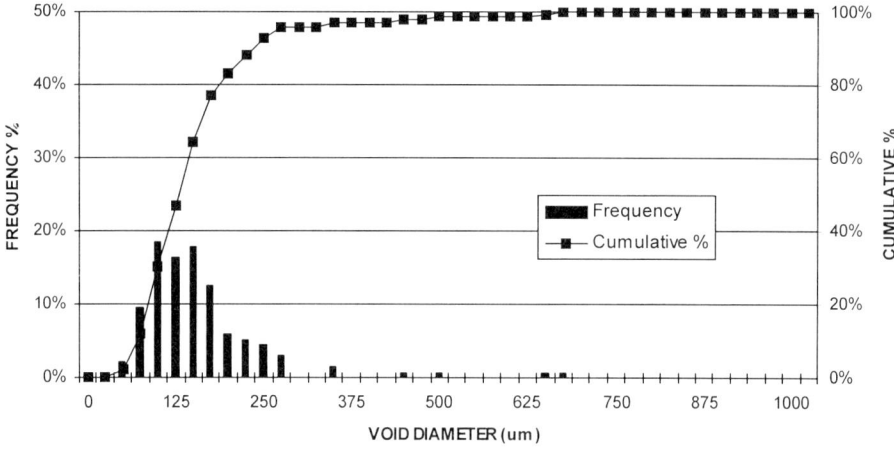

Figure 3 Void distribution of mix B

The air-void size distribution of the two mixes follows a log-normal distribution. Although the composition of the mixes is the same the 28-day compressive strengths and histograms differs. For example the standard deviation and mean void diameter of the two histograms are different. It is therefore necessary to find parameters to explain and quantify relationships between air-void size distribution and structural properties, such as 28-day compressive strength.

Rosin – Rammler Distribution

There are various mathematical functions available for the description of particle size distribution of powdered and cementitious materials. One of these distributions is the Rosin–Rammler distribution function which gained acceptance for use in the cement industry [4].

Rosin and Rammler [4] investigated the particle size distribution of crushed coal and developed a function that describes the distribution as:

$$f(x) = e^{(-bx^n)}$$

They also found that the function does not only apply to crushed coal but also to various other powdered materials. Bennet [4] modified this function to:

$$f(x) = e^{(\frac{x}{x_0})^n}$$

Where:

        $f(x) = RR =$ the percentage retained on sieve size x

x = particle size (μm)

x_0 = the position parameter, it represents the particle size above which 36.8% of the particles are coarser, and gives an indication of the fineness of the material.

n = slope, it indicates the range of distribution of the material.

Taking double logarithm of the above mentioned equation gives:

$$\ln\ln(\frac{1}{RR}) = n(\ln x - \ln x_0)$$

This equation describes a straightline plot where the coordinate system is made up of $\ln x$ on the horizontal-axis and $\ln\ln(\frac{1}{RR})$ on the vertical-axis. n is the slope of the straight line and describes the spread of the particle size distribution. The straight line is intercepted on the horizontal axis at a value describing the particle size x_0 which is a measure of the degree of fineness of the material [4].

Application of Rosin-Rammler Distribution

Provision was made for statistical outliers by not taking the five- percent smallest diameters and five percent biggest diameters into account. The two-sided 95% confidence interval was used to plot the cumulative percentage oversize distribution of air-voids for all the mixes (see Figure 4).

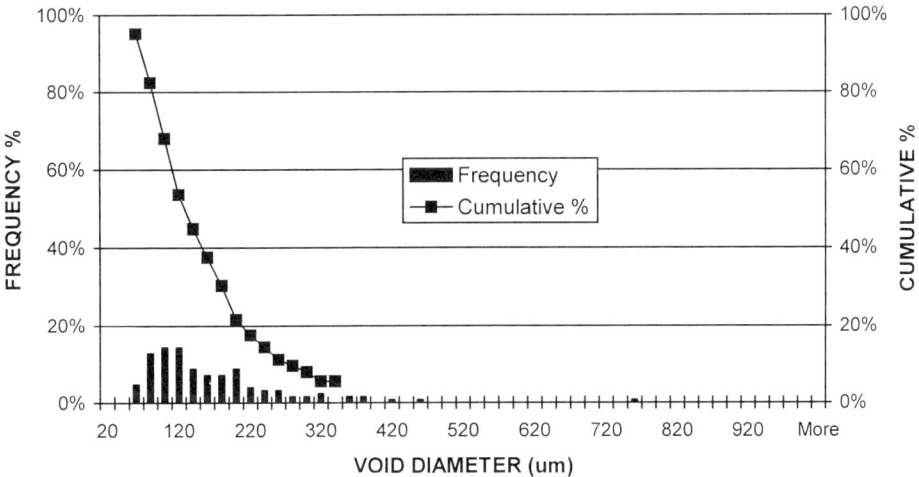

Figure 4 Cumulative percentage oversize voids

An exponential trendline was added to the graph and the equation and R-square value displayed on the graph (see Figure 5). The R-square value for all the mixes was between 0.98 and 0.998 and therefore the equation $RR = e^{-bx}$ determined by the trendline for every mix is a true representation of the cumulative % oversize distribution.

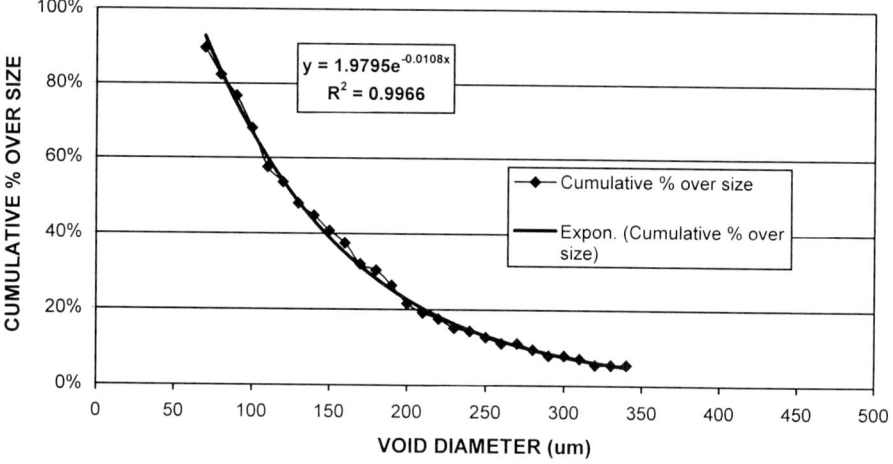

Figure 5 Exponential fit

The values obtained from these equations are used to plot the modified Rosin-Rammler distribution graph $\ln\ln(1/RR)$ versus $\ln x$, with RR the cumulative % oversize and x the air-void diameter (see Figure 6). A linear trendline and equation is also added to these graphs. The slope and interception with the horizontal axis of the line is taken as the n value and $\ln x_0$ value respectively of the modified Rosin-Rammler function. The n value represents the range of the air-void diameter distribution and the x_0 value (position parameter) indicates the air-void diameter (μm) which 36.8% of the air-void diameters are greater and indicates the void size.

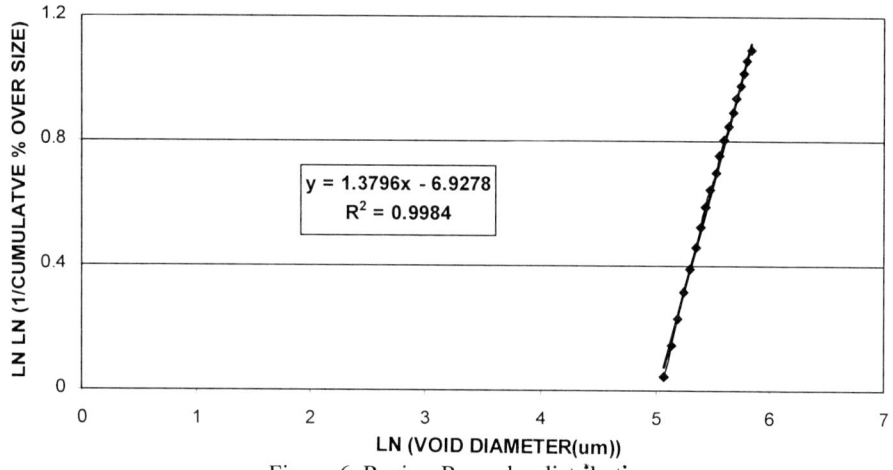

Figure 6 Rosin - Rammler distribution

RESULTS

A summary of the modified Rosin -Rammler distributions as fitted for the six mixtures can be seen in Figure 7. This graph clearly indicates that the air-void diameter (or position parameter - x_0 value) decreases as the dry densities increase but the effect of density on the range of void size distribution (or slope - n value) can not be determined from this graph. Therefore the slope (n) and the intercept ($\ln x_0$) was plotted as a function of density in Figure 8. From Figure 8 it can be seen that a relation exists not only between dry density and the intercept ($\ln x_0$) but also between the dry density and the slope (n). The slope increases with an increase in density, indicating a smaller range of void size distribution at higher densities. The effect of void size and distribution on compressive strength can be seen in Figure 9 and 10.

Figure 9 indicates that an exponential relation exists between void size and compressive strength. The air-void diameters increase as the compressive strength decrease. It seems as if the position parameter's graph incline asymtotical towards 150um. Figure 10 shows that the slope increases with an increase in compressive strength indicating a narrower air-void size distribution for a higher compressive strength.

It is therefore possible that the mixtures with higher compressive strengths contain voids that are more uniform in size. At lower strengths the voids seem to merge; thus forming larger voids, resulting in a wider distribution in void sizes.

More tests will have to be conducted to establish whether an asymtotical trend towards a certain void size exists. The effect of spacing and shape of voids still has to be established.

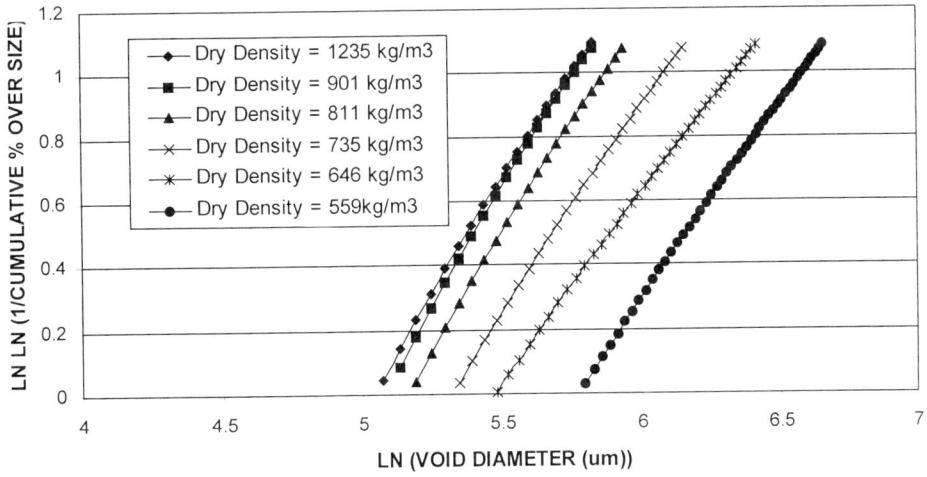

Figure 7 Summary of Rosin - Rammler distributions

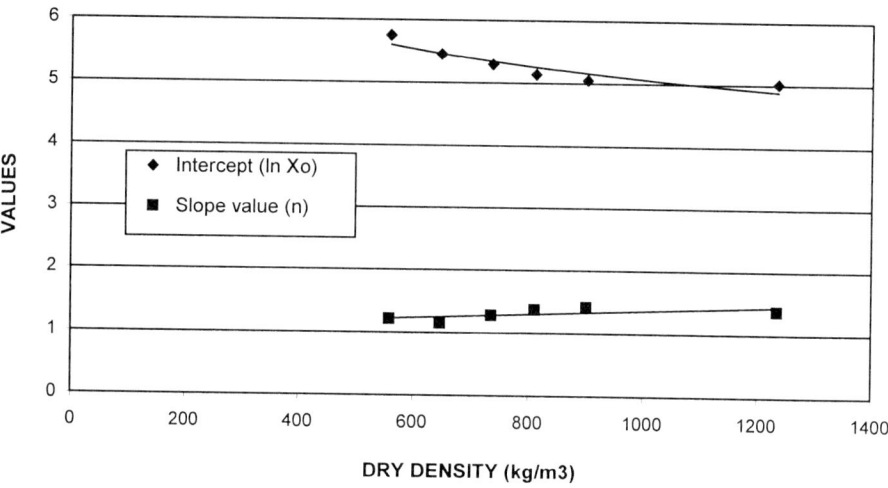

Figure 8 Air void distribution as a function of dry density

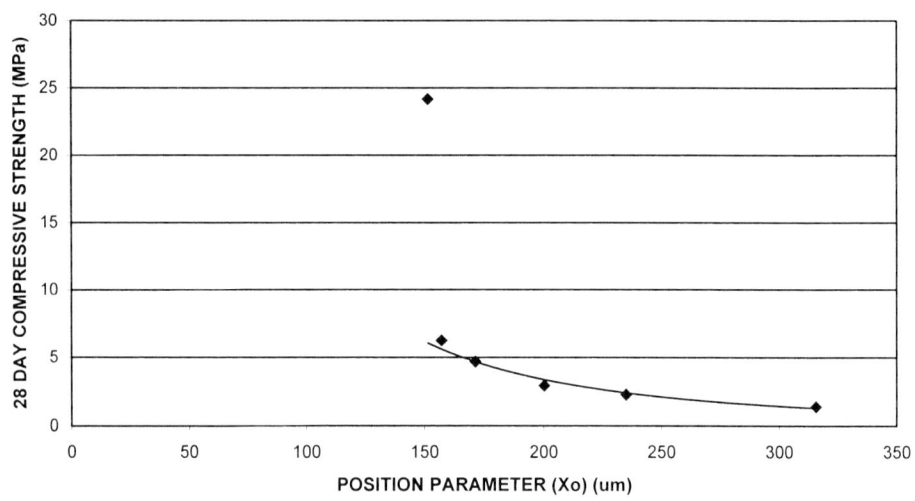

Figure 9 Compressive strength as a function of position parameter

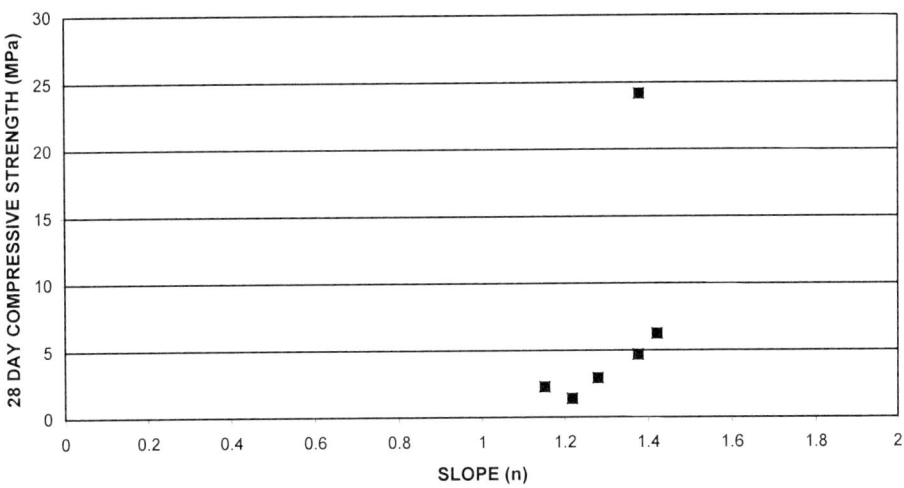

Figure 10 Compressive strength as a function of slope

CONCLUSION

The air-void size distribution of foamed concrete mixes was determined by using image analysis. It was found that although the composition and air content of some mixes are the same the 28-day compressive strengths and histograms differ.

Therefore parameters were established to explain and quantify relationships between air-void size distribution and 28-day compressive strength.

The Rosin-Rammler mathematical function for the description of particle size distribution was fitted to the air-void size distribution to establish the slope (n) and position parameter (x_0). The slope (n) parameter represents the range of the air-void size distribution and the position parameter (x_0) a representative air-void size for every mix.

From this preliminary investigation it was found that a relation exists between these parameters and the dry density of foamed concrete, the slope (n) increases with an increase in density and the position parameter (x_0) decreases with an increase in density.

It was also established that an exponential relation exists between air-void size and compressive strength as well as a narrower air-void size distribution for higher compressive strengths. Work is currently been conducted on the effect of shape and space factors.

REFERENCES

1. NEVILLE, A, M, AND BROOKS, J, J. Concrete Technology, 1987, pp 95.

2. KEARSLEY, E, P. The use of foamcrete for affordable development in third world countries. Proceedings conference on concrete in service of mankind (Congress on appropriate concrete technology), Dundee, UK, 1996. E & F N Spon, London, UK, 1996, pp 233–243.

3. WAINWRIGHT, P, J, AND OLORUNSOGO, F, T. Effects of PSD of GGBS on some durability properties of slag cement mortars. Journal of the South African Institution of Civil Engineering, 1999,Volume 41, First Quarter, pp 9-17.

4. OLORUNSOGO, F, T. Effect of particle size distribution of ground granulated blastfurnace slag on some properties of slag cement mortar, Phd 1990, Leeds.

5. ASTM 1990. Standard Test Method for Microscopical determination of Parameters of the Air-Void System in Hardened Concrete. ASTM Annual Book, Vol 04.02.

6. PrEN 480-II:1993. Admixtures for concrete, mortar and grout – Test methods - Determination of air void characteristics in hardened concrete. European Standard.

7. CAHILL, J, DOLAN, J, C, AND INWARD, P, W. The identification and measurement of entrained air in concrete using image analysis. Petrography of cementitious materials,ASTM STP 1215, Sharon M. Dehayes and David Stark, Eds., American Society for Testing and Materials, Philadelphia, 1994, pp 111-124.

8. NASSER, K, W, AND SINGH, B, P. A new method and apparatus for determining the air-void parameters in fresh and hardened concrete. Adam Neville Symposium on Concrete Technology 1995.

9. VISAGIE, M. The effect of microstructure on the properties of foamed concrete. Proceedings of the young concrete engineers' and technologists conference, Midrand, SA, October 1997, pp 101-109.

10. Optimate 5.21 Utilities menu.

THE INFLUENCE OF THE MIX DESIGN ON THE PROPERTIES OF MICRO-CELLULAR CONCRETE

L De Rose

SIP Project Managers (Pty) Ltd

J Morris

University of the Witwatersrand

South Africa

ABSTRACT. While exploring the possibilities of using lightweight cellular concrete as ceiling insulation for low cost housing in South Africa, it became evident that the mix design would play a crucial role in the success or failure of the project.

To achieve both adequate strength and acceptable insulation values the influence of variations in the mix design, involving OPC, RHPC, Fly Ash and Lime together with separately generated foam, was investigated. This investigation included the compressive strength, transverse strength, shrinkage and thermal conductivity of the different mix designs. Extensive use was made of statistical processes to evaluate the results of the experimental work.

The conclusion was reached that a micro-cellular product to meet the physical requirements that had been set as targets could be achieved.

Keywords: Lightweight cellular concrete, Insulation, Foamed concrete, OPC, RHPC, Fly ash, Pre-formed foam, Sub 500kg/m^3 density.

Laura De Rose MSc Graduate of the Department of Building and Quantity Surveying at the University of the Witwatersrand, Johannesburg. Currently Assistant Project Manager in the employ of SIP Project Managers (PTY) Ltd. in Cape Town, South Africa.

John Morris Dr Es Sc Leuven in Mineral Chemistry. Formally Chief Director of National Building Research Institute (now known as the Division of Building Technology) of South Africa. He has spent the greater part of his professional career researching aspects of the performance of building material. He is currently Professor of Building Science in the Department of Building and Quantity Surveying at the University of the Witwatersrand, Johannesburg.

INTRODUCTION

The main driving force behind this research was to try and find a way of insulating low cost houses and therefore reduce the heat that is being lost to the atmosphere, in order to decrease the energy consumption in winter.

The Need for Energy Conservation

An unfortunate feature of low cost housing that exist at present, in South Africa, is its lack of thermal insulation. This problem results in extensive heat loss in the winter months, which naturally causes an increase in energy consumption for that period. According to Holm[1] et al, most existing low-income houses are energy inefficient and existing pollution levels are detrimental to health. They go on to say that an increase in the housing stock will raise pollution levels and worsen the negative impact on the local and global environment.

Methods of Energy Conservation in Housing

Besides passive thermal designing, the only way to conserve energy and prevent heat loss in houses is by insulation. Most low cost houses do not have ceilings and this is, ironically among the most significant factors that affect indoor temperatures in a house. The ultimate aim behind this research was to come up with an optimum mix design for a material that could be used to produce a cost effective thermally insulating ceiling tile.

MICRO-CELLULAR CONCRETE

An Introduction

The term "Micro-cellular Concrete" (MCC), for the purposes of this research, refers to a lightweight concrete, made up of cement (Ordinary Portland Cement (OPC) or Rapid Hardening Portland Cement (RHPC)) slurry, Fly Ash and Lime mixed together with a separately prepared foam. The foam is produced by passing the foaming agent (diluted with water) through a specialised foam generator. The difference between the MCC produced for this research and normal aerated concrete is that the latter usually contains sand, whereas, the former has no aggregates included in the mix.

This Research Project

To achieve both adequate strength and acceptable insulation values the influence of variations in the mix design were investigated. The main objectives were to try to understand what effects these variations would have on the physical properties (i.e. thermal conductivity and drying shrinkage) and the mechanical properties (i.e. compressive and transverse strength) of MCC. The research examined the relationship between the mix design and the properties, in relation to density; w:c ratio; type of cement used; and FA and Lime content.

In order to facilitate the comparisons of the different mix designs at 28 days, multiple regression analysis was used to develop a statistical model for each test. These models helped to correct for the unavoidable differences, which occurred in the densities and w:c ratios.

Methodology

Test specimens of 32 different mixes were produced for six different types of tests (namely, wet and dry compressive strengths; transverse strengths for both bars and tiles; drying shrinkage; and, thermal conductivity). Each mix was cast over a range of densities between 250 and 550 kg/m³.

Once tested the results of the tests, as well as the mix design parameters were used as input data in a Data Analysis Tool called Multiple Regression [2].

The objective for this was to formulate a performance equation, based on each of the inputs, from which predicted values could be calculated and used to draw up all the necessary charts relevant to the interpretation of the results.

LABORATORY STUDIES

Materials

Cement

For the research two different types of cement were used, namely OPC and RHPC.

Fly ash (FA)

FA was used as a replacement for cement in the mixes. It is the by-product of coal-fired power stations and possesses pozzolanic properties. It is used in concrete because of the financial savings that its use offers; its influence on the heat of hydration of the concrete during setting; and the way in which it improves the workability of fresh concrete. South Africa is a significant producer of FA, Kruger[3] reported that this country had stockpiled 350 million tones of FA, and it was projected that this figure would grow by 23 million tones annually.

Lime (Ca(OH)$_2$)

Lime was used as an additive in the mix, because it accelerates the hardening of FA. The lime used for the purposes of this research was CLC Building Lime (plastic pressure hydrated building lime, Type A2P).

Foam

Mechanically produced foam creates and protects the desired content of air in the matrix. It is produced by introducing an emulsion of protein (Keratin) diluted 1:40 with water into a purpose designed generator. The foam is blended into the prepared mortar matrix by conventional concrete mixing equipment.

Mix Design

There were 32 different mixes cast, and some were repeated at different densities, therefore the total number of mixes cast was 50. Lime was used as an additive to the mix, whereas FA was used as a replacement for the cement. Table 1 shows the dry ingredients that went into all the mixes.

Table 1 Dry ingredients incorporated in the mixes

CEMENT	FLY ASH REPLACEMENT	LIME ADDITIVE
100 %	0 %	0 %
90 %	10 %	2.5 %
85 %	15 %	5 %
80 %	20 %	10 %

The entire mix design process for this kind of material is complex. Due to the complexity and lack of knowledge concerning the design of this material, a pragmatic approach was adopted to try to achieve the required results. A mix was designed, cast and tested, and then the physical properties of the product were measure in order to establish how far from the target that the resultant product was, in terms of density and w:c ratio. This knowledge was, then related back when designing the next mix. The main objective, when designing the mixes, was to keep the w:c ratio within a specific range and to make sure that all the different combinations of dry ingredients were designed at least three times to provide a range of densities.

The calculation of the mix design is broken down into the following steps: -

Step 1: Firstly, there is the design of the cementitious slurry, which in this case has up to three components (cement, FA, water).

In order to determine what mass of dry ingredients to incorporate into the mix the desired final oven dried density is used as a rough guideline. There is no sand or stone incorporated in the mixes; therefore, the oven dried density of the MCC will depend mainly on the mass of the binder volume. 400 kg/m³ was used as a first approximation.

To calculate the water requirement for the mix an initial w:c ratio of 0.32 was used for all mixes. The reason the ratio was so low initially is that there is an additional amount of water added to the mix at a later stage in the mixing process. This additional water is introduced into the mix by way of the foam and "extra water" that is added to the slurry. The "extra water" is added to achieve a desired consistency, based on visual observation, to facilitate proper mixing of the slurry and the foam. As each mix differs in the quantities of dry ingredients that affect workability, each mix has a different amount of water added. The final w:c ratio was calculated by including the initial water, the foam water content and the "extra water".

The masses of the ingredients were converted to volumes to determine how much foam needed to be added to make up 1 m³ of mix. The relative densities used were: -

- OPC = 3.15 kg/l
- RHPC = 3.14 kg/l
- FA = 2.2 kg/l
- Water = 1.0 kg/l

Step 2: Secondly, there was the design of the foam, which included water and the foaming agent. The volume of foam necessary for each mix was the amount of foam needed to make up the remainder of the 1000*l* once all the dry ingredients had been added together.

Step 3: Thirdly, the combination of the first two materials in suitable proportions was required. It was in this phase that the extra water is added to the slurry before combining the two. The visual observation is based on past knowledge and experience obtained while mixing similar mixes [4].

Production

Storage

All materials were stored in a room which was kept relatively warm (25-28° C) and dry (relative humidity approximately 40%). All dry ingredients were stored above the ground on wooden planks for periods of time not exceeding two months.

Mixing

The equipment used for the mixing process included a foam generator; a 125 litre tilting drum mixer, modified to accept a lid to facilitate proper mixing; and electronic balances.

Up to four mixes were cast on a specific day. Once all the dry ingredients were mixed together, the water was added and allowed to mix for 20 minutes. The density of the foam was then checked to ensure a target density of 70 – 90 kg/m³. The density of the foam is a parameter which is difficult to keep constant, it varies according to the ambient temperature, and even though there is a device on the generator that allows adjustments of the density, to achieve foam of the exact same density each time is difficult.

The foam is then pumped directly into the mixer and a lid was used to close the opening of the mixer. This was necessary in order to keep the contents of the mixer from spilling out when the mixer was tilted at an angle greater than 40°, in order to facilitate proper mixing. This lid was removed periodically to check the density of the mix as well as to scrape the walls of the mixer. The slurry and the foam were mixed together for approximately 10 minutes.

It was important not to over or under mix the slurry and foam. Over mixing would have broken down the air cells of the foam resulting in an increase in the mixture's density. Under mixing would have resulted in an inconsistent mix. On average, the mixing time for each mixing cycle was approximately 30 minutes.

There were a number of problems encountered in the mixing process mainly: -

1. Variations in mixing time, as mentioned above, due to the fact that it was sometimes difficult to prevent the dry ingredients from coagulating, which would result in cement balls forming within the mix affecting the homogeneity of the mix;

2. Variations in the water requirement: - each mix required a different amount of "extra water" to achieve the same consistency. The reason for this was that each mix differed as to the amount of FA and lime added. FA increased the workability of the mix and therefore reduced the amount of water required, and, lime decreased the workability of the mix and therefore increased the amount of water needed;

3. Variations in the density of the foam: - The foam, which was added to the mix, was made up of 40 parts water and 1 part solution. This additional water introduced into the mix had to be included in the calculation for the w:c ratio. The amount of additional water was calculated using the density of the foam. The problem presents itself because the density of the foam was a parameter, which was difficult to keep constant. It varied according to the ambient temperature, and although there was a device on the generator that allowed variations of densities, to achieve foam of the exact same density each time was difficult.

The last two problems were alleviated by creating a statistical model using multiple regression analysis.

Casting

There were six tests performed, the samples needed for each, and the curing methods adopted are shown in Table 2.

Table 2 Test to be performed with samples required

TEST	SAMPLE	No	CURING METHOD
Compressive Strength (wet)	100mm cubes	3	Moist cured entire 28 days
Compressive Strength (dry)	100mm cubes	3	Moist cured 24 days oven dried 4 days
Transverse Strength (bars)	75*75*285mm bars	3	Air cured for 28 days
Transverse Strength (tiles)	600*600*45mm tiles	3	Moist cured entire 28 days
Thermal Conductivity	300*300*45mm tiles	3	Moist cured entire 28 days
Drying Shrinkage	75*75*285mm bars	3	Air cured for 28 days

The mixture is transported from the mixer using buckets, then poured into the moulds, covered with plastic, and only demoulded after 48 hours because it was anticipated that insufficient strength would have been reached after only 24 hours.

Curing

The three basic curing methods, described below, were used to comply with the specific tests performed.

1. Moist Curing: The samples were kept wet and maintained at a temperature of 22-25° C for the entire curing period. This was achieved by standing samples in a tray of water, covered with plastic sheeting. Bath curing was not suitable as the samples floated.

2. Air Curing: The samples were placed on suitable racks in a drying room at 22-25° C, and left to cure in open air.

3. Wet and Oven Dried Curing: Initially the same as (1) above, but only for 24 days, after which they are oven dried at 50° C until constant mass was reached (approximately four days).

Testing

Compressive strength

"The measured compressive strength of a given concrete depends on the intrinsic properties of the concrete and on the way in which it is tested. It is therefore essential that methods of determining compressive strength should be standardised."[5]

Each specimen was tested immediately after it was removed from the water (or oven). Surface water, grit and projecting fins were removed, and the mass (to an accuracy of at least 1%) of each specimen was determined before testing it.

The bearing surfaces of the platens of the compression-testing machine (Avery Dennison) were wiped clean, and a specimen was positioned in the machine so that the load was applied to opposite as-cast faces of the specimen.

The compression load was applied without shock and increased at a uniform rate of 15 MPa/minute (starting at zero) until the specimen failed. The maximum load applied was then recorded and a mean was calculated from the three specimens representing each mix.

Once crushed the remains of each cube were placed in drying trays in the oven and oven dried in order to calculate the oven dried density.

The calculation of the compressive strength was carried out using the stress equation: -

$$\text{Stress (MPa)} = P / A$$

Where
P = maximum load (N)
A = area of cube face (mm²)

The results of the compressive testing were calculated working to three decimal places. The reason for this was that very low strengths were expected and the results were compared relative to each other only.

Transverse (flexural) strength

The bars were tested dry, whilst the tiles were tested immediately after they had been removed from the curing chamber and while they were still wet. Surface water, grit and fins were removed, and the mass of each specimen, to an accuracy of at least 1%, was determined before testing.

The rollers of the compression machine (Avery Dennison) were wiped clean, and the specimen placed centrally on the supporting rollers. The load was applied to the center of the bar from above, through a loading roller. The bar was supported below by two rollers placed 12,5% from each edge, i.e. the distance between the supports was 75% of the length of the specimen (see Figure 1).

It was ensured that all loading and supporting rollers were evenly in contact with the specimen and the load applied without shock. The load was increased continuously at a constant rate of 15 MPa/minute until the specimen failed.

Three samples of each mix were tested to obtain a mean and the transverse strength calculated using the equation: -

$$\text{Transverse strength (MPa)} = 3(P \times L) / 2(b \times h^2)$$

Where P = breaking load (N)
 L = distance between the supporting rollers (mm)
 b = breadth of the beam (mm)
 h = height of the beam (mm)

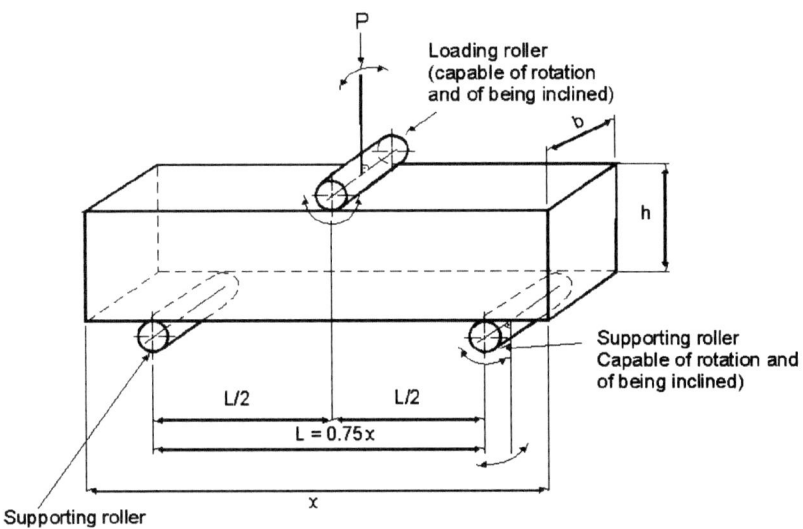

Figure 1 Indication of how bars were loaded for transverse testing

Drying shrinkage

These test specimens were cast and tested using the moulds and length comparator as specified in ASTM C 490 – 74: "Apparatus for use in measurement of length change of hardened cement paste, mortar and concrete".

The specimens were removed from their moulds after 48 hours. An initial measurement for length was taken within ½ an hour of removing the specimen from the mould. Thereafter the specimens were placed in the drying room and measurements were taken at set intervals up until 28 days, making sure that the specimen was oriented in the same direction in relation to the comparator as before.

Before any reading was taken, however, the device was checked against a reference bar. This reference bar was made of a steel alloy that has a very low coefficient of thermal expansion.

In order to take the readings the anvils on either end of the specimen were carefully placed into the slots provided by the comparator. The specimen was then rotated and the minimum reading was recorded. The measurement was read off a dial micrometer, which was accurate to within 0.002mm.

To calculate the results the linear shrinkage of each specimen was based on the initial measurement taken after demoulding and this was then expressed as a percentage.

Thermal conductivity

"The coefficient of thermal conductivity of a material is the rate of flow of heat per unit area per unit temperature gradient when the heat flow is at right angles to the faces of a thin parallel-sided slab of the material under steady state conditions."[6]

The testing was done using a Rapid-K Thermal Conductivity Instrument. Three representative samples of each mix were tested and a mean was calculated.

Before testing, the slabs were oven dried at a temperature of 50° C until constant mass (approximately five days). This was done to eliminate any moisture retained in the slabs, as it would have had an effect on the conductivity results. The reason the temperature of the oven was kept so low was to keep the water within the samples from reaching boiling point. If boiling point had been reached the vapor pressure within the cells would have caused the cells to burst. This would have resulted in cracking of the sample, creating a path for heat to flow through and making thermal testing useless.

As the slabs were taken out of the oven, they were wrapped in plastic bags to prevent any moisture absorption from the atmosphere and this plastic covering was only removed minutes before testing.

The sample was placed between the two plates of the instrument and upper and lower temperature limits were chosen at 45° C and 0° C respectively. (These temperatures were chosen as they could reasonably be expected to be the maximum and minimum temperature variance that would be experienced in the attic of a South African household.)

Heat was allowed to flow between the two plates until the system stabilised. The maximum time allowed for the samples to stabilise was three hours. On stabilisation, the thermal conductivity was calculated using the Fourier heat flow equation.

$$\lambda = (Q \times D) / (A \times \Delta T)$$

Where λ = thermal conductivity
   Q = time rate of heat flow
   D = sample thickness
   A = cross-sectional area
   ΔT = temperature difference across the sample

RESULTS

Results Measured

Statistical model

Once all the tests had been carried out, the mix designs, along with all the results obtained, were used as input data in a Data Analysis Tool called Multiple Regression. This was done to arrive at an equation for performance, for each test, based on each of the inputs.

The need for a statistical model (and an equation of performance) arose primarily because the densities and w:c ratios for each mix were different. As mentioned earlier the main reasons for these differences can be attributed to the fact that there were differences in the water requirements and the density of the foam.

As a result of these differences it makes comparing, say, the strength of one mix with another impossible, because the density and w:c ratios both have a direct influence on strength. A model was needed to correct for the densities and w:c ratios when making comparisons into the effects of mix design.

When applying the model, use can only be made of variables that are within the range of those values used to calculate the model; i.e. the models will not work on values taken at the extremes.

From these performance equations, predicted values could be calculated and used to draw up all the necessary charts relevant to the interpretation of the results.

Summary Interpretation of the Results

Compressive strength (see Figure 2)

- The performance equations for wet and dry cubes were very similar.
- Lime had no effect on compressive strength.
- Strength decreased with increasing w:c ratio up until approximately 0.45, after which it increased with increasing w:c ratio.

- Strength increased with increasing density.
- RHPC was stronger than OPC at 28 days.
- Oven dried cubes were stronger than saturated cubes.
- Strength increased with increasing FA content.
- OPC benefited more than RHPC, from an increase in the FA content.

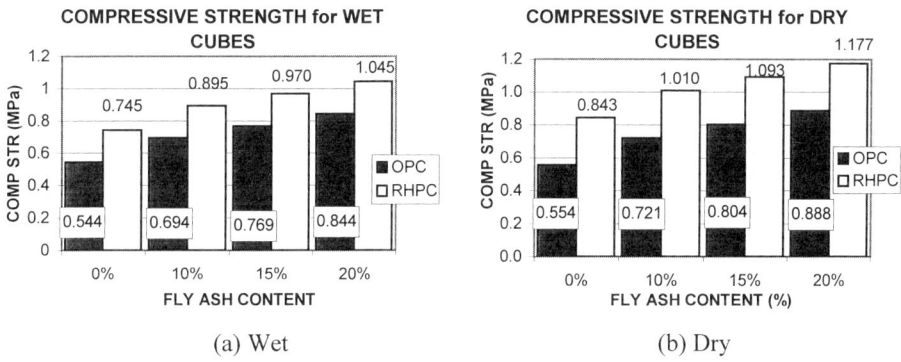

(a) Wet (b) Dry

Figure 2 Compressive strength figures at fixed w:c ratio and fixed density

Transverse / flexural strength (see Figure 3)

- FA had no effect on the transverse strength.
- Strength decreased as w:c ratio increased.
- Strength increased with increasing density.
- RHPC had a higher strength than OPC, at 28 days.
- Minimum strength was achieved at ~5% lime for bars and ~2.5% lime for tiles. Same lime concentrations yielded the greatest advantage when using RHPC as opposed to OPC.
- Tiles were stronger than bars due to wet curing.

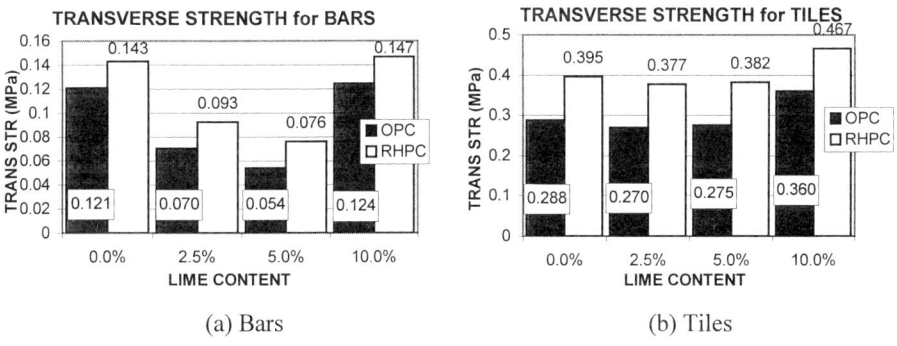

(a) Bars (b) Tiles

Figure 3 Transverse strength at fixed w:c ratio and fixed density

Drying shrinkage (see Figure 4)

- Both lime and FA had an influence on drying shrinkage.
- Drying shrinkage was a function of time.
- At 28 days, a w:c ratio of 0,45 yielded the minimum shrinkage.
- Shrinkage decreased with increasing density.
- RHPC yielded approximately 5.5% higher shrinkage than OPC.
- 4-6% lime yielded highest shrinkage, which was also the concentration of lime which yielded the lowest transverse strength.
- Shrinkage decreased at a decreasing rate with increasing FA, within the range examined.

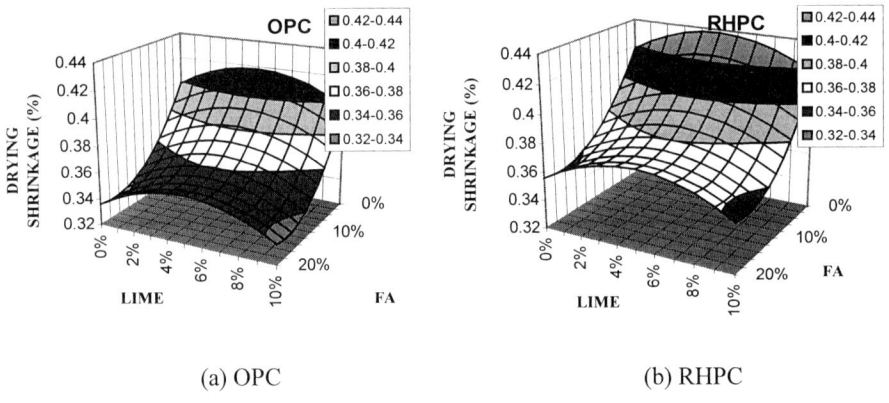

(a) OPC (b) RHPC

Figure 4 Drying shrinkage at fixed w:c ratio and fixed density

Thermal conductivity (K-Value) (see Figure 5)

- Both lime and FA had an influence on thermal conductivity.
- An increase in density increased the k-value within the limits explored, this could be explained using the logic that there is effectively more solid per given volume. This increase in solid seems to provide more material along which the heat can be transferred via conduction.
- RHPC increased the k-value by approximately 5.5% over the values for OPC.
- An increase in lime increased the k-value, which lines up with the relationship between density and k-values within these limits.
- 10% FA yields the best (i.e. lowest) k-value.

INTERPRETATION AND RECOMMENDATIONS

The aspects of mix design that were investigated were type of cement used, FA content and lime content. RHPC yielded higher strength and shrinkage for all tests (as opposed to OPC) at 28 days, but this factor would probably even out if the tests were taken later.

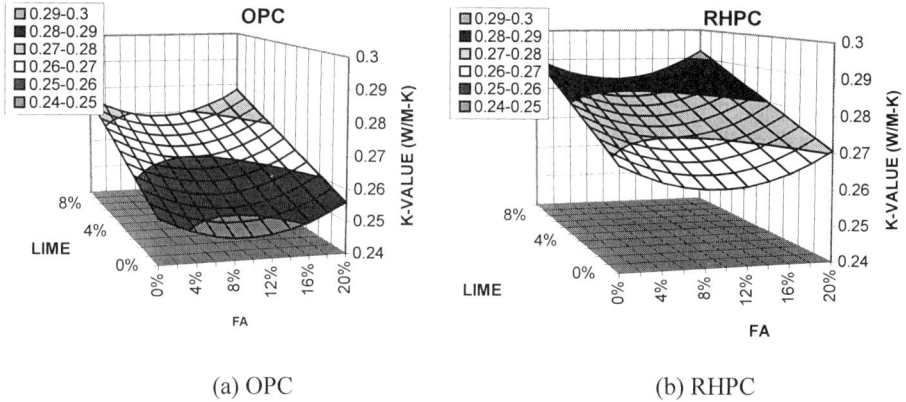

(a) OPC (b) RHPC

Figure 5 Thermal conductivity at fixed w:c ratio and fixed density

A FA content of between 6-8% yielded the lowest thermal conductivity and strength increased with increasing FA. Within the range 0 to 20%; 10% lime yielded the highest strength, but zero lime gave the best thermal conductivity figures; within the range 0 to 10%.

The authors believe that this investigation into the implications of mix design on the potential usefulness of Micro Cellular Concrete (MCC) for the production of insulating ceiling panels has contributed to the debate. The strengths achieved and the k-values attained confirm that the physical properties of such insulating ceiling tiles are within the range required to be able to contribute to the improvement of human comfort in low cost houses. This will effectively lead to a reduced expenditure on energy to ensure comfort during the cold winter nights experienced on the South African highveld.

REFERENCES

1. HOLM, D, TRIELOFF, K, VAN ASWEGEN, D F, AND VAN WYK, S. Towards a Policy for Passive Thermal Design in Low Cost Housing, Department of Mineral and Energy Affairs, Combined Report No.:EO9307/8, South Africa, August 1994

2. Microsoft Excel for Windows 95, Version 7.0

3. KRUGER, R A. Ten Years of Research into Fly Ash Utilization, Proceedings, First National Syposium, South African Coal Ash Association, Pretoria, 1990.

4. DE ROSE, L. The Viability of Cellular Lightweight Concrete Tiles as Thermally Insulating Ceiling Boards, Undergraduate Discourse, Department of Building and Quantity Surveying, University of the Witwatersrand, October, 1995.

5. ADDIS, B J. Strength of Hardened Concrete, Fulton's Concrete Technology, Portland Cement Institute, 7th Edition, The Natal Witness Printing and Publishing Co. (Pty) Ltd.: South Africa, 1994

6. MUNCASTER, R. A - Level Physics, 2nd Edition, Stanley Thornes Ltd.: England, 1985

MOISTURE FIXATION AND TRANSFER IN CLAYEY CELLULAR CONCRETE - RELATION WITH THERMAL CHARACTERISTICS

L Marmoret A Bouguerra
A t'Kint de Roodenbecke
University of Rennes
O Douzanet M Queneudec
University of Picardie Jules Verne
France

ABSTRACT. Moisture is an important factor in the thermal behaviour of a building material. Moisture exchanges in a porous medium can occur by either adsorption or capillarity. This article focuses on the clayey cellular concretes which have been developed. A thermomechanical study has enabled selecting a formulation which can be used in non-bearing elements and has demonstrated a sizeable storage capacity. Permeability to liquid, vapour and gaseous phases, along with capillary pressure, have been determined as a function of moisture content. Adsorption and desorption phenomena are also studied for relative moistures ranging from 15% to 97%. In each test case, the temperature has been held at 20°C. The hygroscopic behaviour is characterised and its relationship with both thermal conductivity and specific heat capacity is analysed.

Keywords: Moisture and thermal transfer, Cellular concrete.

Dr L Marmoret is a post doctoral searcher in Civil Engineering, University of Rennes. He is specialised in the study of heat and mass transfer in building materials.

Dr A Bouguerra is a post doctoral searcher in Civil Engineering, University of Rennes. His main research interest concerns heat and mass transfer in building materials.

Dr O Douzane is a lecturer in Civil Engineering, University of Picardie Jules Verne, Amiens. His major centre of interest is thermal behaviour of building materials.

A t'Kint de Roodenbeke is a reader in building equipments. His major centre of interest is indoor environments.

Professor M Queneudec, Head of the « Laboratoire des Transferts et Réactivité dans les Milieux Condensés », University of Picardie Jules Verne, Amiens. She is specialised in the design and characterisation of building materials.

INTRODUCTION

In developing countries, population growth constitutes one of the most significant current transformations and has given rise to a serious shortage of housing supply. One response to the economic constraints experienced could be the utilisation of local resources in construction processes and systems. Among these resources, clay is a preferred material due both to the expanse of its deposits and to its specific qualities. Furthermore, the aggregate industry has incited, in some regions, the development of large quantities of clayey co-products. Moreover, considerable market potential exists in the area of soil rehabilitation and environmental restoration. It is from the standpoint of this triple objective that proposals have been forwarded to enhance the use of such clayey products in the form of lightweight concrete [1]. One of the techniques used in this respect is the creation of a cellular structure by means of a chemical reaction in the mass.

As is the case with all porous materials, clayey cellular concrete can at times entrap significant quantities of water. With this material, water gets entrapped in either the vapour state or the liquid state and then migrates. Transfer phenomena are governed both by the characteristics of the porous medium and by hygrothermal conditions. These phenomena are capable of considerably modifying the material's thermal properties.

STUDIED MATERIAL

The process of lightening clay-cement pastes has previously been described [1]. It involves a complex reaction that depends on both the ambient conditions and the composition of the paste.

The clayey base material is a co-product obtained from utilising alluvial sand. It is mainly composed of kaolinite [2]. Various samples extracted have displayed a percentage of quartz of less than 5%. The laser gap-grading analysis is shown in Figure 1. The cement used is a CPA CEM I 52.5 whose chemical analysis is presented in Table 1; Figure 1 provides the results of the laser gap-grading analysis.

The choice of the material to be studied in this work has been made on the basis of a compromise between mechanical and thermal properties. It corresponds to a cement/clay ratio of 25%, a water/clay ratio of 85% and an aluminium/clay ratio of 0.1%. In the dry state, the apparent volumic mass is equal to 850 kg/m^3; the compressive strength is 2 N/mm^2. The thermal conductivity is near 0.2 W/m°C, and the porosity is approximately 60%.

Table1 Chemical analysis of the Portland cement CPA CEM I 52.5

CaO	MgO	Fe_2O_3	SiO_2	Al_2O_3	SO_3	PF (1000°C)
64.8	0.9	4.3	21.3	3.7	2.7	1.2

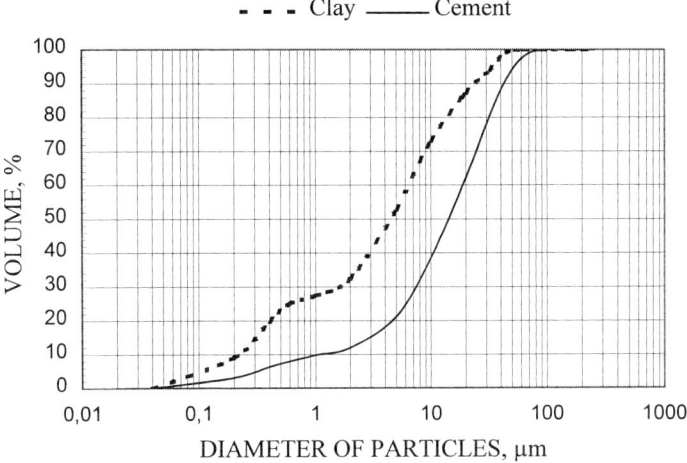

Figure 1 Laser gap-grading analysis of both the cement and the clayey material

EXPERIMENTAL RESULTS AND ANALYSIS

Permeability

The measurement of the permeability to vapour was conducted by the "normalised dish" method. We deduce the coefficient of diffusion K_{av} in the air from the permeability in the vapour phase K_v and then the volumic mass in the gaseous phase ρ_g by the relationship:

$$K_v = \rho_g \cdot K_{av}$$

The process employed to determine both the permeability to gas K_g and the permeability to liquid K_l has already been developed by Nicolas [3]. These coefficients change with temperature [4,5]; however, the amount of this change remains relatively small within the range of temperatures applied herein. These coefficients also vary with the material's water content. Figure 2 shows this evolution for water contents between 0.014 and 0.040, inclusive, as obtained after stabilisation in an atmosphere at 20°C and with a relative humidity varying from 15% to 97%. The permeability in the liquid phase is to be multiplied by a factor of 1000 between the two extreme water contents being studied. The evolution observed is very fast for low values of water content. In the vapour phase, this evolution is less pronounced; the permeability to gas decreases as water content increases.

Isothermal of Sorption - Desorption

The isothermal of sorption and desorption allows representing the water entrapped by both adsorption and capillary condensation [6,7,8]. These curves display a hysteresis phenomenon which can be explained by both the difference in evolution of the liquid-vapour interface and the pore geometry.

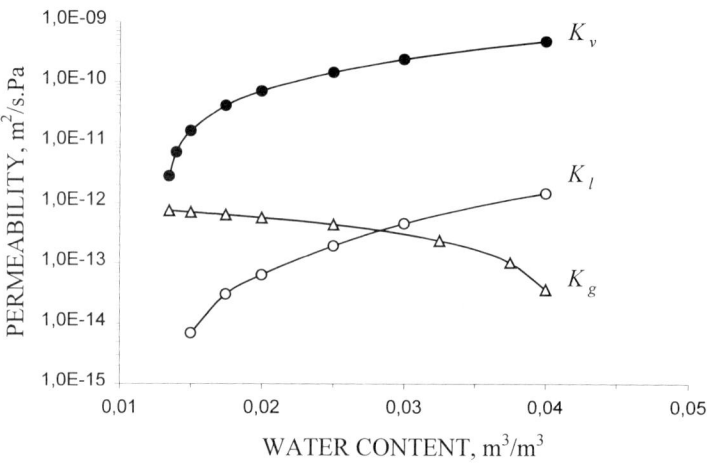

Figure 2 Permeability measurements

Experimental results are presented in Figure 3. Tests of the theoretical representation of the entrapment of humidity have previously been carried out. The entrapment of water vapour at the macroscopic level has been translated by Nicolas [3] by means of introducing, into the free energy of water, the function $h(\beta_l)$, an interaction function. The resultant curve is therefore obtained without experiencing the hysteresis phenomenon, a situation which corresponds to the ideal case of humidifying the porous medium at an infinitely slow rate; hence, at each instant, a state of thermodynamic balance is attained. It can thus be concluded that the interaction $h(\beta)$ becomes infinite as the porous medium nears either the dry state ($\beta_l = 0$) or the saturated state ($\beta_l = n$).

Figure 3 reveals that the interaction function makes it possible to appropriately translate the hydric exchanges occurring in the studied area; this translation becomes even more pertinent since the hysteresis phenomenon is only very slight. Moreover, for a relative humidity of 97%, the volumic water content at equilibrium is near 8%; this value is distinctly different from the material's total porosity.

Capillary Pressure

Whenever both liquid water and a gaseous phase are present in a material, they get split inside the pores in accordance solely with the laws of capillarity (in the absence of external forces, such as gravity, etc.). For a given water content, a difference can thus be established between the pressure determined in the two phases, which is referred to as the capillary pressure. An expression for the pressures that serves to define the capillary pressure is as follows:

$$P_c = -\frac{\rho_l.R.T}{M_v} LnHR(\beta_l) + \frac{\rho_l.R.T}{M_v} h(\beta_l)$$

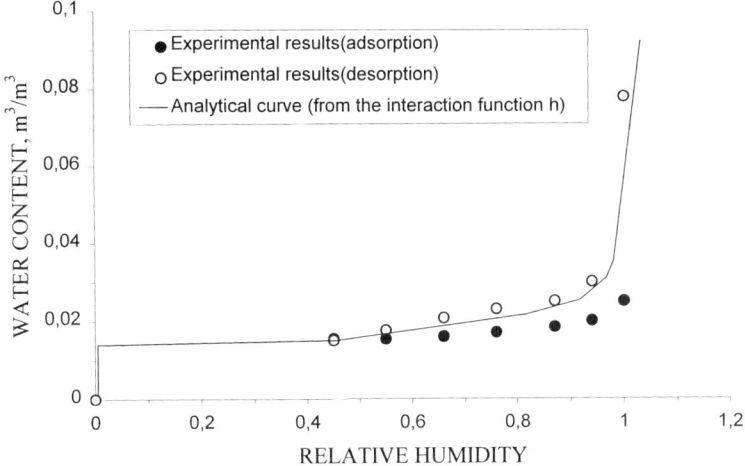

Figure 3 Isothermal of sorption and desorption

The capillary pressure commonly used P_k (called the "corrected capillary pressure") in macroscopic models of humidity transfer in porous media is defined by:

$$P_k = -\frac{\rho_1 \cdot R \cdot T}{M_v}.$$

The correction is even smaller when the interaction function approaches zero. Figure 4 presents the capillary pressure P_c and the corrected capillary pressure P_k for the studied material. A minimal water content β_{lei}, the so-called "irreducible water content", can be obtained for a relative humidity of close to zero. A rapid drop in capillary pressure is observed once the water content has reached a low level. This phenomenon represents the invasion of the porous medium by the liquid phase which has come into contact with the medium.

Evolution in Thermal Characteristics as a Function of Both Temperature and Water Content

Heat transfer through a non-saturated porous material occurs not only by conduction (in both air and water), convection and radiation, but also by diffusion of the water vapour through the air of pores. This distribution process, resulting from the temperature gradient, gives rise to additional energy exchanges during the presence of evaporation-condensation phenomena on the walls of pores; it is coupled with a displacement of the humidity in the vapour phase from hot zones to cold zones and in the liquid phase in the opposite direction of that of the capillarity. This entire process is known as the "thermomigration phenomenon". Thermal characteristics hence are no longer identical at all points. Under such conditions, thermal characteristics can only be referred to in conjunction with the redistribution of humidity.

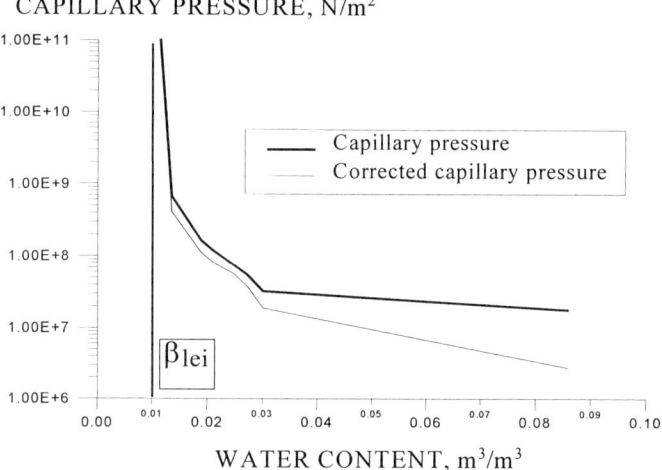

Figure 4 Capillary pressure

So as not to disrupt the determination of the thermal characteristics, we have elected to use a transitory method [9].

Thermal conductivity

By neglecting the effects of thermomigration, we can define an apparent thermal conductivity [10] in the presence of humidity as a function of the thermal conductivity of the dry porous medium λ^*, thanks to the following:

$$\lambda^*_{app} = \lambda^* + \rho_s . a_{mv} . \delta_v . L$$

where ρ_s is the apparent volumic mass of the solid, a_{mv} the transfer coefficient of the vapour phase in the porous medium, and δ_v the coefficient of mass transfer in the vapour phase under the effect of the thermal gradient. L is the latent heat of the vaporisation of water, expressed in J/kg.

Experimental points are presented in Figure 5. A very sharp, but practically regular, variation in the thermal conductivity of the water contents studied can be observed. This variation turns out to be just about doubled between the dry state and the study's maximal water content. For a water content $\beta_1 = 0.028$, which corresponds to a relative humidity of the enclosed test area of 85% (i.e. in a relatively humid medium), the thermal conductivity is nearly stabilised at room temperature.

The curve drawn with a solid line corresponds to the Bories expression [10]. A good level of agreement with experimental values has been obtained.

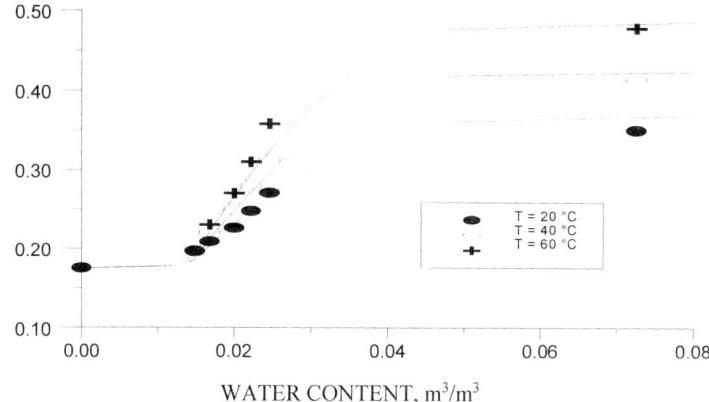

Figure 5 Influence of both water content and temperature on the thermal conductivity

Specific heat capacity

Figure 6 provides experimental values for the specific heat capacity as a function of both humidity and temperature.

The equivalent specific heat capacity C* of the humid material has been modelled by summing the enthalpies of the various phases in accordance with the following relationship:

$$(\rho.C)^* = \sum_{i=s,v,l} \beta_i.\rho_i.C_i = \rho_s.C_s + \beta_v.\rho_v.C_v + \beta_l.\rho_l.C_l$$

This expression is represented by the curve drawn with a solid line in Figure 6. Here again, a good level of agreement with the experimental points can be observed.

CONCLUSION

The work presented herein has been oriented towards the overall objective of enhancing the use of clayey fines in the form of an insulating lightweight concrete.

Emphasis has been placed on a porous material which exchanges humidity with the ambient medium. Such exchanges do act to modify the material's thermal behaviour.

However, results have shown that under normal operating conditions (H.R. = 65%), the quantity of absorbed water remains small; hence, thermal performances are hardly affected at all.

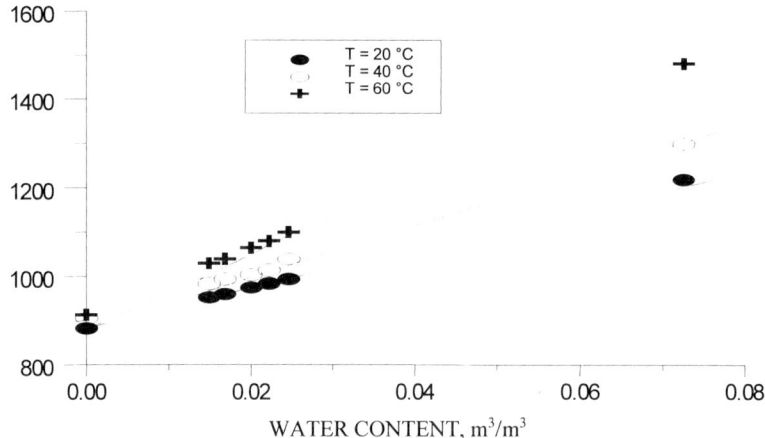

Figure 6 Influence of both water content and temperature on the specific heat capacity

REFERENCES

1. Al RIM, K., Etude de l'influence de différents facteurs d'allégement des matériaux argileux : le béton argileux léger. Généralisation à d'autres fines de roches. Thèse de Doctorat, Université de Rennes 1, 1995, pp 193.

2. ESTEOULE-CHOUX, J., Contribution à l'étude des argiles du Massif Armoricain. Argiles des altérations et argiles des bassins sédimentaires tertiaires. Thèse de Doctorat ès Sciences, Université de Rennes, 1967, pp 319.

3. NICOLAS, P., Modélisation mathématique et numérique des transferts d'humidité en milieu poreux, Thèse de Doctorat, Université de Paris VI (diffusée par le Laboratoire Central des Ponts et Chaussées), 1992.

4. DE VRIES, D.A., Thermal Conductivity of granular materials, Annexe 1952, 1, Bull. Inst. Intern. du Froid, 1952, pp. 115-131.

5. KRISCHER, O., ROHNALTER, H., Warmeleitung und Dampfdiffusion in feuchte gutern, V.D.I. Forchungsheft, 1940, p. 42.

6. COUASNET, Y., Contribution à l'étude du transfert de vapeur d'eau en régime permanent et non stationnaire dans les matériaux poreux hygroscopiques, Cahiers du Centre Scientifique et Technique du Bâtiment, n° 2349. 1989.

7. BRUNAUER, S., EMMET, P.H., TELLER, E., Adsorption of gases in multimolecular layer, J. Am. Chem. Soc., Vol. 60, 1938, pp. 309-319.

8. BADMANN, R., STOCKHAUSEN, N. and SETZER, M.J., The statistical thickness and the chemical potential of adsorbed water films, J. Coll. Int. Sci., Vol. 82, N° 2, 1981, pp. 534-542.

9. BOUGUERRA, A., Contributions à l'étude d'un procédé de valorisation de déchets argileux: comportement hygrothermique des matériaux élaborés, Thèse de Doctorat, I.N.S.A. de Lyon, 1997, pp 212.

10. BORIES, S., Transfert de chaleur et de masse dans les matériaux. Analyse critique des différents modèles mathématiques utilisés, Séminaire "L'humidité dans le bâtiment" Saint Rémy lès Chevreuse, 23-25 Novembre 1982.

CONFIGURATION OF PORES ON THE MECHANICAL AND THERMAL CHARACTERISTICS OF CELLULAR CONCRETE THROUGH A HOMOGENIZATION METHOD

Z Malou
R Cabrillac
University of Cergy Pontoise
France

ABSTRACT. This study deals with the modelization of the mechanical and thermal behaviour of an anisotropic porous material made up of ellipsoidal pores uniaxially oriented according to the geometric parameters that characterize porosity. To do so we used a homogeneization technique based on the finite element method and the coupling of porosity parameters was taken into account. This approach made possible both the introduction of the quantitative and qualitative aspect of porosity and the transcription of mechanical and thermal anisotropy due to geometric anisotropy. We thus determined the homogeneized elasticity moduli and thermal conductivities of the material both parallel and perpendicular to the flattening of the pores according to their shape coefficient and porosity. The results show the advantage of anisotropy in the case of building materials such as cellular concretes. Indeed we noticed that with equal porosity a material made up of ellipsoidal pores is more rigid parallelly to the flattening of pores and more insulating perpendicularly than a material made up of spherical pores.

Keywords: Cellular concretes, Optimization, Homogeneization, Porosity, Mechanical characteristics, Thermal characteristics, Anisotropy.

Doctor Zahir Malou has made his doctor's thesis in the Civil Engineering Laboratory of the University of Cergy Pontoise, France.

Professor Richard Cabrillac is Director of the University Institute of Civil Engineering Science at the University of Cergy Pontoise, France and Head of the Civil Engineering Laboratory of the same University.

INTRODUCTION

Currently cellular concretes on the market are materials made up of spherical pores evenly distributed with isotropic mechanical and thermal behaviour. To get such porosity configuration requires a highly accurate adjustment of the expansion phenomenon and thus of the composition parameters [1,2]. Experimental studies on cellular concretes [2,3] have shown that without specific adjustment anisotropic materials both on a mechanical and thermal level are naturally obtained. The anisotropic behaviour of the material is due to the geometric configuration of the porosity (shape, distribution and direction of the pores). Indeed the pores present uniaxial flattening perpendicular to the direction of expansion. This anisotropy phenomenon may have advantages for the optimization of cellular concrete performances be they autoclaved or not. Compared to isotropic materials made up of spherical pores anisotropy may indeed, with equal porosity, allow for the improvement mechanical performances of the material in the direction parallel to the flattening of the pores and of the thermal performances in a perpendicular direction [4,5,6].

This study deals with the mechanical and thermal modelization of a porous material according to porosity and geometric configuration of this one through a homogeneization technique based on the finite element method [1,7,8].

GEOMETRIC MODELIZATION OF THE MATERIAL

The mechanical and thermal modelization of a heterogeneous material requires in the first place a geometric idealization of heterogeneities which is necessary to simplify the mathematical formulation of the problem and to facilitate digital processing. In the case of highly heterogeneous materials the existence of a periodic structure of the material is assumed and the latter is thus assimilated to the juxtaposition of an infinite number of identical cells. The reference cell is called representative elementary volume (REV). This volume is defined by a basic period containing a heterogeneity. It must take into account the geometric and mechanical and thermal characteristics of heterogeneity as well as those of the material matrix. In the case of porous materials heterogeneity is shown by a void and the peridiocity of the material leads us to formulate geometric hypotheses on the pore and the pore-containing basic period.

The field of porosities that may be contemplated directly depends on those hypotheses and implicitely defines the field of validity of mechanical and thermal models. Limit porosity (PL) is defined [1,2,5,6] as porosity corresponding to the case when the pore is tangent in one or several points to the perimeter of the basic period that contains it. It depends on the type of distribution and on the shape of the pores.

In the case of simple cubic distribution (modelization 3D) and in the case of distribution following a square basic period (modelization 2D) containing respectively ellipsoidal or elliptical pores, limit porosity depends on the shape coefficient of the pores n= a/b (big axis - small axis ratio). It is maximal when the latter equals 1.(Figure.1).

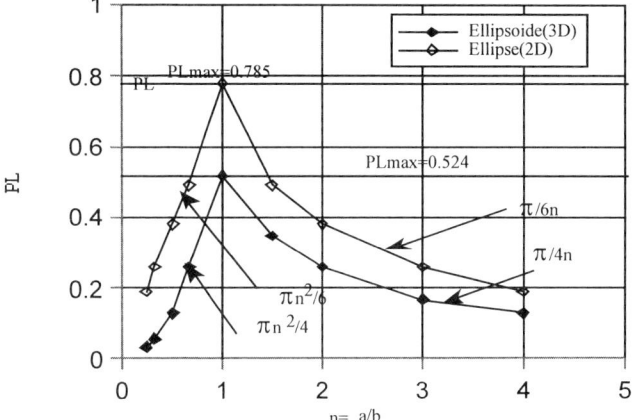

Figure 1 Comparison ellipsoid simple cubic network – ellipse square basic period network

MECHANICAL MODELIZATION

Homogeneization by Finite Element Method

The principle of homogeneization methods consists in substituting heterogeneous material for a homogeneous equivalent material [7, 8] and to determine the equivalent mechanical characteristics. The resolution of the problem consists in putting on the REV an average macroscopic strain (D) and imposing a displacement field on the limit.

$$U_i = D_{ij} y_j$$

Then we solve on the volume (V) of the REV the following elasticity problem with limit conditions on the displacements.

$$\begin{cases} \text{div}_y \sigma = 0 \text{ in } V \\ \sigma_{ij}(u) = a_{ijkh} e_{kh} \text{ in } V \\ U_i = D_{ij} y_i \text{ on the limit of } V \end{cases} \text{ with } e_{ij}(u) = \frac{1}{2}\left(\frac{\partial U_i}{\partial y_j} + \frac{\partial U_j}{\partial y_i}\right) \text{ in } V$$

The determination of the equivalent rigidity coefficients is defined as follows:

$$\langle \sigma_{11} \rangle = \langle a_{11kh} \varepsilon_{kh}(u) \rangle = C_{11kh}^{hom} D_{kh} = C_{1111}^{hom}$$
$$\langle \sigma_{22} \rangle = C_{2211}^{hom}$$
$$\langle \sigma_{12} \rangle = C_{1211}^{hom}$$

In 2D case the equivalent behaviour by the theory of effective moduli is given by the average [1]:

$$\langle \sigma_{ij} \rangle = \frac{1}{|S|} \int_S \sigma_{ij}(y) dy$$

The equivalent homogeneous material (Figure 2) is orthotropic and the homogeneized rigidity coefficients in the direction of orthotropy are written as follows:

$$C_{1111}^{hom} = \left(\frac{E_1^{hom}}{1 - \nu_{12}^{hom} \nu_{21}^{hom}} \right) = C11 \qquad C_{2222}^{hom} = \left(\frac{E_2^{hom}}{1 - \nu_{12}^{hom} \nu_{21}^{hom}} \right) = C22$$

The mechanical anisotropy is characterized by $\quad \dfrac{C_{2222}^{hom}}{C_{1111}^{hom}} = \dfrac{E_2^{hom}}{E_1^{hom}}$

The load put on the REV can be seen in Figure 3

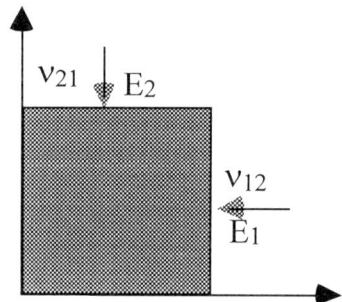

Figure 2 Equivalent homogeneous material

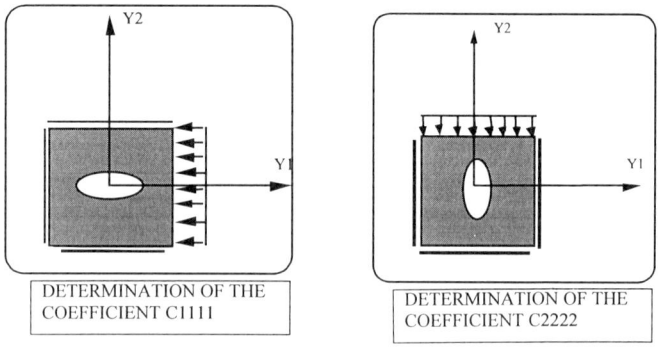

Figure 3 Mechanical load put on the REV

Results and Remarks

Figure 4 shows the evolution of rigidity coefficients in relation to porosity for all geometric models taken into account in the study. It must be noted that the curves are each time limited owing to porosity limits that corresponds to the geometric configuration of the targeted porosity. It may also be noted that there are two distinct envelopes of grading. One corresponds to the rigidities parallel to the flattening of the voids (C11), the other to those perpendicular to it (C22).

It must also be noted that the difference of rigidities between the parallel and perpendicular direction to the flattening of voids becomes wider and wider as the shape coefficient increases. The two envelopes of grading are separated by the curve which corresponds to the circular void. One notes that with equal porosity the material is more rigid when parallel to the flattening of the voids than when perpendicular to it.

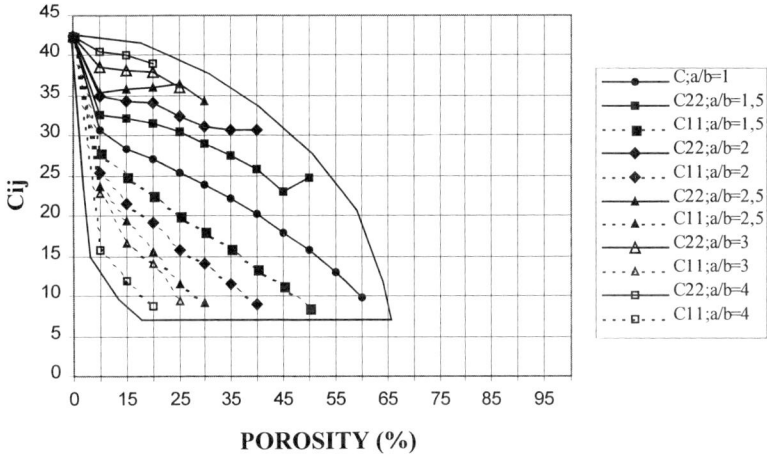

Figure 4 Evolution of rigidity coefficients according to porosity

Figure 5 shows the evolution of the rigidity ratios both parallel (C22) and perpendicular (C11) to the flattening of the voids. The ratio expresses mechanical anisotropy due to geometric anisotropy. It may be noted that the C22/C11 ratio increases when porosity increases for a constant shape coefficient. The ratio also increases when the shape coefficient increases for a constant porosity. Maximal mechanical anisotropy is obtained for a 20% porosity and the ratio C22/C11 reaches a value about 4.

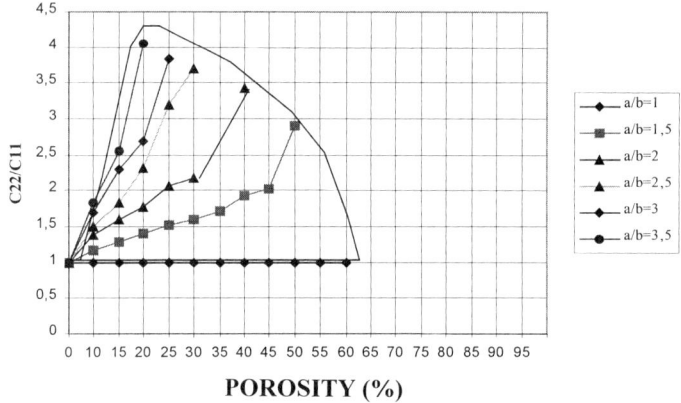

Figure 5 Evolution of mechanical anisotropy according to porosity

Figure 6 expresses the evolution in relation to the porosity of the envelop curves of theoretical anisotropy ratios expressed in terms of elasticity moduli and of the experimental anisotropy ratios expressed in terms of compressive strengths [1]. One notes a similar evolution of the experimental and theoretical curves expressing the material anisotropy. This clearly confirms, on a qualitative level, the conclusions drawn from modelization and the optimization potentialities of cellular concretes on a mechanical level by playing on the shape and the orientation of pores.

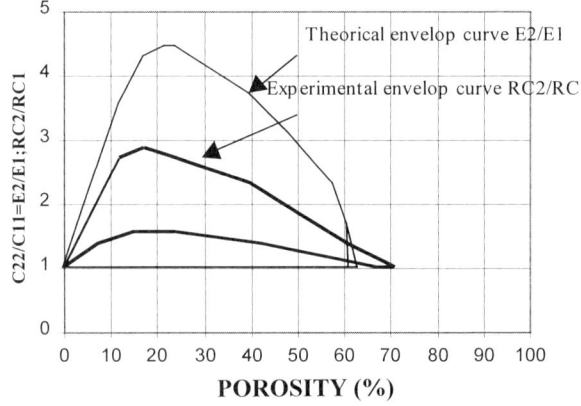

Figure 6 Evolution of the anisotropy - Experimental and theoretical results

THERMAL MODELIZATION

Homogeneization by Finite Element

The thermal analysis [1] is based on the homogeneization of the thermal characteristics of the heterogeneous material constituted by the matrix of thermal conductivity (λ) and the void with a weak conductivity which we consider as nil to simplify the mesh. The principle is similar to that of the homogeneisation of mechanical properties of material. The heterogeneous medium is defined by the REV constituted by a square basic period containing a circular void or elliptical void oriented uniaxially. The digital résolution is made on the REV. The heat diffusion Fourier's law giving the flux of heat in function of the gradient of temperature is written $\phi=\lambda\text{grad T}$.

We defined the homogeneized macroscopic characteristics :

$$\langle\phi(y)\rangle=\lambda^{hom}\langle\text{grad}T(y)\rangle$$
$$\langle\phi(y)\rangle=\lambda^{hom}G$$

G : average macroscopic imposed gradient of temperature

$$\langle\phi(x,y)\rangle v = 1/v \int \phi(x,y)dv$$

For the digital resolution using the finite element method we have applied between the two opposite limits of the REV a gradient of temperature and a nil flux of heat on the others limits (Figure 7). The digital treatment has permitted to calculate the average of heat flux in the REV and from Fourier's law we deduced the equivalent thermal conductivity. The digital treatments have been made on different REV and the choice of these is the same as for mechanical modelization.

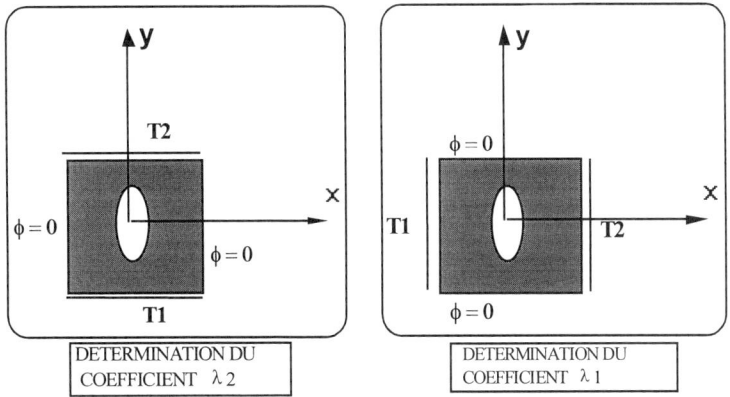

Figure 7 Thermal load put on the REV

Results and Remarks

Figure 8 shows the evolution of thermal conductivities in relation to porosity for all geometric models taken into account in the study. It must be noted that the curves are each time limited owing to porosity limits that correspond to the geometric configuration of the targeted porosity. It may also be noted that there are two distinct envelopes of grading. One corresponds to the conductivities parallel to the flattening of the voids($\lambda 1$), the other to those perpendicular to it ($\lambda 2$). It must also be noted that the difference of conductivities between the parallel and perpendicular direction to the flattening of voids becomes wider and wider as the shape coefficient increases. The two envelopes of grading are separated by the curve which corresponds to the circular void. It is to be noted that with equal porosity the material is more conductive when parallel to the flattening of the voids than when perpendicular to it.

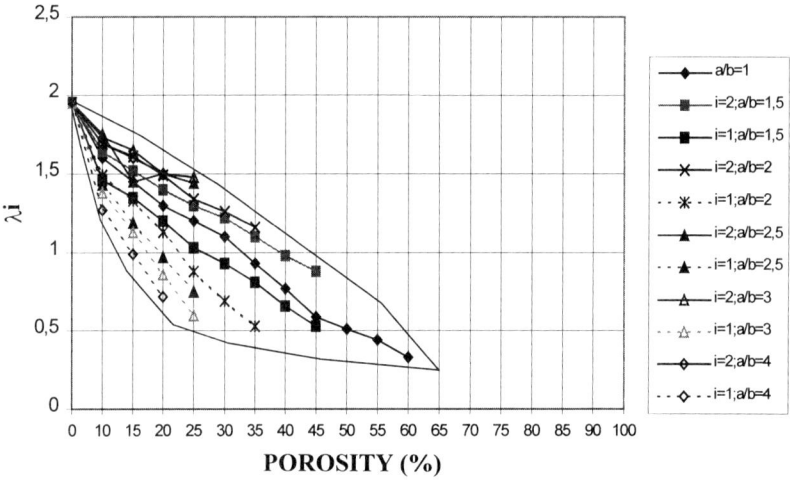

Figure 8 Evolution of thermal conductivities according to porosity

Figure 9 shows the evolution of the ratios $\lambda 2/\lambda 1$ of conductivities in the parallel and perpendicular direction to the flattening of the voids. The ratio expresses thermal anisotropy due to geometric anisotropy. It may be noted that the ratio $\lambda 2/\lambda 1$ increases when porosity increases for a constant shape coefficient. Likewise, the ratio also increases when the shape coefficient increases for a constant porosity. Maximal thermal anisotropy is obtained for a 20% porosity and the ratio $\lambda 2/\lambda 1$ is about 2,5. We can note that the thermal anisotropy is less important than mechanical anisotropy (Figure10). The phenomena are similar from a mechanical and thermal point of view but concerning perpendicular directions.

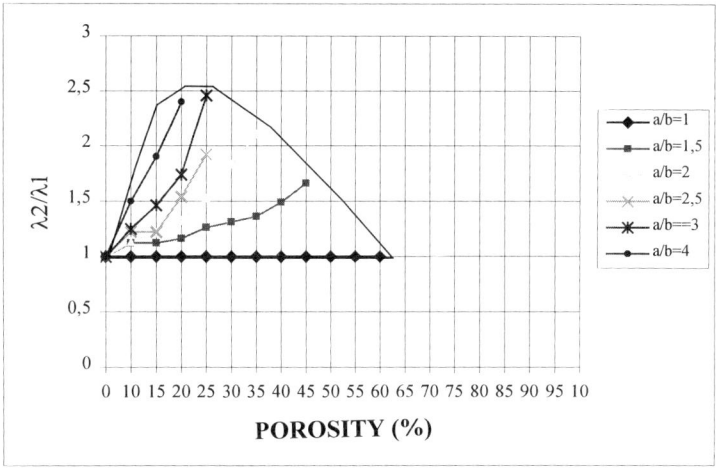

Figure 9 Evolution of thermal anisotropy according to porosity

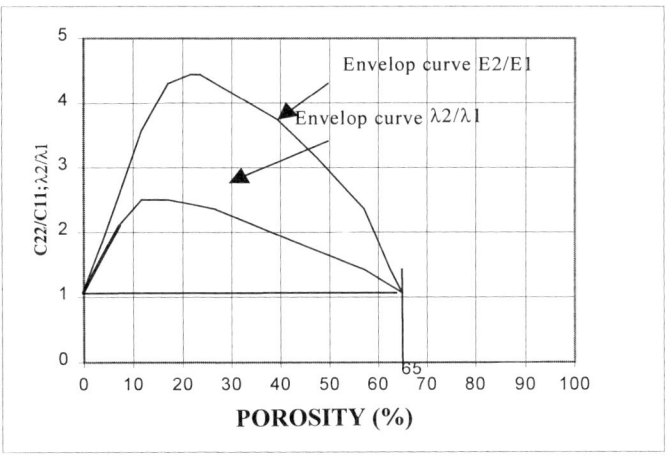

Figure 10 Mechanical and thermal anisotropy

CONCLUSIONS

During this work we have made a modelization using homogeneization by finite element which permits us to determine the mechanical and thermal characteristics of porous material constituted by voids of anisotropic shape and taking into account the coupling of porosity parameters. The homogeneization that we have used has permitted us to introduce at the same time the quantitative and qualitative aspect of porosity which is not the case of the different existing models[1, 4, 5, 6]. This study has shown that with equal porosity a material constituted by ellipsoidal voids can be more rigid than a material consitituted by spherical voids and can be less thermally conductive. These theoretical results have a good qualitative correlation with experimental results on the mechanical point of view and we have noted the duality existing between the evolution of the mechanical and thermal properties in relation to porosity and its configuration. This modelization has permitted us to evaluate the mechanical and thermal anisotropy phenomenon due to geometrical anisotropy. This study has shown the advantage of this phenomenon in the case of building materials like cellular concretes to improve their performances in the vertical direction corresponding to the direction of the loads and their thermal performances in the horizontal direction corresponding to the direction of thermal flux.

REFERENCES

1. MALOU, Z. "Etude de l'optimisation des propriétés mécaniques et thermiques des matériaux poreux de type bétons cellulaires". Thèse de Doctorat ,Université de CERGY PONTOISE , Octobre 1997.

2. CABRILLAC, R. "De la connaissance des matériaux à l'ingénierie des matériaux de construction". Synthèse Habilitation à Diriger des Recherches, UP. X. NANTERRE, Avril 1991.

3. MALOU, Z, CABRILLAC, R, AGGOUN, S, AND ROCHELLE, C. "Influence des paramètres de composition sur la porosité et les caractéristiques mécaniques des bétons cellulaires". Annales de l'Institut Technique du Bâtiment et des Travaux Publics, Mai 1995.

4. BASCOUL, A, CABRILLAC, R, AND MASO, J C. "Influence de la forme et de l'orientation des pores sur le comportement mécanique des bétons cellulaires". Congrès International RILEM, VERSAILLES, Septembre 1987.

5. CABRILLAC, R, PERRIN, B."Influence de la forme et de l'orientation des pores sur la conductivité thermique des milieux poreux". Congrès International RILEM, VERSAILLES, Septembre 1987.

6. CABRILLAC, R, AND SICARD, J. "Modélisation thermique des milieux poreux et prise en compte du couplage des paramètres de porosité pour l'étude de l'influence de la forme des pores". RILEM, Materials and Structures, Mai 1996.

7. SUQUET, P. "Plasticité et homogénéisation". Thèse de Doctorat d'Etat, Université PARIS VI, 1982.

8. LENE, F, "Contribution à l'étude des matériaux composites et de leur endommagement". Thèse de Doctorat d'Etat, Université , PARIS VI, 1984.

MIXTURE DESIGN OPTIMISATION OF CELLULAR CONCRETE

A Rodriguez   M Pedraza
Cemex
Mexico

J Luciano   D Constantiner
Master Builders
United States of America

ABSTRACT. An experiment was designed to determine the relationship between mixture composition and compressive strength of cellular concrete, which was foamed in the field. The independent variables studied were aggregate-to-cement ratio, water-to-cement ratio, percentage of lightweight fine aggregate, and unit weight. A predictive model relating the independent variables to compressive strength was developed using statistical methods. The compressive strength predictive model, coupled with cost information, was input into a nonlinear optimization routine. This spreadsheet model was used as an interactive tool to illustrate the relationship between optimum mixture composition, cost, strength and unit weight. The results demonstrate that small changes in the desired properties can substantially change the required mixture proportions and significantly reduce the concrete cost.

Keywords: Cellular concrete, Foamed concrete, Mixture optimization, Cost optimization.

Arturo Rodriguez is a production manager in charge of several Cemex ready-mix plants in Mexico City.

Marco Pedraza is a concrete production manager for Cemex in Monterrey, Mexico. For several years he was manager of new products at the Cemex R&D facility in Monterrey, Mexico. He specializes in troubleshooting production problems and the introduction of new products such as cellular concrete and CLSM.

John Luciano is a statistical consultant for Master Builders in Cleveland, Ohio, USA. He has worked extensively with optimization of concrete, admixtures, and related construction materials.

Daniel Constantiner is s staff scientist for Master Builders in Cleveland, Ohio, USA.

INTRODUCTION

Experimental design and mixture optimizations have been used for several years in different settings to improve product development. Examples can easily be found in statistical publications for the chemical and physical sciences [1,2]. These statistical methods have been used to study and model concrete properties and behavior [3,4]. These statistical methods are equally powerful tools when the problem at hand is the design of concrete mixtures. They are particularly useful when the concrete is required to perform under unusual conditions, such as hot weather [5], or high strength with modulus of elasticity requirements [6,7].

This study was run to determine a way to produce cellular concrete with challenging strength and unit weight requirements using local aggregates. Generally foamed concrete can be readily produced, but due to the stringent requirements a detailed evaluation was warranted. Two options were available to produce cellular concrete with desirable compressive strength for a given unit weight. The first option was to make cellular concrete with relatively expensive lightweight aggregates and preformed chemical foam. The other option was simply to use only the preformed chemical foam.

The primary objective was to develop a tool to facilitate the selection of materials and concrete properties. The first step was to model the behavior of cellular concrete as a function of the design variables and the second, to optimize this behavior with respect to the raw material cost. The three primary criteria used in the optimization were cost, strength and unit weight of the concrete.

EXPERIMENTAL DESIGN

A modified four variable central composite response surface design was used to select mixture composition and target unit weights. The design modification to accommodate the difficulty and expense associated with actual field testing was to generate 1800 kg/m^3 and 1200 kg/m^3 data from the *same* truck when possible. In addition, cylinders were cast when the unit weight was not close enough to the specified target unit weight. Though these points do not fit the experimental design, they provide valuable information to assess the validity of the statistical models generated from the data.

This type of design was chosen for many reasons. The experimental layout allows many candidate models to fit the data. In particular, it is possible to go beyond simple linear effects since a full second order model can be fit to the data. In short, this design contains the flexibility to address the needs of this study.

Design Variables

The variables undergoing intentional change, and their respective levels, are listed in Table 1.

Table 1 Intentionally changed variables and respective levels

	VERY LOW VALUE	LOW VALUE	CENTER VALUE	HIGH VALUE	VERY HIGH VALUE
Aggregate:Cement Ratio	1.4	1.7	2.0	2.3	2.6
Water:Cement Ratio	0.35	0.40	0.45	0.50	0.55
Lightweight Sand %	0	15	30	45	60
Unit Weight, kg/m^3		1200	1500	1800	1950

The aggregate:cement ratio (A/C) is defined as the combined weight of saturated surface-dried (SSD) lightweight and regular sand relative to the weight of cement. The lightweight sand percentage is also defined on a weight basis. The "very low" unit weight value was not run because of the expected lack of strength. Lower unit weight data will tend to have a greater variance from the target water-to-cement ratio than higher unit weight data, because the foaming system delivers a significant amount of water. Consequently, the foaming time can also be used as a predictor variable

Quantities used to typically describe concrete mixture proportions are different from the design variables listed above. However, once values are selected for the design variables, other quantities such as cement, total aggregate, air, and water contents are fixed and can be calculated algebraically.

EXPERIMENTAL SETUP

Concrete mixtures were produced in a ready-mix batching plant. The ready-mix trucks (9 m^3 capacity) were batched with a minimum volume of 2 m^3 before the concrete was foamed. The batched material was converted to cellular concrete by using a foaming system based on water pressure [8]. The final volume of concrete in the truck varied depending on the amount of foam required to reach the target unit weight. A total of 22 trucks were evaluated over four days.

DATA ANALYSIS

Sources of Variability

The mechanism that relates compressive strength to the experimental conditions is complex and unknown. The analytical goal is to develop a model that approximates this unknown relationship by using the known sources of variation as building blocks. These sources of variation include the design variables (e.g., A/C, w/c, lightweight sand replacement percentage, and the unit weight), environmental variables (e.g., temperature), covariables such as foaming time, amount of foaming admixture, and experimental error (e.g., batching and testing).

Model Building

The greatest value of the experimental design is to allow the effects of one design variable to be estimated independently of another design variable. Regression (least squares) analysis is the tool of choice to build predictive models because of its power and flexibility.

RESULTS

Figure 1 shows the nonlinear relationship between 28-day compressive strength and the unit weight of the concrete. The data points are arranged in a complex "S shaped" fashion. Figure 2 shows the relationship between 28-day strength and the calculated air content. A simple quadratic approximation to the data is contained within this graph.

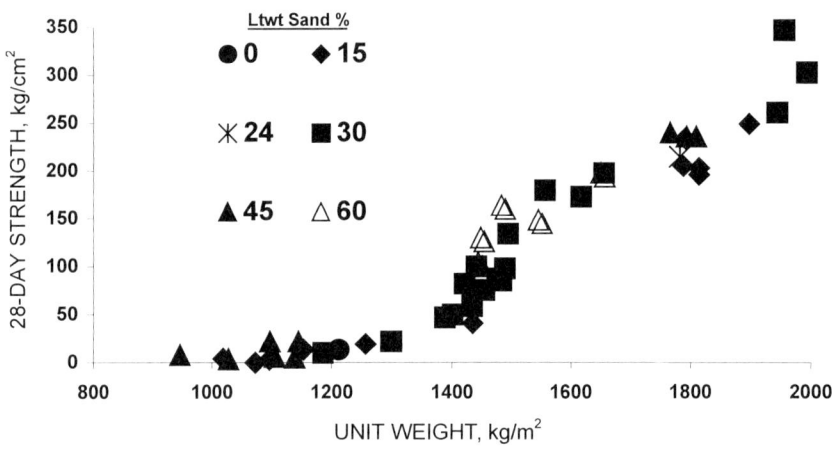

Figure 1 Compressive strength of cellular concrete made with different levels of lightweight sand and various unit weights

Figure 3 shows the strong linear relationship between the unit weight and the calculated air content. The legend shows that the deviation from linearity is mostly due to the lightweight sand percentage in the mix. A simple line would underestimate mixtures with lower percentages of lightweight sand and overestimate mixtures with higher percentages of lightweight sand. A predictive equation to account for the lightweight sand percentage is given in the graph.

The statistical model to estimate 28-day strength (kg/cm^2) for the evaluated materials is given by Equation 1:

$$f'_c 28 = 102.4 - 1.5\, x_2 + 11.5\, x_3 + 100.6\, x_4 - 9.0\, x_2 x_4 + 6.4\, x_3^2 + 12.6\, x_4^2 \quad (1)$$

where $x_2 = (w/c - 0.45)/0.05$,
$x_3 = $ (Lightweight Sand % $-30)/15$, and
$x_4 = $ (Unit Weight $-1500)/300$.

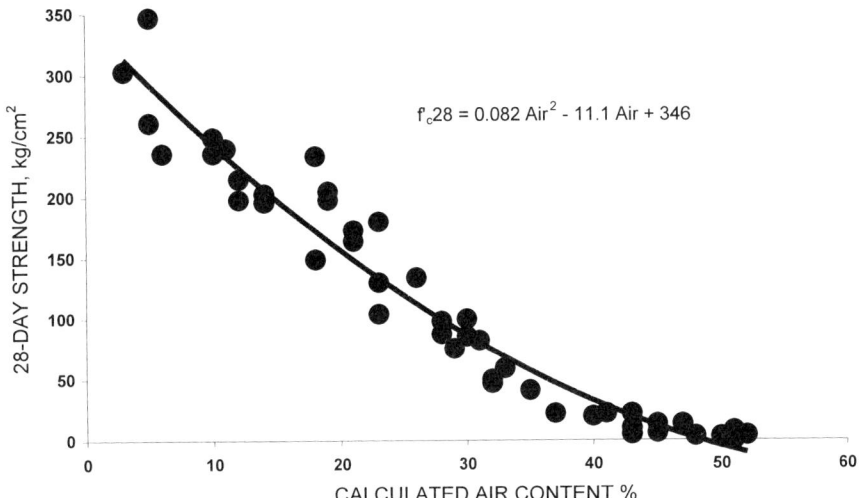

Figure 2 Compressive strength as a function of the calculated air content

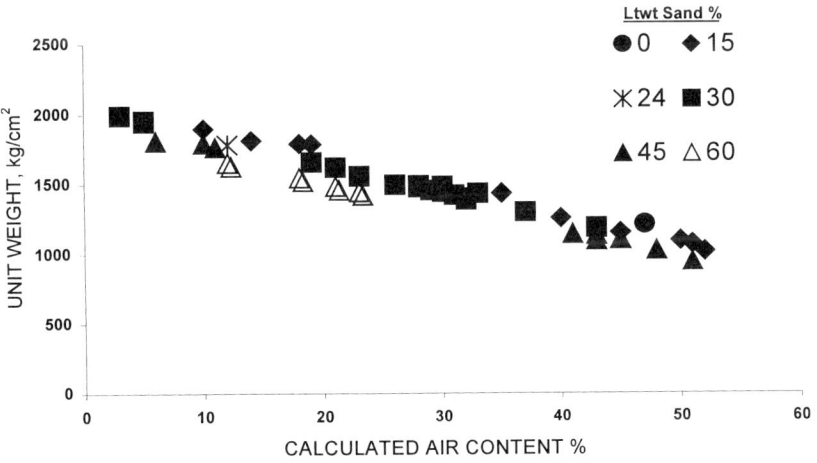

Figure 3 Plot illustrating the relationship between unit weight and calculated air content

Equation 1 is an empirical model to approximate performance within an experimental region when values of d are less than or equal to 2 where $d = \sqrt{x_1^2 + x_2^2 + x_3^2 + x_4^2}$ and x_1=(A/C-2)/0.3. Figure 4 is a scatter plot comparing the actual to the predicted 28-day strengths. This graph shows that the predicted 28-day strengths are unbiased and provide a reasonable estimate for the corresponding actual strengths.

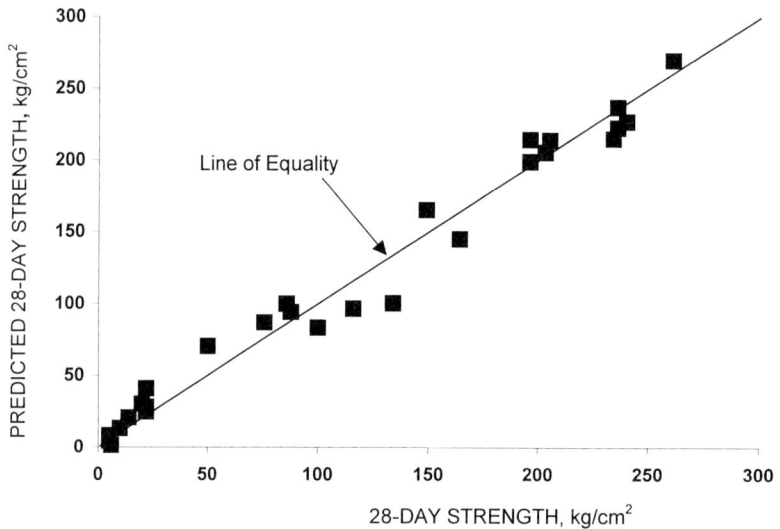

Figure 4 Scatter plot illustrating the relationship between the predicted and actual compressive strength

OPTIMIZATION

An optimization model was constructed using the empirical relationship for 28-day strength and cost information. This optimization model has been developed using a non-linear optimization routine that is available in Microsoft Excel called "Solver". Solver iteratively searches for optimal solutions subject to input constraints (i.e., within the experimental region) that minimize cost while maintaining other parameters relatively constant (such as 28-day strength and unit weight). The result is a single solution that meets certain predefined parameters. To obtain a global or broader picture, it is easy to run the optimization routine different times changing the predefined constraints. Examples of this exercise are shown in Tables 1 to 3.

The first line in Table 1 shows the cost-minimized optimization results when the criteria used are: maximum unit weight of 1700 kg/m³, and a minimum strength of 200 kg/cm² at 28 days. The desired properties can be obtained with the use of a high amount of lightweight sand and a relatively low amount of air. If a cost of 100% was assigned to this mix, then other mixtures can be compared on the basis of cost. The second and third lines of Table 2 show relatively similar mixtures where the predefined constraint has been expanded to include a higher concrete unit weight. In these cases, the mixtures that should develop the required 200 kg/cm² are significantly different than the original. It is important to note the significant drop in concrete cost (16%) when the unit weight is increased by only 50 kg/m³ (less than 3%).

Similar examples are shown in Tables 3 and 4 for concrete expected to deliver compressive strengths of 150 and 100 kg/cm², respectively. Table 3 shows a similar trend as in Table 2; large reduction in cost while increasing the unit weight only by 25 kg/m³. But it also shows that further increase in the unit weight does not provide a significant cost advantage. Table 4 shows similar results when the criteria are changed to allow a maximum weight of 1400 kg/m³ and a minimum strength of 100 kg/cm². Again, it is necessary to use a high amount of lightweight sand into the mix to reduce the amount of air. There is a cost benefit to increasing the unit weight upper limit to 1450 kg/m³, although the cost advantage is not as significant as in the previous two cases.

Table 2 Mixture optimization results for cellular concrete with strength higher than 200 kg/cm²

DESIRED STRENGTH, (Kg/cm²)	UNIT WEIGHT, (kg/m³)	SUGGESTED MIXTURE PROPORTIONS (kg/m³)					RELATIVE COST
		Air, %	Cement	Ltwt Sand*	Reg Sand#	Water	
200	1700	14.2	447	485	586	182	100%
200	1750	21.9	496	65	986	203	84%
200	1800	20.5	481	58	1010	216	82%

* Ltwt sand – lightweight sand; # Reg sand – normal weight sand

Table 3 Optimization results for cellular concrete with strength higher than 150 kg/cm²

DESIRED STRENGTH, (Kg/cm²)	UNIT WEIGHT, (kg/m³)	SUGGESTED MIXTURE PROPORTIONS (kg/m³)					RELATIVE COST
		Air, %	Cement	Ltwt Sand*	Reg Sand#	Water	
150	1600	20.5	405	406	619	170	100%
150	1625	28.0	463	19	947	196	85%
150	1650	26.9	444	32	959	201	84%

Table 4 Mixture optimization results for cellular concrete with strength higher than 100 kg/cm²

DESIRED STRENGTH, (Kg/cm²)	UNIT WEIGHT, (kg/m³)	SUGGESTED MIXTURE PROPORTIONS (kg/m³)					RELATIVE COST
		Air, %	Cement	Ltwt Sand*	Reg Sand#	Water	
100	1400	26.6	366	435	419	180	100%
100	1450	26.0	364	386	522	178	97%
100	1500	28.6	373	246	719	162	91%

CONCLUSIONS

- Mixture optimization that takes into consideration the cost and the performance of the concrete can be used to estimate cost implications of changing the performance requirements of a concrete mixture.

- Slightly changing the performance requirements can substantially change the required mixture proportions and produce a significant cost advantage.

REFERENCES

1. MONTGOMERY, D. C., Design and Analysis of Experiments, 3rd Ed., John Wiley & Sons, New York, 1991.

2. BOX, G.E.P. and DRAPER, N., Empirical Model-Building and Response Surfaces, John Wiley & Sons, New York, 1991.

3. GAO, D., HEIMANN, R. B., and ALEXANDER, S. D. B., Box-Behnken Design Applied to Study the Strengthening of Aluminate Concrete Modified by a Superabsorbent Polymer/Clay Composite, Advances in Cement Research, Vol. 9, No. 35, pp. 93-97, July 1997.

4. NEHDI, M., MINDESS, S., and AÏTCIN, P-C., Statistical Modelling of the Microfiller Effect on the Rheology of Composite Cement Pastes, Advances in Cement Research, Vol. 9, No. 33, pp. 37-46, Jan. 1997.

5. AL-GAHTANI, H.J, ABBASI, A. G. F., and AL-AMOUDI, O.S.B, "Concrete Mixture Design for Hot Weather: Experimental and Statistical Analyses", Magazine of Concrete Research, Vol. 50, No.2, pp. 95-105, June 1998.

6. LUCIANO, J.J., and BOBROWSKI, G. S., "Using Statistical Methods to Optimize High-Strength Concrete Performance", Transportation Research Record, Vol. 1284, pp. 60-68, 1990.

7. LUCIANO, J.J., NMAI, C.K., and DELGADO, J.R., "A Novel Approach to Developing High-Strength Concrete", Concrete International, pp. 25-29, May 1991.

8. NMAI, C.K., MCNEAL F, and MARTIN, D., "New Foaming Agent for Controlled Low-Strength Materials Applications", Concrete International, Vol. 19, No. 4, pp. 44-47, April 1997.

JUST FOAMED CONCRETE – AN OVERVIEW

E P Kearsley
University of Pretoria
South Africa

ABSTRACT. Foamed concrete is manufactured by adding a foaming agent to a cement based mortar. The foaming agent can be added to the mixture and foam formed through vigorous mixing or the foaming agent can be aerated before being added to the mixture (pre-foaming). The foamed concrete discussed in this paper has been manufactured using pre-foaming. Over the last fifty years the use of foamed concrete in the construction industry has been almost exclusively limited to non-structural applications such as void filling, thermal insulation, acoustic dampening, trench filling and building blocks. Although foamed concrete has been widely used it is perceived to be weak and non-durable. Tests conducted at the University of Pretoria in South Africa during the past ten years indicate that the short and long term properties of foamed concrete is such that the material could well be used for structural elements. In South Africa large volumes of pulverized fuel ash (pfa) is available at relatively low cost and the effect of using high percentages of pfa under different curing regimes has been investigated. Tests have been conducted on the use of steel reinforced foamed concrete in structural applications. The behavior of foamed concrete elements has been compared to that of normal concrete elements of the same compressive strength. The use of fibers in foamed concrete to improve shear behavior is discussed.

Keywords: Foamed concrete, Pulverized-fuel ash (PFA), Curing, Durability, Fibers, Shear strength, Lightweight concrete, Aerated concrete.

Mrs Elsabe P Kearsley is a Senior Lecturer in Civil Engineering at the University of Pretoria in South Africa. Her main interests are the development of affordable alternative building materials and methods.

INTRODUCTION

Over the last fifty years the use of foamed concrete in the construction industry has been almost exclusively limited to non-structural applications such as void filling, thermal insulation, acoustic dampening, trench filling and building blocks. Although foamed concrete has been widely used it is perceived to be weak and non-durable.

Cost is a major concern in the development of any infrastructure and to address the need of developing communities for affordable infrastructure and permanent residential units, schools and clinics, modern specialized technology should be applied to improve the quality of building materials used in under developed areas. Foamed concrete is a lightweight building material that can be used if the strength is optimized and the cost minimized and therefore the use of waste products, such as ash, in foamed concrete is being investigated.

During development the emphasis should be on energy efficient housing in order to minimize electricity consumption. The good thermal properties of foamed concrete increase the possible use of foamed concrete in energy efficient applications [1]. The relative high drying shrinkage and low mass of foamed concrete makes the material most suitable for manufacturing small pre-cast structural elements that can be used in labor intensive or self-help projects.

Tests conducted at the University of Pretoria in South Africa during the past ten years indicate that the short and long term properties of foamed concrete is such that the material could well be used for structural elements. This paper is based on some of the results obtained from the ongoing research project.

MATERIALS USED

Foamed concrete is produced under controlled conditions from cement, filler, water and a liquid chemical that is diluted with water and aerated to form the foam. The foaming agent used is "Foamtech", consisting of hydrolyzed proteins and manufactured in South Africa. The foaming agent is diluted with water in a ratio of 1:40 and then aerated to a density of 70 kg/m^3.

The cement used in this investigation is rapid hardening Portland cement that can be classified as CEM I 42.5R according to the South African specification SABS ENV 197-1:1992 [2].

In South Africa the production of electricity and liquid fuel from coal causes the accumulation of vast quantities of coal ash. In 1990 it was estimated that approximately 350 million tons of coal ash was stored in South Africa and that this amount was growing by 23 million tons annually [3]. A large proportion of this ash can be used at low cost as a cement extender. SABS 1491: Part II, the South African standard for the use of ash as cement extender [4], limits the particles with diameters exceeding 45 µm to a maximum of 12.5%, which increases the cost of producing classified ash. Unclassified ash, where approximately 40% of the particles have diameters exceeding 45 µm, has been used as cement extender and filler in the mixtures discussed in this paper.

PROPERTIES OF FOAMED CONCRETE

Foamed concrete is more sensitive to water demand than normal concrete. Water demand is normally not a problem but if the water in the mixture is not sufficient for the initial reaction of the cement, the cement withdraws water from the foam, causing rapid degeneration of the foam. If too much water is added segregation takes place, causing a variation in density [5]. The density of foamed concrete is a function of the percentage foam added to the mixture and the maximum compressive strength of the material decreases as the density decreases (and the foam content increases) as indicated in Figure 1.

Figure 1 contains a graph of 28 day compressive strength as a function of dry density for mixtures with different percentages of ash replacement. All the mixtures contained water, cement and a pre-foamed protein foam. The first set of mixtures contained no ash and for each set of mixtures a larger percentage of the cement was replaced with unclassified ash. The ash/cement ratios of 1, 2, 3 and 4 as indicated reflects 50%, 66.7%, 75% and 80% cement replacement (per weight). For all the mixtures the water/cementitious ratio was kept constant at approximately 0.3. From this graph it can be seen that 50% of the cement can be replaced with unclassified ash without any reduction in 28 day compressive strength. Higher percentages replacement results in reduced strength and the effect of high ash content on the 28 day compressive strength seems to increase with increased density. The dry densities as indicated in Figure 1 were determined by oven drying a cube at 105 °C until no further reduction in strength took place.

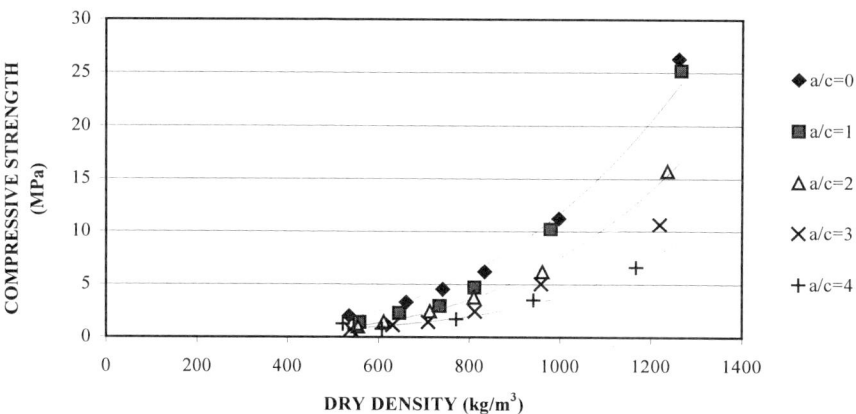

Figure 1 Effect of ash replacement on compressive strength.

The dry densities as indicated in Figure 1 can however not be used in calculating the exact composition of foamed concrete mixtures. A so-called "casting density" is required for mix design and casting control purposes. The relation between casting and dry density of the above-mentioned mixtures is indicated in Figure 2. For casting densities between 700 and 1500 kg/m^3 there seems to be a linear relation between the casting and the dry density and this relation can best be explained by the following equation:

$$\gamma_{dry} = 0.868\gamma_{cast} - 55.07 \qquad (1)$$

Where:

γ_{dry} = dry density (kg/m³)
γ_{cast} = casting density (kg/m³)

This equation explains 99.35% of the variability in dry density. If foamed concrete with a specific dry density is required the equation as indicated above can be used to determine what the casting density would need to be. When designing foamed concrete a target casting density is determined and the water/cement and sand/cement ratios are chosen [6]. Using these ratios and the relative densities of the materials the mass of the cement and the volume of foam that should be added to obtain the required density can be determined.

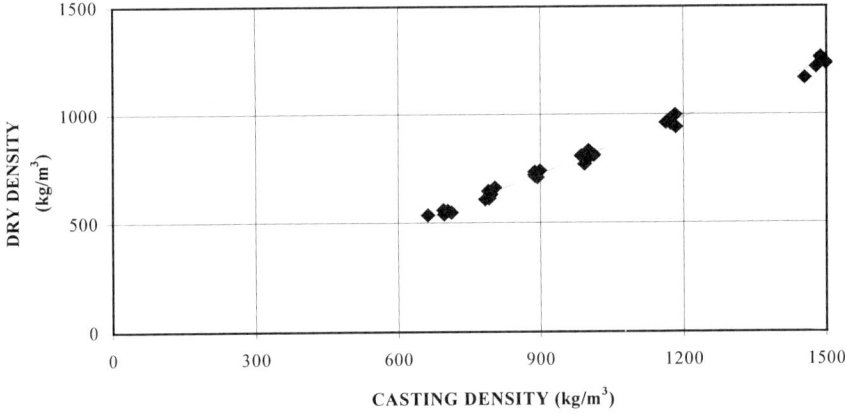

Figure 2 Dry density as a function of casting density

EFFECT OF CURING

The compressive strengths as indicated in Figure 1 was obtained for samples that were wrapped in plastic after demoulding, and then placed in a constant temperature room at 22 °C and approximately 65% relative humidity. The compressive strengths obtained after 28 days for mixtures with dry densities in the region of 1 250 kg/m³ indicate that foamed concrete with similar densities could be used in structural elements.

If structural elements are manufactured in a pre-cast factory it is of utmost importance that the required strength is reached in the shortest viable time. The influence of pfa on the strength development of concrete has been studied extensively and it has been concluded that the rate of strength development is related to the mix proportions, properties and type of pfa, temperature and curing conditions and the hydration of the cementitious materials. In general terms at normal temperatures the use of coal ash results in reduced early strengths and increased long term strengths [7]. When concrete made with normal Portland cement is cured at temperatures in excess of 30 °C, an increase in strength at early ages is followed by a marked decrease in strength at later ages. In contrast however pfa shows strength gains as a result of heating. The fact that no reduction in 28 day strength was observed for foamed concrete mixtures with up to 50% cement replacement, cured in air at 22 °C (see Figure 1), indicates that elevated curing temperatures, could result in increased strength for mixtures with high pfa contents.

To establish the effect of elevated curing temperature on the strength development of foamed concrete, cubes manufactured from mixtures with a casting density of 1 500 kg/m^3 were cured in water at different temperatures. A mixture with a casting density of 1 500 kg/m^3 will have a dry density in the region of 1 250 kg/m^3.

In Figure 3 the compressive strength of a mixture with an ash/cement ratio of 1 (50% cement replacement) is plotted as a function of time for different curing temperatures. The temperatures as indicated are 22 °C, 40 °C, 50 °C, 60 °C and 70 °C. These results indicate that it is possible to obtain compressive strengths as high as 40 MPa within 3 days for foamed concrete with dry densities as low as 1 250 kg/m^3. From the results obtained the optimum temperature for the highest ultimate strength seems to be 40 °C, which yielded an average compressive strength of 62 MPa after 56 days.

Although these mixtures have high compressive strengths they will not be cost effective and the cement content will have to be reduced, not only to reduce the cost but also to reduce shrinkage [5]. One way of reducing the cement content is to increase the percentage cement replacement and therefore the effect of the curing temperature on the strength gain of a mixture with a casting density of 1 500 kg/m^3 and an ash/cement ratio of 3 (75% cement replacement) was investigated.

The compressive strength of this mixture is plotted as a function of time in Figure 4. The compressive strength at all the curing temperatures is noticeably lower than for the mixture with 50% cement replacement, but this difference seems to be the largest at low curing temperature. The higher ash content seems to result in a decrease in the rate of strength development even at high temperatures. After 28 days the mixtures containing 75% pfa that were cured at high temperature have virtually the same strength than the mixtures containing only 50% ash. The optimum curing temperature again seems to be somewhere between 40 °C and 50 °C, but with such a high ash content the early age strength development at these temperatures is very slow as can be seen in the 3 and 7 day compressive strengths in Figure 4.

For early age strength the curing temperature at 75% cement replacement will have to be increased to at least 60 °C, if less than 7 days curing is required. Although the ultimate strengths are in the region of only 40 MPa in stead of 60 MPa where 50% cement replacement took place, the strengths are still such that the material could be used for structural applications. The fact that up to 50% of the concrete own weight can be eliminated can however lead to savings the cost of the super- and substructure of any building constructed using foamed concrete.

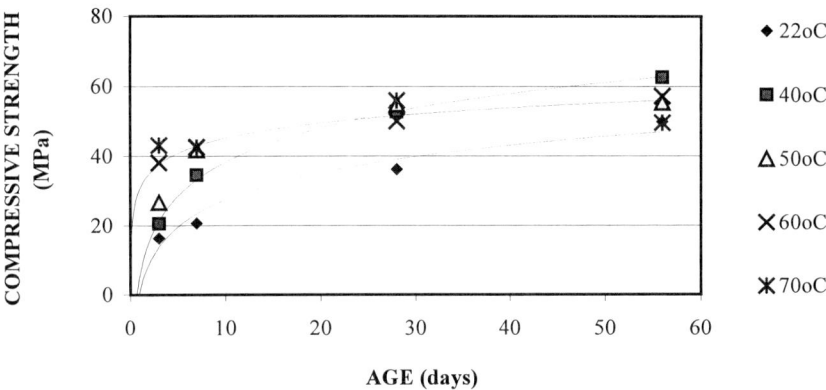

Figure 3 Strength development for 50% cement replacement

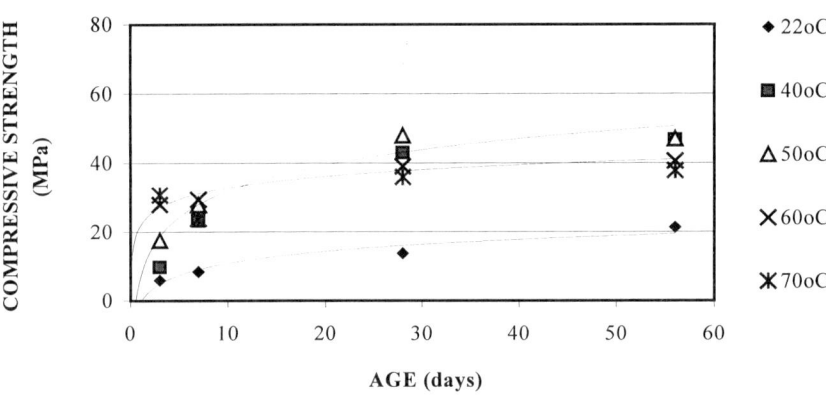

Figure 4 Strength development for 75% cement replacement

STRUCTURAL BEHAVIOUR

As the compressive strength of foamed concrete at higher densities falls within the same range than the characteristic compressive strengths as used in structural design codes such as BS 8110, it was decided to compare the behavior of foamed concrete beams to that of identically reinforced normal concrete beams. If the behavior does not differ much it should be possible to use normal concrete design codes for designing foamed concrete structural elements.

Beams with a compressive strength of 15 MPa were cast using foamed concrete and normal concrete. Each beam was reinforced using two 10 mm bars and shear stirrups at 150 mm centers. These beams were tested using a third-point-loading configuration and the results of these tests can be seen in Figure 5 [8]. The modulus of elasticity as measured for foamed concrete is approximately 50% of the value measured for normal concrete with comparable compressive strength. This reduction in modulus of elasticity indicates that under similar loading the foamed concrete should have significantly higher deflections than normal concrete. From the results as shown in Figure 5 it can be seen that the foamed concrete does deflect noticeably more than the normal concrete but the difference in deflection is relatively small and the stiffness of the reinforcing must be contributing to limit the deflection.

The foamed concrete beam with stirrups at 150 mm centers failed in shear while an identically reinforced normal concrete beam carried a higher load and failed in bending. To obtain a result from the foamed concrete similar to that obtained from the normal concrete the spacing of the shear stirrups had to be decreased. From Figure 5 it can be seen that the behavior of a foamed concrete beam with stirrups at 125 mm centers can be compared to that of a normal concrete beam with stirrups at 150 mm centers [5].

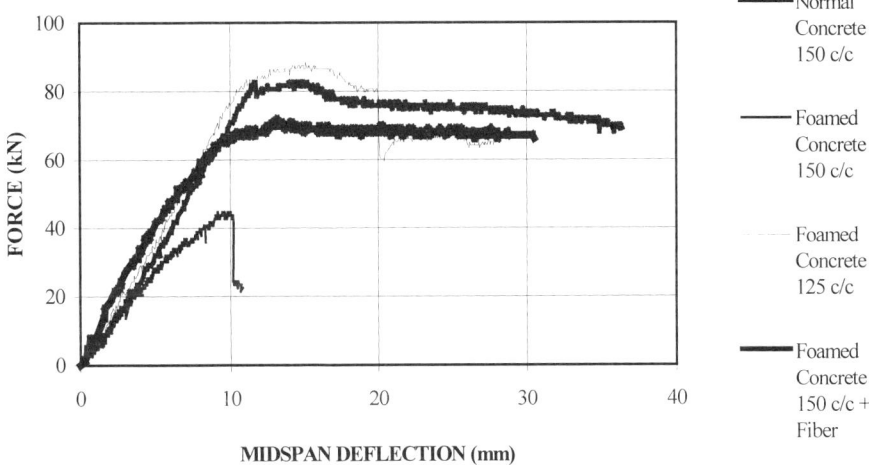

Figure 5 Load deflection behaviour of 15 MPa beams

Aggregate interlocking is said to improve the shear resistance of concrete and the lack of course aggregate in foamed concrete can be the reason for the reduced shear resistance of the material. Some form of interlocking needs to be provided to improve the shear behavior of foamed concrete. The inclusion of steel fibers in normal concrete is known to enhance the flexural and shear strength of the resulting composite material and in beams made of mortars the fibers are said to be generally more effective as they are free from aggregate interference [9]. Steel fibers can not be used in foamed concrete as their high density causes them to settle to the bottom of the mixture while the foamed concrete is still setting. The use of various types, diameters, lengths and volumes of fibers have been investigated [9] and it was decided to add 2 kg/m^3 of 12 mm long chopped polypropylene fibers to the foamed concrete beams to enhance the shear behavior. The fibers did not improve the compressive strength but from the results as indicated in Figure 5 it can be seen that the fibers improved the shear behavior of the foamed concrete sufficiently that the behavior of foamed concrete beams containing fibers does not differ noticeably from normal concrete beams.

The compressive strength of all these beams were only 15 MPa, which is significantly lower than the compressive strength normally used for structural concrete elements and the behavior observed might therefore not be typical of structural elements. 30 MPa beams were cast to determine whether the behavior of structural foamed concrete beams could be improved by the addition of fibers.

The load deflection behavior of the 30 MPa beams is shown in Figure 6. Each of these beams was reinforced with two 10 mm diameter high yield steel bars, and 8 mm mild steel stirrups at 150 mm centers. From this graph it can be seen that the fibers is again required to prevent shear failure. Although the compressive strength of the concrete is double that of the concrete used in Figure 5 the ultimate load carried by the beam is only marginally higher than before. The beams were under-reinforced and bending failure occurred with the steel flowing.

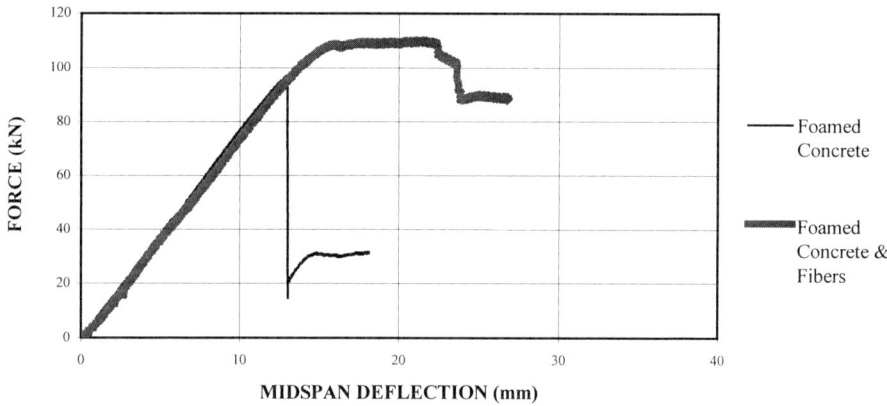

Figure 6 Load deflection behaviour of 30 MPa beams

DURABILITY OF FOAMED CONCRETE

Although foamed concrete has been widely used throughout the world it is perceived to be weak and non-durable. A durable structure will maintain its required strength and serviceability during its service life, and the durability of concrete depends largely on the permeability of the concrete [10]. The permeability of foamed concrete of different densities was determined and compared to that of normal 25 MPa concrete [11]. Oxygen permeability tests were undertaken using a falling head permeameter [12] and the results obtained are indicated in Figure 7. In this graph time is plotted as a function of the log of the ratio of pressure at the beginning of the test (P_0) to the pressure at the end of the test (P). From these results it can be seen that the pressure decline is, as expected, a function of the density of the foamed concrete, with the higher density mixes exhibiting slower rates of decline in pressure. These results indicate that the concrete with a casting density of 1 500 kg/m^3 was less permeable than the normal 25 MPa concrete. If oxygen permeability is taken as an indication of concrete durability, it does seem possible that foamed concrete with relatively high densities could be used as a durable alternative to normal concrete.

Repeating the test on lower density samples resulted in a higher rate of decline indicating that the pressure might damage the porous structure of the material. The falling head permeameter might therefore not be a suitable method for determining the permeability of low density foamed concrete.

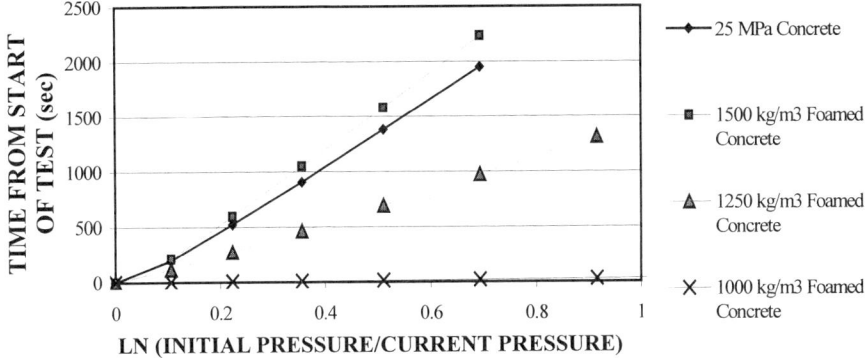

Figure 7 Permeability

CONCLUSIONS

Replacing large volumes of cement with unclassified pfa can reduce the cost of manufacturing foamed concrete. The pfa not only reduces the drying shrinkage of foamed concrete but can also result in increased strength. Curing at high temperatures can improve the early and long term compressive strength of foamed concrete. The strengths obtained from well cured mixtures with high ash contents are such that foamed concrete can be used as a structural material in reinforced foamed concrete.

The shear resistance of foamed concrete is much lower than that of normal concrete due to the lack of aggregate interlocking but the shear strength can be improved by the addition of chopped polypropylene fibers. The behavior of foamed concrete beams containing fibers is similar to that of normal concrete. At this stage of our investigation it seems that the structural design codes used for normal reinforced concrete can be used for foamed concrete elements.

The durability of foamed concrete needs to be investigated before reinforced foamed concrete structures can be built. The cell-like structure of foamed concrete and the possible porosity of the cell walls do not make the foamed concrete less resistant to penetration of moisture than normal concrete. The air voids seem to act as a buffer, preventing rapid moisture penetration. Initial tests indicate that foamed concrete with a casting density of 1 500 kg/m^3 can be less permeable than normal 25 MPa concrete, and accelerated corrosion tests indicate that foamed concrete can be just as durable than normal concrete. These tests still need to be correlated with full scale, long term durability tests.

ACKNOWLEDGEMENTS

The author would like to express appreciation for the research grants made by Grinaker Duraset (Mining) Limited and Alpha Limited to carry out the work reported in this paper.

REFERENCES

1. KEARSLEY, E. P. & MOSTERT, H. F. Foamcrete in Developing Communities. Proceedings of the FIP International Symposium: The concrete Way to Development, Johannesburg, March 1997, pp. 735 – 745.

2. SOUTH AFRICAN BUREAU OF STANDARDS. SABS ENV 197-1:1992: Cement – Composition, specification and conformity criteria, Part I: Common cements. SABS, Pretoria.

3. KRUGER, R. A. Ten years of research into ASH utilisation. Proceedings of the First National Symposium of the South African Coal Ash Association, Pretoria, South Africa. May 1990, pp. 1 – 13.

4. SOUTH AFRICAN BUREAU OF STANDARDS. SABS 1491: Part II – 1989: Portland cement extenders Part II: Fly ash. SABS.

5. KEARSLEY, E. P. The use of foamcrete for affordable development in third world countries. Proceedings Concrete in Service of Mankind Conference (Appropriate Concrete Technology) Dundee, UK, 1996, pp. 233 – 243.

6. ACI COMMITTEE 523. 1975. Guide for Cellular Concretes Above 50 pcf, and for Aggregate concretes Above 50 pcf with Compressive Strength Less Than 2 500 psi. ACI Journal, February 1975, Title no. 72-7, pp. 50-66.

7. SARKAR, S, L & GHOSH, S. N. Progress in Cement and Concrete. Volume 4: Mineral Admixtures in Cement and Concrete. 1993, New Dehli.

8. KEARSLEY, E. P. & MOSTERT, H. F. The Use of Foamcrete in Southern Africa. Proceedings of the ACI International Conference on High Performance Concrete, SP-172, Kuala Lumpur, Malaysia, 1997, pp. 919 – 934.

9. SWAMY, R. N., JONES, R. & CHAIM, T. P. Influence of steel fibers on the Shear resistance of Lightweight Concrete I-beams. ACI Structural Journal, Volume 90, No 1, January – February 1993, pp. 103 – 114.

10. NEVILLE, A. M. 1995. Properties of Concrete. Fourth Edition. Longman Group Limited, Essex, England.

11. KEARLSEY, E. P. & BOOYENS, P. J. Reinforced Foamed Concrete – Can It Be Durable? Concrete/Beton No 91, November 1998, pp. 5 – 9.

12. BALLIM, Y. A Low Cost Falling Head Permeameter for Measuring Concrete Gas Permeability. Concrete Beton, Number 61, 1991, pp. 13 – 18.

THEME FOUR:
UNDERWATER CONCRETE

Keynote Paper

UNDERWATER CONCRETE PLACEMENT: MATERIALS METHODS AND CASE STUDIES

J E McDonald
B Neeley
US Army Engineer Waterways Experiment Station
United States of America

ABSTRACT. Dewatering an area so that construction or repair of hydraulic structures can be accomplished under dry conditions is often difficult and usually expensive. For example, costs to dewater a stilling basin can exceed $1,000,000, and the average cost to dewater is more than 40 percent of the total repair cost. Therefore, underwater concrete placement should be an economical alternative for repair of many hydraulic structures. Consequently, studies were conducted by the US Army Corps of Engineers, as part of the Repair, Evaluation, Maintenance, and Rehabilitation (REMR) research program, to evaluate state-of-the-art methods and materials for repair of concrete underwater. These studies included (a) identification of methods and equipment for underwater cleaning and inspection of concrete surfaces, (b) evaluation of materials and procedures for anchor embedment in hardened concrete under submerged conditions, (c) development of improved materials and techniques for underwater placement of freshly mixed concrete, and (d) development of prefabricated elements for underwater repair. This paper summarizes the results of these laboratory and field studies and includes selected case studies that illustrate the application of this technology in repair of existing structures.

Keywords: Underwater concrete, Repair, Antiwashout admixtures, Tremie concrete, Surface preparation, Anchors, Precast concrete, Prefabricated steel

Mr James E McDonald is the senior research civil engineer in the Concrete and Materials Division, Structures Laboratory, US Army Engineer Waterways Experiment Station (WES), Vicksburg, MS, USA. His main research interest is development of improved technology for evaluation and repair of concrete. Mr. McDonald has published widely and serves on many technical committees. He is the current President of the International Concrete Repair Institute.

Mr Billy Neeley is a research civil engineer in the Concrete and Materials Division, Structures Laboratory, US Army Engineer Waterways Experiment Station (WES), Vicksburg, MS, USA. He has over 16 years of experience in research and application of specialized areas of concrete, including underwater concrete, high-strength concrete, fiber-reinforced concrete, heavyweight concrete, lightweight concrete, and mass concrete. He has authored or co-authored over 30 technical reports and papers.

INTRODUCTION

The U.S. Army Corps of Engineers owns and operates more than 600 civil works projects throughout the United States. About one-half of the hydraulic structures (dams, navigation lock chambers, powerhouses, etc.) associated with these projects were constructed prior to 1940. Therefore, it is not surprising that the concrete in these structures has experienced varying degrees of deterioration.

Dewatering a hydraulic structure so that repairs can be made under dry conditions is usually difficult and always expensive. For example, in repair of the Libby Dam stilling basin, the cost of dewatering alone exceeded $1,000,000. Historically, the average cost of dewatering is more than 40 percent of the total repair cost. Also, dewatering often disrupts project operations and can have potentially adverse impacts on the environment. Therefore, underwater concrete placement should be an attractive alternative for repair of many hydraulic structures.

The tremie is a well established underwater concrete placement procedure; however, it is most often used in new construction to place thick sections involving large quantities of concrete. In these cases, the end of the tremie pipe can be maintained in the mass of fresh concrete to minimize loss of cement. However, in repairs the sections are generally much thinner, making it difficult to maintain a tremie seal. Consequently, studies were conducted as part of the Repair, Evaluation, Maintenance, and Rehabilitation (REMR) research program to develop improved materials and methods for underwater repair of concrete. These studies included (a) identification of methods and equipment for underwater cleaning and inspection of concrete surfaces, (b) evaluation of materials and procedures for anchor embedment in hardened concrete under submerged conditions, (c) development of improved materials and techniques for underwater placement of freshly mixed concrete, and (d) development of prefabricated elements for underwater repair. Results of these studies are summarized in the following.

UNDERWATER INSPECTION

A through and logical evaluation of the current condition of the concrete in a structure is the first step of any repair project. A variety of procedures and equipment are available for conducting underwater surveys [1]. Manned diving, including scuba and surface-supplied air systems, is the traditional method for performing underwater inspections. This method offers a number of advantages: they are (a) applicable to a wide variety of structures, (b) flexible inspection procedures, (c) simple (especially the scuba diver in shallow water applications), and in most cases, (d) relatively inexpensive. Also, a variety of commercially available instruments for testing concrete above water have been modified for underwater use by divers. These instruments include a rebound hammer to provide data on concrete surface hardness, a magnetic reinforcing steel locator to locate and measure the amount of concrete cover over the reinforcement, and direct and indirect ultrasonic pulse velocity systems which can be used to determine the general condition of concrete based on sound velocity measurements [2].

Underwater vehicles can compensate for the limitations inherent in diver systems because they can function at extreme depths, remain underwater for long durations, and repeatedly perform the same mission without sacrifice in quality.

Also, they can be operated in environments where water temperatures, currents, and tidal conditions preclude the use of divers. Remotely operated vehicles (ROVs) include video systems for real-time or slow-scan viewing and have some degree of mobility. Most ROVs are capable of accommodating various attachments for grasping, cleaning, and performing other inspection chores. Specially designed ROVs can accommodate and operate nondestructive testing equipment. ROVs have been used to inspect a variety of hydraulic structures, including dams, intake towers, tunnels and conduits and breakwaters and jetties.

A high-resolution acoustic mapping system has been successfully used to rapidly and accurately survey erosion damage in stilling basins at a number of dams. Survey results are in the form of real-time strip charts showing the absolute relief for each run, three-dimensional surface relief plots showing composite data from all the survey runs in a given area, contour maps of selected areas, and printouts of the individual data points. The system is designed to operate in water depths of 1.5 to 12 m and produce accuracies of $\pm$ 50 mm vertically and $\pm$ 0.3 m laterally [3]. The major limitation of the system is that it can only be used in relatively calm water. Wave action causing a roll angle of more than 5 deg will automatically shut down the system.

UNDERWATER REPAIRS

Surface Preparation

Proper surface preparation is essential for any significant bond to occur between the repair material and the existing concrete substrate. All marine growth, sediments, debris, and deteriorated concrete must be removed prior to placement of the repair material. Equipment and methods specifically designed for underwater excavation and debris removal are available. Also, a wide variety of underwater cleaning tools and methodologies have been designed specifically for cleaning the submerged portions of underwater structures.

Underwater Excavation

The three primary methods for excavation of accumulated materials, such as mud, sand, clay, and cobbles, include: air lifting, dredging, and jetting. Selection of the best method for excavation depends on several factors: the nature of the material to excavated; the vertical and horizontal distances the material must be moved; the quantity of the material to be excavated; and, the environment (water depth, current, and wave action) [4]. Air lifts are appropriate for removal of most types of sediment material in water depths of 8 to 23 m. Jetting and dredging techniques, or combinations thereof, are not limited by water depth.

Debris Removal

The primary types of debris that accumulate in hydraulic structures, such as stilling basins, include cobbles, sediment, and reinforcing steel. Cobbles and sediment can be removed with one of the excavation techniques discussed in the previous section. Removal of exposed reinforcing steel often requires underwater cutting of the steel. The most common techniques for underwater steel cutting can be categorized as either mechanical and thermal. Equipment used for mechanical cutting includes portable, hydraulically-powered shears and bandsaws.

Three thermal techniques are recommended for underwater cutting: oxygen-arc cutting, shielded metal arc cutting and gas cutting. Oxygen cutting is the preferred technique for US Navy Underwater Construction Team diver operations [4]. The technology exists for underwater abrasive-jet cutting systems, and commercially available equipment is continually evolving.

Cleaning Tools

There are three general types of cleaning tools: hand tools, powered hand tools, and, self-propelled cleaning vehicles [4]. Hand tools include conventional devices such as scrapers, chisels, and wire brushes. Because of their low cleaning efficiency, hand tools should be used only where there is light fouling or spot cleaning is to be done in limited areas. Powered hand tools include rotary brushes, abrasive discs, and waterjet systems. A high-pressure waterjet is the best tool to use in obstructed or limited access areas. A high-pressure, high-flow system can be used to remove most types of moderate to heavy fouling. A high-pressure, low-flow system may be required to clean an area that is difficult or impossible to reach with a high-flow system because of the retrojet. Self-propelled cleaning vehicles are large brush systems that travel along the work surface on wheels. A self-propelled vehicle can be used to quickly and effectively remove light to moderate marine and freshwater fouling on large and accessible concrete surfaces. For areas that are not large enough to justify the use of a self-propelled vehicle, hydraulically powered hand tools, such as rotary cutters, can efficiently remove all fouling from concrete surfaces.

Anchors

Repairs are often anchored to the existing concrete substrate with dowels. Anchors are particularly necessary in areas where it is difficult to keep the concrete surface clean until the repair is placed. Since most of the anchors used in concrete repair are post-installed, anchors can be classified as either grouted or expansion systems. Expansion anchors are designed to be inserted into predrilled holes and then expanded by either torquing the nut, hammering the anchor or expanding into an undercut in the concrete. These anchors transfer the tension load from the anchor to the concrete through friction or keying against the side of the drill hole. Grouted anchors include headed or headless bolts, threaded rods, and deformed reinforcing bars. They are embedded in predrilled holes with either cementitious or polymer materials. Cementitious materials include portland-cement grouts, with or without sand, and other commercially available premixed grouts. Polymer materials are generally two-component compounds of polyesters, vinylesters, or epoxies. These resins are available in four forms: tubes or "sausages," glass capsules, plastic cartridges, or bulk.

Anchor Embedment Materials

The effectiveness of neat portland-cement grout, epoxy resin, and prepackaged polyester resin in embedding anchors in hardened concrete was evaluated under a variety of wet and dry installation and curing conditions [5]. Pullout tests were conducted at eight different ages ranging from 1 day to 32 months. Creep and durability tests were also conducted.

Beyond 1 day, all pullout strengths were approximately equal to the ultimate strength of the reinforcing-bar anchor when the anchors were installed under dry conditions, regardless of the type of embedment material or curing conditions. With the exception of the anchors embedded in polyester resin under submerged conditions, pullout strengths were essentially equal to the ultimate strength of the anchor when the anchors were installed under wet or submerged conditions. The overall average pullout strength of anchors embedded in polyester resin under submerged conditions was 35 percent less than the strength of similar anchors installed and cured under dry conditions. The largest reductions in pullout strength, approximately 50 percent, occurred at ages of 6 and 16 months. Although the epoxy resin performed well in these tests when placed in wet holes, it should be noted that the manufacturer does not recommend placement under submerged conditions.

Creep tests were conducted by subjecting pullout specimens to a sustained load of 60 percent of the anchor-yield strength and periodically measuring anchor slippage at the end of the specimen opposite the loaded end. After 6 months under load, anchors embedded in portland-cement grout and epoxy resin, that were installed and tested under wet conditions exhibited low anchor slippage, averaging 0.07 and 0.08 mm, respectively, or 2 to 4 times higher than results under dry conditions. Anchors embedded in polyester resin, installed and cured under submerged conditions, exhibited significant slippage; in fact, in one case the anchor pulled completely out of the concrete after 14 days under load. After 6 months under load, the two remaining specimens exhibited an average anchor slippage of 2.1 mm, approximately 30 times higher than anchors embedded in portland-cement grout under the same conditions.

Long-term durability of the embedment materials was evaluated by periodic compressive strength tests on 51-mm cubes stored both submerged and in laboratory air. After 32 months, the average compressive strength of polyester-resin and epoxy-resin specimens stored in water was 37 and 26 percent less, respectively, than that of companion specimens stored in air. The strength of portland-cement grout cubes stored in water averaged 5 percent higher than that of companion specimens stored in air during the same period.

The performance of anchors embedded in vinylester resin, prepackaged in glass capsules, was also evaluated under dry and submerged conditions [6]. Pullout tests were conducted at four different ages ranging from 1 to 28 days. The tensile capacity of anchors embedded under submerged conditions was approximately one-third that of similar anchors embedded in dry holes.

Anchor Installation Procedures

The reduced tensile capacity of anchors embedded in concrete under submerged conditions with prepackaged polyester resin and vinylester resin cartridges is primarily attributed to the anchor installation procedure. Resin extruded from dry holes during anchor installation was very cohesive, and a significant effort was required to obtain the full embedment depth. In comparison, anchor installation required significantly less effort under submerged conditions. Also, the extruded resin was much more fluid under wet conditions, and the creamy color contrasted with the black resin extruded under dry conditions. Although insertion of the adhesive capsule or cartridge into the drill hole displaces the majority of the water in the hole, water will remain between the walls of the adhesive container and the drill hole. Insertion of the anchor traps this water in the drill hole and causes it to become mixed with the adhesive, resulting in an anchor with reduced tensile capacity. Consequently, an anchor-installation procedure was developed that eliminated the problem of resin and water mixing in the drill hole [7].

In the revised installation procedure, a small volume of adhesive was injected into the bottom of the drill hole in bulk form prior to insertion of the adhesive capsule. This injection was easily accomplished with disposable, paired plastic cartridges which contained the vinylester resin and a hardener. The cartridges were inserted into a tool similar to a caulking gun which automatically dispensed the proper material proportions through a static mixing tube directly into the drill hole. Once the injection was completed, insertion of a prepackaged vinylester-resin capsule displaced the remainder of the water in the drill hole prior to anchor insertion and spinning.

Anchors installed with the revised procedure exhibited essentially the same tensile capacity under dry and submerged conditions. At 25-mm displacement, the tensile capacity of vertical anchors installed with the revised procedure under submerged conditions averaged more than three times greater than that of similar anchors installed with the original procedure. The ultimate tensile capacity of anchors installed under submerged conditions was near the yield load of the anchors. Also, the difference in tensile capacity between horizontal anchors installed under dry and submerged conditions was less than 2 percent.

Additional Tests of Embedment Materials

Epoxy resins were not prepackaged in sausage-type cartridges because insertion and spinning of the anchor did not provide adequate mixing. However, with development of the coaxial or paired disposable cartridges with static mixing tubes, a number of suppliers began marketing epoxies for anchor embedment under submerged conditions. Also, some suppliers contended that insertion of the prepackaged capsule in the second step of the two-step installation procedure can be eliminated by injecting additional epoxy. Consequently, laboratory tests were initiated to evaluate these claims. Under dry conditions, pullout strengths for the injected epoxies were similar to that for the two-step procedure; however, tests on anchors installed under submerged conditions, resulted in drastic reductions in pullout strengths for the epoxies [8]. The epoxies also performed very poorly in creep tests under submerged conditions. In fact, the anchors embedded in one of the epoxies pulled completely out of the concrete within 1 day following loading.

During the course of these tests, a cementitious grout in sausage form was introduced in the US, specifically for underwater applications. According to the supplier, the cartridge contains a fast-setting, nonshrink cementitious compound encased in a unique envelope, which when immersed in water will allow controlled wetting of the contents, forming a thixotropic grout. Pullout strengths for submerged anchors installed with the cement capsules were essentially the same as that for the two-step installation. Anchors embedded with the cement capsule also exhibited very low creep similar to that for the two-step procedure.

Based on these test results, the current Corps of Engineers guidance is use the two-step anchor installation procedure when prepackaged polyester resin or vinylester resin is to be used as an embedment material for short (less than 380-mm embedment length) steel anchors in hardened concrete under submerged conditions.

Similar anchors embedded in neat portland-cement grout exhibit excellent performance when the grout is allowed to cure for a minimum of 3 days prior to loading. If in a given situation, divers find it easier to work with capsules, then use of the cementitious capsules is acceptable.

Repair Materials

Cast-in-place concrete and prefabricated elements of concrete and steel have been used successfully in underwater repair of hydraulic structures. Each material has inherent advantages and limitations which should be considered in design of a repair for specific project conditions.

Cast-in-Place Concrete Materials

Successful underwater concrete placement requires that the fresh concrete be protected from the water until it is in place and begins to stiffen so that the cement and other fines cannot wash away from the aggregates. This protection can be achieved through proper use of placing equipment, such as tremies and pumps [9,10]. Also, the quality of the in-place concrete can be enhanced by the addition of an antiwashout mixture (AWA) which increases the cohesiveness of the concrete [11-13]. Basic considerations for proportioning and placing underwater concrete mixtures are detailed in a recent Corps of Engineers report [14].

Concrete mixtures for underwater placement are typically highly workable and cohesive. A minimum slump of 150 to 230 mm is generally required [9]. The degree of workability and cohesiveness can vary somewhat depending upon the type of placing equipment being used and the physical dimensions of the placement area. For example, massive and confined placements, such as cofferdams, or bridge piers, can be completed with a conventional tremie concrete mixture that is less workable (flowable) and cohesive than a mixture for a typical repair where the concrete is placed in relatively thin sections with large surface areas. Concrete mixtures proportioned for underwater placement generally have a minimum cementitious content of 356 kg/m^3 and a maximum W/C + P ratio of 0.45 [9]. Underwater concrete mixtures proportioned for maximum flowability and cohesiveness usually have cementitious contents approaching 415 kg/m^3 and w/(c+m) of 0.40 or less. The high cementitious contents and low w/(c+m) are necessary for cohesiveness and flowability even if high strength concrete is not structurally necessary. If the underwater concrete placement area is sufficiently large, generation of heat from the high cementitious content can result in thermal stresses sufficient to cause cracking. To minimize the generation of heat, pozzolans and/or ground-granulated blast-furnace slag (GGBFS) are commonly used as part of the cementitious materials. In fact, a properly proportioned mass underwater concrete mixture can have a significant portion of the total cementitious material being comprised of pozzolan and GGBFS. Pozzolans and GGBFS also improve flow characteristics. Silica fume should be used to enhance the hardened properties of concrete subjected to abrasion-erosion. The addition of silica fume will also increase the cohesiveness of the fresh concrete mixture and minimize bleeding. The final selection of concrete mixture proportions should be based on the results of test placements in a placement box or pit that can be dewatered after the placement.

AWA should be used to enhance the cohesiveness of concrete that must flow laterally in thin lifts for a substantial distance. The two basic types of AWA are cellulose and gum. Both materials are effective in increasing the cohesiveness of concrete and preventing an excessive amount of fines from washing away from the aggregates when the concrete comes in contact with water [11,12]. However, there are some important differences in the manner that these admixtures interact with other concrete admixtures, especially high-range water-reducing admixtures. Cellulose-based AWA's are incompatible with sulfonated-napthlene-based high-range water-reducing admixtures. Also, excessive air can be generated when certain water-reducing or high-range water-reducing admixtures are used in combination with cellulose based AWA's [12]. This reaction can be determined only through a trial batch of concrete containing the proposed admixtures. If excessive air is present, an air-detraining agent can be used to reduce the air content to acceptable levels. Gum-based AWA's can be used with any type of water-reducing or high-range water-reducing admixture. An air-entraining admixture will be required to entrain the desired amount of air in the concrete mixture. AWA's are sold by most major concrete admixture suppliers. Although these admixtures are not required in every underwater concrete placement, their use does enhance the quality of the inplace concrete. An evaluation should be made prior to every underwater concrete placement to determine whether an AWA is necessary for a successful placement. If an AWA is not necessary to place the concrete, a further evaluation should be made to determine whether the enhanced quality of the inplace concrete would justify the additional cost of the admixture. Chemical admixtures should not be used to compensate for poor mixture proportions, poor materials quality, or poor construction practices.

Fluid concrete mixtures described in the preceding are typically proportioned for high mobility and self-compaction with moderate resistance to water erosion. Such mixtures may not be particularly suitable for repair of thin scour holes (< 0.3 m deep) if the limited depth requires that the concrete be dropped freely in water during placement. If free-fall of the concrete through water in necessary for a particular placement, low-slump and highly washout-resistant concrete mixtures have been proportioned that can be cast underwater then spread and compacted in place [11]. These mixtures generally have a higher cementitious content, a much higher AWA dosage, and a much lower slump compared to the self-leveling, self-compacting mixtures. If thin scour holes can be filled with minimal free-fall of the concrete through water, highly cohesive, yet flowable mixtures containing high dosages of AWA may be suitable.

Cast-in-Place Concrete Methods

The tremie method has been successfully used for many years to place concrete underwater [9] Tremie pipes must be long enough to reach from above water to the location underwater where the concrete is to be deposited. Concrete flows through the tremie, the lower end of which is embedded in a mound of the fresh concrete so that all subsequent concrete flows into the mound and is not exposed directly to the surrounding water. Tremie concrete mixtures must be fully protected from exposure to water until in place. AWA's can be used in tremie concrete but are not necessary, although their use will enhance the cohesiveness and flowability of the concrete. If an AWA is used, embedment of the tremie in the fresh concrete is still desirable, but some exposure of the fresh concrete to water may be permitted. Free-fall exposure of the concrete through water is not recommended under any conditions. Usually the tremie is maintained in a vertical position.

Inclined tremies have also been successfully used with self-leveling concrete to place flat slabs underwater [11]. The tremie pipe should be inclined at about 45 deg and the discharge point should be close to the edge of the casting area. Discharge of the concrete should proceed without interruption and the bottom of the pipe should be kept within the freshly-placed concrete. Self-leveling concrete has been used to form smooth, flat surfaces (an average slope of 1V:190H) and flow at least 4 m with no significant reduction in quality. The average densities and strengths of cores taken following placement were only about 1 and 5 percent lower, respectively, than control cylinders which were cast above water [15].

The Hydrovalve method and the Kajima Double Tube tremie method [10] are each variations of the traditional tremie. Each uses a flexible hose that collapses under hydrostatic pressure and thus carries a controlled amount of concrete down the hose in slugs. This slow and contained movement of the concrete helps to prevent segregation and is particularly useful in placing conventional tremie concrete. These methods are reliable, inexpensive, and can be used by any contractor with personnel experienced in working underwater.

In recent years, pumping concrete underwater has become more common. There are fewer transfer points for the concrete, the problems associated with gravity feed are eliminated, and the use of a boom permits easier movement to different locations during placement. These advantages can be important when concrete is being placed in thin layers, as is the case in many repair situations. Pneumatic valves attached to the end of a concrete pump line permit better control of concrete flow through the lines and even allow termination of the flow to protect the concrete within the lines and to prevent excessive fouling of the water while the boom is being moved. Some units incorporate a level detector to monitor the concrete placement. Pumping is considered to be an effective method for placing concrete underwater [10]. However, the surging characteristics inherent to a concrete pump can increase the opportunity for washout. It is important to maintain the discharge end of the pump line submerged in the freshly placed concrete to minimize washout.

Preplaced-aggregate concrete is an effective method for repairing large void areas underwater. The coarse aggregate is enclosed in forms, and grout is then injected from the bottom of the preplaced aggregate. To prevent loss of fines and cement at the top of the repair, venting forms have a permeable fabric next to the concrete, backed with a wire mesh, and supported by a stronger backing of perforated steel and plywood. The pressure generated by the grout beneath the forms necessitates doweling to hold down the forms. The grout must have a high fluidity to ensure complete filling of voids. Use of an antiwashout admixture will lessen the need for the protective top form.

Stiff, highly washout-resistant concrete mixtures can be placed underwater with bottom-dumping skips, tilting pallets, or bottom-dumping buckets [15]. The latter can be useful for placing discrete slugs of concrete for repairing small, isolated areas, and a bottom-dumping skip is appropriate for placing concrete layers of uniform thickness. Placement devices should be closed and sealed to minimize water erosion caused by turbulent water flow as the device is lowered in water. Quick-release bottom gates should be provided to permit discharge of the concrete as a continuous mass. Free-fall discharge through water should be limited to 380 mm. Repetitive passes with a concrete or steel roller is recommended for spreading and compaction of the concrete.

Prefabricated Elements

Precast concrete panels and modular sections have been successfully used in underwater repair of lock walls, stilling basins, and dams [16]. Prefabricated steel panels have been used underwater as temporary and permanent top forms for preplaced-aggregate concrete. Also, prefabricated steel modules have been used in underwater repair of erosion damage. Research indicates that it is feasible to use prefabricated steel, precast concrete, or composite steel-concrete panels in underwater repair of stilling basins [17]. Each material has inherent advantages, and the following factors should be considered in designing panels for a specific project.

Abrasion resistance is a primary concern in selection of materials for repair of erosion damage caused by waterborne debris. Tests results [18] indicate that the widths and depths of abrasion of concrete mixtures with an average 28-day compressive strength of 75 MPa were 1.6 and 2.3 times higher, respectively, compared to abrasion-resistant steel. However, when the depths of abrasion were compared as a percentage of the thickness for typical panels of each material, it was concluded that the durability of the high-strength concrete was equivalent to that of the abrasion-resistant steel.

Uplift forces caused by high-velocity water flowing over the panel surface is a concern in repair of hydraulic structures, particularly stilling basins. The Old River Low Sill Control Structure demonstrated the effect of uplift on steel panels [19]. Thirty modules, 7.3 m long and ranging in width from 0.9 to 6.7 m, were prefabricated from 13-mm thick steel plate for the stilling basin repair. After the modules were installed and anchored underwater, the voids between the steel plate and the existing concrete slab were filled with grout. Additional anchors were installed in holes drilled through the grouted modules and into the basin slab to a depth of 0.9 m. An underwater inspection of the basin 8 months after the repairs showed that 7 of the 30 modules had lost portions of their steel plate, ranging from 20 to 100 percent of the surface area. A number of anchor bolts were found broken flush with the surface plate or the grout or pulled completely out of the substrate. A second inspection, approximately 2 years after repair, revealed that additional steel plate had been ripped from four of the modules previously damaged, and an additional nine modules had sustained damage. There were only a few remnants of the steel plates when the basin was dewatered 11 years after the repairs. Precast concrete panels, with a minimum thickness of about 100 mm, should be less susceptible to these problems. The increased panel stiffness, damping, and bond to the infill concrete will reduce uplift problems.

The design of the anchor system should ensure that the prefabricated panels are adequately anchored to the existing concrete to resist the uplift forces and vibrations created by flowing water. Welding of anchor systems as nearly flush with the prefabricated steel plate surface as possible appears more desirable than raised bolted connections. Preformed, recessed holes for anchors were easily incorporated during precasting of the concrete panels used in underwater repairs at Gavins Point Dam [16]. The ability of the anchor system, including any embedment material, to perform satisfactorily under the exposure conditions, particularly creep and fatigue, should be evaluated during design of the repair.

Joint details are important design considerations because once a panel fails and is displaced, adjacent panels are more susceptible to failure as a result of their exposed edges. Consequently, each panel should be designed as an individual repair unit and should not rely on adjacent panels for protection.

The vulnerability of joints can be reduced by providing stiffened and recessed panel edges. The joint between the existing concrete and the leading edge of the upstream panels should be designed to provide a smooth transition to the repair section. A general design philosophy should be to minimize the number of joints by using the largest panels practical.

Panels are usually installed with a barge-mounted crane. Therefore, panel weight can be important, depending on the lifting capacity of available cranes. For panels of a given area, the weight of concrete panels in air will be about four times the weight of steel panels. Panel supports are bottom-installed platforms or seats which may be required to ensure that panels are placed at the desired elevation and are properly aligned. Several concepts for panel supports have been proposed [17], some of which include a means to attach and anchor the panel to resist uplift forces. The potential of these concepts should be evaluated based on specific project requirements.

The physical dimensions of the placement area dictate whether portland-cement grout or concrete is used as the infill material. Grout should be used in those applications that require the infill to flow substantial distances in very thin lifts, whereas pumped concrete or preplaced-aggregate concrete should be used to fill larger voids.

CASE STUDIES

Red Rock Dam

An underwater concrete mixture containing AWA was effectively used for repair of erosion damage at Red Rock Dam [20]. Approximately 75 m^3 was placed by pumping through a 100-mm diameter line. The water depth was 7.5 m and a minimum discharge of 8.5 m^3/s was being released through the dam during repair. A diver controlled the end of the pump line, keeping it embedded in the mass of newly discharged concrete and moving it around to completely fill the repair area. The effects of the AWA used were apparent; even though the concrete had a slump of 230 mm, it was very cohesive. The concrete pumped very well and, according to the diver, self-levelled within a few minutes following placement. The diver also reported that the concrete remained cohesive and exhibited very little loss of fines on the few occasions when the end of the pump line kicked out of the concrete. The total cost of the repair was $128,000. In comparison, estimated costs to dewater alone for a conventional repair ranged from $1/2 to $3/4 million. Subsequent diver inspections indicate that the concrete continues to exhibit good performance.

Soo Locks

A concrete pier immediately downstream of the locks was originally constructed on welded steel cells filled with stone and sand. During more than fifty years in service, several cells sustained damage and loss of fill that endangered the pier. Underwater repairs included welding new steels sheets over distorted sections, reinforcement of split joints, and placement of a concrete footing along the base of the cellular structure [21]. A local concrete producer in northern Michigan supplied the concrete that was proportioned with antiwashout, water-reducing, and accelerating admixtures. The concrete was placed by pumping through a 150-mm diameter pump line, that was insulated against the low water temperature.

The locks continued normal operations during placement of the 230 m³ of concrete. Two divers that controlled the pump line reported that there was virtually no washout of fines and that significant turbulence created by passing ships had little effect on the freshly-placed concrete. The aggregate remained coated with cement paste and the concrete set quickly as required.

St. Lambert Lock

A vessel hit the upstream bullnose of the lock during a severe wind storm in November 1992 and the impact caused significant damage. Emergency repairs were made to allow continued operations until the end of St. Lawrence Seaway shipping season in December. Permanent repairs had to be completed during the 2-month winter shutdown.

Substantial spalling at lower elevations necessitated complete removal and replacement underwater. Approximately 130 m³ of concrete containing AWA was placed underwater by pumping [22]. The specified concrete slump was 180 ± 20 mm which was somewhat lower than conventional concrete typically used for underwater repair. At the beginning of each placing operation, a polystyrene bag was attached to the end of the pump line. After the pump line was partially filled with concrete, it was lowered into the water to approximately 100 mm above the receiving surface. Subsequent concrete pumping broke the bottom seal and formed a mound around the discharge point. The pump line was fitted with a metal tube for the final 1.2 m to facilitate its insertion into freshly-cast concrete. A rope was attached at the bottom of the pump line to enable a diver to manipulate the line. Whenever possible, the line was embedded approximately 0.5 m into the fresh concrete. Given the near zero visibility conditions underwater, it was impossible for divers to maintain a continuous seal, thus subjecting the concrete to some free drop in the water. However, the selection of a highly cohesive concrete with AWA reduced the risk of washout under very difficult placing conditions. The river water temperature was approximately 1 C during concrete placement.

Rapid concrete placement without the need to erect a cofferdam for dewatering allowed the repair to be completed prior to opening of the navigation season. The additional cost of the AWA was estimated at $90/m³ (CAN) in 1993 or about $10,000 for the project; however, the estimated savings over a conventional repair was almost $200,000.

Wilbur D. Mills Dam

Flooding along the Arkansas River in 1982, caused barges to break loose from a fleeting area upstream of the dam. The barges blocked 12 of the 16 gates of the dam and created upstream cross currents and downstream surging through the remaining 4 gates. These conditions resulted in severe scouring both upstream and downstream of the dam. The immediate repair efforts included barge removal and placement of approximately 31,800 tons of riprap stone in the downstream scour holes. Work on a permanent repair to the downstream scour problem was initiated in October 1990. The existing stilling basin was extended 64-m downstream by sinking barges in the desired position, filling the sunken barges with riprap, and underwater placement of concrete to fill voids within the riprap [23].

Extensive underwater foundation preparations were required to properly slope and prepare the riverbed before the barges could be placed. Once the desired grade and slope were achieved, a bed of leveling stone 0.3 to 0.6 m thick was placed. Each of the 26 hopper barges was partially filled with 690 m^3 of concrete and floated into a slot between two lay barges. Cables suspended from girders lying across their decks were used to lower each 270-ton barge into position. The 11 by 59-m barges were placed on a 1V:6H downstream slope in 3 to 17 m of water. After each hopper barge was sunk, its cargo hole was filled with rock, and an underwater concrete mixture containing AWA was used to fill the voids in the rock-filled barge. This same concrete mixture was used to fill the spaces between the existing stilling basin upstream of the barges and spaces between the sunken barges. The 230-mm slump concrete was pumped for distances up to 460 m. Cores taken from the repair indicate that the concrete fully penetrated all voids within the riprap and between barges. The average compressive strength of the concrete cores was 31 MPa. The contract for the permanent repair was $17.6 million, which was much less than the cost of other possible solutions.

Santa Fe Railroad Bridge

Flooding along the Mississippi River in 1993 scoured away large portions of the riverbed near the Santa Fe Railroad Bridge at Ft. Madison, Iowa. The high water and swift currents left the footing of one concrete pier 4.6 m above the river bottom with the footing supported only by the original 128 wooden piles. The first step in the repair was to install sheet piling around the perimeter of the footing. Fifteen steel H-piles were then driven into bedrock between the sheet piling and the existing footing. Once they were embedded in concrete, the steel pile would relieve the pressure on the weakened wooden piles [21].

The underwater concrete mixture was similar to that previously used successfully in repair of the stilling basin at Red Rock Dam. AWA was added at a dosage of 780 mL/100 kg for the initial 76 m^3 of concrete and then reduced to 520 mL/100 kg for the remainder of the placement. The concrete was placed by pumping through a 150-mm diameter line. The discharge end of the line was positioned approximately 300 mm from the river bottom during placement of the concrete with the high dosage of AWA. The remaining concrete was pumped into the plastic concrete previously placed which further minimized the potential for washout. The water within the sheet piling remained clear throughout the placement even when the pump line was repositioned. A total of 717 m^3 of concrete treated with the AWA were placed on the project. Early strength test were requested because both railroad tracks were scheduled to reopen four days after the concrete placement. Results of compressive strength tests averaged 27 and 42 MPa at 4 and 28 days, respectively.

Florida Keys Reef

A section of coral reef frequented by sight-seeing boats and recreational divers was damaged by ship impact. The impact formed a shallow crater and destroyed approximately 315 m^2 of the living surface of the reef located in 1.8 to 3 m of water. Restoration of the reef was essential because erosion from wave action was undermining the living coral around the rim of the crater causing the depression to become larger with time. Coral requires a firm, textured foundation to re-establish itself within a damaged area; therefore a plan was developed to install textured precast concrete panels in the crater and to infill the interstitial spaces between the panels and between the panels and the reef with concrete placed underwater [24].

The cast-in-place concrete was necessary to keep the precast panels in position and to secure the living coral at the perimeter of the crater thus minimizing further undermining.

A number of factors were considered in development of the underwater concrete placement procedure, including (a) placement would occur in an environmentally sensitive area and dispersion of cement particles into the seawater must be minimized, (b) concrete would be placed in very thin layers ranging from approximately 1 m down to 75 mm thick, and (c) stone and coral would be inserted into the fresh concrete in some areas to provide a textured surface for the living coral to anchor itself. An underwater concrete mixture was proportioned with a water-cementitious ratio of 0.40 for a compressive strength of 41 MPa at 28 days. A 1-m^3 batch of concrete contained 20, 41, and 272 kg of silica fume, fly ash, and cement, respectively. The mixture also contained AWA and a large dosage of high-range water-reducing admixture was used to maintain a concrete slump of more than 254 mm. The concrete produced on barge using a mobile continuous screw-type mixer and was pumped through a line suspended by a crane boom to a diver that controlled the concrete placement. The concrete was largely self-leveling and was successfully placed between the precast units and around the perimeter of the crater.

Gavins Point Dam

The powerhouse and the tailrace slab are founded on excavated chalk and shale. The draft tube portals are part of a massive monolithic concrete placement which is part of the powerhouse foundation. The free-floating tailrace slab is 0.3-m-thick concrete. Rebound of the foundation over the years resulted in an offset and spalling at the interface between the draft tube portal and the tailrace slab. Underwater inspections also identified concrete erosion and spalling in a retaining wall downstream of the powerhouse where it interfaced with the tailrace slab.

The consensus was that a repair in the dry would be better; however, the cost of building a cofferdam and the lengthy power plant outage during construction were unacceptable. It was decided that the repairs would be made underwater by divers working at a depth of approximately 15 m [16]. The contractor used several diving crews so work could continue 24 hr a day and each step of the repair was reviewed on land before being done underwater

After removal of debris and cleaning of concrete surfaces, the spalled areas and voids were filled with preplaced aggregate and covered by an anchored form (steel plate or precast concrete panel). Forms were necessary because grout was injected into the aggregate under pressure. A steel plate was used as a form for the repair made at the draft tube portal. However, the offset between the south retaining wall foundation and the tailrace slab was much larger, approximately 0.3 m. To provide for this large offset and for some additional tailrace slab rebound, 0.4-m-thick precast concrete panels were used as stay-in-place forms instead of steel plates.

Following form installation, water in the aggregate voids was displaced by injecting prepackaged cementitious grout through holes in the forms. The contract allowed for the power units to be shut down for 14 consecutive days for this repair; however, underwater construction was completed 3-1/2 days ahead of schedule.

CONCLUSIONS

Laboratory and field experience summarized herein shows that underwater concrete placement can be accomplished with the same degree of reliability as in-the-dry construction. However, it is technically demanding and often involves complex construction scheduling and logistics. Underwater concreting operations must be executed properly with appropriate mixture proportions and placement procedures to be successful and avoid adverse impact on project cost and schedule.

ACKNOWLEDGMENT

The research studies summarized herein were authorized by Headquarters, US Army Corps of Engineers, as part of the Repair, Evaluation, Maintenance, and Rehabilitation (REMR) Research Program. Program Manager for REMR was Mr. William F. McCleese, US Army Engineer Waterways Experiment Station (WES). The studies were conducted at WES under the general supervision of Mr. Bryant Mather, Director, Structures Laboratory, and Dr. Paul F. Mlakar, Chief, Concrete and Materials Division. At the time of preparation of this paper, the Commander of WES was COL Robin R. Cababa, EN. The permission of the Chief of Engineers to publish this paper is gratefully acknowledged.

REFERENCES

1. HEADQUARTERS, US ARMY CORPS OF ENGINEERS. Evaluation and Repair of Concrete Structures. Engineer Manual 1110-2-2002, Washington, DC, June, 1995, 184 pp.

2. SMITH, A P. New Tools and Techniques for the Underwater Inspection of Waterfront Structures. OTC 5390, 19th Annual Offshore Technology Conference, Houston, TX, April, 1987.

3. THORNTON, H T, JR., AND ALEXANDER, A M. Development of Nondestructive Testing Systems for In-Situ Evaluation of Concrete Structures. Technical Report REMR-CS-10, US Army Engineer Waterways Experiment Station, Vicksburg, MS, Dec., 1987, 167 pp.

4. KEENEY, C A. Procedures and Devices for Underwater Cleaning of Civil Works Structures. Technical Report REMR-CS-8, US Army Engineer Waterways Experiment Station, Vicksburg, MS, Nov., 1987, 54 pp.

5. BEST, J F, AND MCDONALD, J E. Evaluation of Polyester Resin, Epoxy, and Cement Grouts for Embedding Reinforcing Steel Bars in Hardened Concrete. Technical Report REMR-CS-23, US Army Engineer Waterways Experiment Station, Vicksburg, MS, Jan., 1990, 69 pp.

6. MCDONALD, J E. Evaluation of Vinylester Resin for Anchor Embedment in Concrete. Technical Report REMR-CS-20, US Army Engineer Waterways Experiment Station, Vicksburg, MS, Feb., 1989, 53 pp.

7. MCDONALD, J E. Anchor Embedment in Hardened Concrete Under Submerged Conditions. Technical Report REMR-CS-33, US Army Engineer Waterways Experiment Station, Vicksburg, MS, Oct., 1990, 42 pp.

8. MCDONALD, W E. Evaluation of Grouting Materials for Anchor Embedments in Hardened Concrete. Technical Report REMR-CS-56, US Army Engineer Waterways Experiment Station, Vicksburg, MS, Feb., 1998, 164 pp.

9. ACI COMMITTEE 304. Guide for Measuring, Mixing, Transporting, and Placing Concrete. ACI Manual of Concrete Practice, American Concrete Institute, Detroit, MI, 1997, 49 pp.

10. GERWICK, B C. Review of the State of the Art for Underwater Repair Using Abrasion-Resistant Concrete. Technical Report REMR-CS-19, US Army Engineer Waterways Experiment Station, Vicksburg, MS, Sep., 1988, 42 pp.

11. KHAYAT, K H. Underwater Repair of Concrete Damaged by Abrasion-Erosion. Technical Report REMR-CS-37, US Army Engineer Waterways Experiment Station, Vicksburg, MS, Dec., 1991, 348 pp.

12. NEELEY, B D. Evaluation of Concrete Mixtures for Use in Underwater Repairs. Technical Report REMR-CS-18, US Army Engineer Waterways Experiment Station, Vicksburg, MS, Apr., 1988, 130 pp.

13. NEELEY, B D, SAUCIER, K L, AND THORNTON, H T, Jr. Laboratory Evaluation of Concrete Mixtures and Techniques for Underwater Repairs. Technical Report REMR-CS-34, US Army Engineer Waterways Experiment Station, Vicksburg, MS, Nov., 1990, 92 pp.

14. YAO, S X, BERNER, D E, AND GERWICK, B C. An Assessment of Underwater Concrete Technologies for "In-the-Wet" Construction of Navigation Structures. Technical Report INP-SL-102, US Army Engineer Waterways Experiment Station, Vicksburg, MS, Oct., 1998, 83 pp.

15. KHAYAT, K H, GERWICK, B C, AND HESTER, W T. Self-leveling and Stiff Consolidated Concretes for Casting High-Performance Flat Slabs in Water. Concrete International, Vol 15, No. 8, American Concrete Institute, Detroit, MI, Aug., 1993, pp. 36-43.

16. MCDONALD, J E, AND CURTIS, N F. Applications of Precast Concrete in Repair and Replacement of Civil Works Structures. Technical Report REMR-CS-49, US Army Engineer Waterways Experiment Station, Vicksburg, MS, Jun., 1995, 298 pp.

17. RAIL, R D, AND HAYNES, H H. Underwater Stilling Basin Repair Techniques Using Precast or Prefabricated Elements. Technical Report REMR-CS-38, US Army Engineer Waterways Experiment Station, Vicksburg, MS, Dec., 1991, 97 pp.

18. SIMONS, B P. Abrasion Testing for Suspended Sediment Loads. Concrete International, Vol 14, No. 3, American Concrete Institute, Detroit, MI, Mar., 1992, pp. 58-61.

19. MCDONALD, J E. Maintenance and Preservation of Concrete Structures; Repair of Erosion-Damaged Structures. Technical Report C-78-4, Report 2, US Army Engineer Waterways Experiment Station, Vicksburg, MS, Apr., 1980, 306 pp.

20. NEELEY, B D AND WICKERSHAM, J. Repair of Red Rock Dam. Concrete International, Vol 11, No. 10, American Concrete Institute, Detroit, MI, Oct., 1989, pp. 36-39.

21. BURY, M A, NMAI, C K, AMEKUEDI, G, AND BURY, J. Unique Applications of a Cellulose-Based Antiwashout Admixture. Master Builders, Inc., Beachwood, OH, Nov., 1997, 10 pp.

22. KHAYAT, K H, AND GAUDREAULT, M. Underwater Repair of the St. Lambert Lock. Vol 19, No. 3, American Concrete Institute, Detroit, MI, Mar., 1997, pp. 36-40.

23. ANON, Corps Finds Solution in Sinking Barges. Construction News, Vol 59, No. 13, Southam Business Communications, Little Rock, AR, Jul., 1992, pp. 8-10.

24. BERNER, D E. Tremie Concrete Restoration of a Florida Keys Reef. News, Ben C. Gerwick, Inc., San Francisco, CA, May, 1996, pp. 6.

UNDERWATER CONCRETE CONSTRUCTION AND REPAIR

T J Collins
Collins Engineers Inc
United States of America

ABSTRACT. Concrete structural components located in water serve as the foundation elements of many structures. In the United States, periodic underwater inspections are mandated for all bridges and many other owners of waterfront structures routinely conduct underwater inspections. After these inspections owners have been faced with problems of accomplishing repairs to them. Many owners, and consultants have not had the in-house expertise to design and manage repairs underwater. To provide bridge owners the necessary design and management skills, the U.S. Department of Transportation retained Collins Engineers to evaluate and report on available underwater construction and repair techniques and conduct seminars throughout the U.S. Factors to be addressed in selecting underwater repair methods include environmental considerations such as currents, depth, pollution and regulatory mandates; material considerations such as durability and workability; unformed and formed repairs with rigid, flexible and fabric systems; and material placement methods. Many repairs can be more economically accomplished using divers, and employing divers effectively requires an understanding of the capabilities and limitations of the diver, and an understanding of dive safety regulations. Underwater construction and repair also involves contracting considerations not normally necessary for above water work. Contract documents require special considerations because there are few standard details, and specifications must allow for the expectation that underwater conditions may vary greatly. This paper addresses some of these special concerns in this type of work

Keywords: Diving, Underwater, Underwater inspection, Underwater construction, Underwater repair, Engineer-diver.

Thomas J Collins S E, P E is the President of Collins Engineers, Inc. Chicago, Illinois, USA, a consulting engineering firm specializing in underwater inspection, design of repairs and design of new bridge and waterfront structures. Currently, Mr. Collins is Principal Instructor for the U.S. Department of Transportation's Demonstration Project 98, a two-day course on Underwater Bridge Evaluation and Repair, a program which is being presented to government engineering agencies throughout the USA.

INTRODUCTION

Owners of bridges, piers and wharves, dams, intake and outfall conduits and other waterfront and in-water structures have extensive and expensive facilities located underwater. Some owners and engineers have realized the importance of proper inspection and maintenance of these underwater facilities; others have adopted an "out of sight, out of mind" approach. Experience, however, has shown that a well planned underwater inspection and maintenance program can help ensure safe operation of those facilities and reduce overall operating and maintenance costs.

In the United States, the U.S. Navy and bridge owners have been most active in underwater inspection and repair. The U.S. Navy has had a worldwide underwater facility assessment program since about 1980. The program consists of underwater inspection and assessment of all their facilities by registered professional engineer-divers at intervals of six years or less. The Navy program includes subsequent design of remedial measures by the same team of engineers who conducted the underwater inspection, and timely inspection of repairs as they are accomplished.

Since 1988, bridge owners in the U.S. have been required to inspect their bridges underwater at intervals of five years or less. Now, following at least two rounds of inspections completed to date, these bridge owners generally have a good idea of the underwater condition of their bridges These bridge inspections have resulted in the recognition of the need for underwater repairs for many bridges. In coastal states, there has generally been an adequate underwater engineering expertise, either as a consequence of waterfront structures or offshore structures, but this expertise was often not available or familiar to highway bridge engineers. In inland states, there was an even greater lack of specialized underwater expertise in the bridge community.

To meet the need for underwater engineering and construction expertise, the U. S. Department of Transportation's Federal Highway Administration retained Collins Engineer, Inc. to develop a two-day seminar on Underwater Evaluation and Repair of Bridge Components and present it to bridge and waterfront engineers throughout the U.S. In addition, the project team provided specialized technical assistance to those owners on an as-needed basis. This paper presents some of the special concerns, related to concrete structures, which were covered in the project.

WET VS DRY CONSTRUCTION

Almost any construction technique that can be used above water can also be used underwater, but it will probably take longer and cost more, and the technique must be adapted for underwater work. There are a number of products and materials that have been formulated or developed specifically for underwater work. In deciding whether to perform the work in the dry, in a cofferdam or behind a dike, or in the wet, i.e., using a diver, the engineer must evaluate the overall cost of the work, considering time, and equipment needs, and the capability of a diver to accomplish the specific task.

The engineer must determine first, if the work can be accomplished by a diver. The diver is best suited for performing work that is of a more gross than fine nature. Because of the equipment the diver must wear, precise movements are difficult. Therefore, as much of the work as possible, such as assembly of components, should be performed above water.

Environmental conditions may make underwater inspection or repair difficult. Depth and currents are the two greatest natural constraints, but pollution considerations may also affect the planning for the work. Divers are limited in the amount of actual work time, i.e., bottom time, that they can achieve each day. As the depth of the work increases, less time is available to accomplish the work without special measures, such as in-water or surface decompression. For example, at a depth of approximately 20 meters, a diver will probably only be able to work for two to three hours per day. This could mean that additional divers are needed in order to keep the above water support workers fully productive.

High currents can affect diving and topside operations, as well as cofferdam construction. In currents greater than about 0.5 meters per second, divers must be shielded or tethered in place, and it is difficult to handle tools, forms and materials. It is also difficult to manoeuvre boats and barges around construction in high currents. Sites with consistently high currents will probably require cofferdams or similar measures. However, piling may be difficult to position and drive in currents greater than about 1 meter per second.

Even accounting for the higher costs of a diving operation, it may be preferable to the cost of a cofferdam. A cofferdam may also not be acceptable because of the reduction in cross sectional area of the waterway caused by the structure. The reduction may not be acceptable from a hydraulic standpoint or it may interfere with waterway traffic. It should be noted that the sheets must be driven some distance from the edge of the existing structure in order to use heavy equipment to excavate the space between the structure and the sheets to place a seal in the bottom of the cofferdam prior to dewatering.

Pollution concerns fall in two categories, the effect of the water on the diver and the effect of construction activities on the waterway. Waters suspected of pollution should be tested and, if necessary divers can take special precautions such as using dry suits and helmets, and washing before removing their suits. Materials removed from the water, such as sediment and debris, may require disposal as hazardous or contaminated waste. Construction and repair activities must be planned to prevent pollution and contingency plans should be in place to handle pollution caused by fuel spills, lubricating and hydraulic oil spills, form oil release, and mortar and admixture chemical release.

Fish and wildlife related regulations may also impose significant restrictions on the time when work may be performed. Construction activities may be restricted by fish habitats and spawning, the presence of endangered species, or the effect of repair materials on aquatic life. Special testing, such as water quality sampling, may be needed before and during the work.

UNDERWATER REPAIRS

Concrete, including Portland cement, polymer and epoxy mortars, is the most common repair material for underwater repairs, whether the original material is concrete or not. Concrete is used to repair conventionally reinforced concrete, prestressed concrete, steel and timber substructure elements in water.

For underwater repairs, durability and workability are usually more important than strength. These properties can be enhanced with proper mix design and the addition of admixtures. Special mix designs are required for marine repairs because of the harsh environment and difficult placement methods.

Hand Patching

Hand patching is the simplest and least costly repair method, and it requires the simplest diving operation with minimal support equipment. It is often used to repair broken corners and cover exposed reinforcing steel on corners of piles. Materials that are readily available and commonly used include: Portland cement mortar, hydraulic mortar, and neat epoxy mortar. Though the coefficient of expansion of epoxies is generally three to four times higher than that of cement mortars, epoxies formulated for underwater, not just moisture tolerant, can perform satisfactorily in most situations. Epoxies do create a vapor barrier, however, that can lead to debonding for large patches extending above water.

Formed Repairs

For larger areas, formed repairs are commonly used both above and below water. Form selection is usually left to the contractor who will consider forces acting on the form, ease of installation and overall cost. Because hydrostatic pressure acts on at least one face of the form underwater, form pressures are generally reduced, allowing for use of lighter forms with fewer anchors or ties. Stay-in-place forms may be economical underwater, where they would not be economical above water, because of the high cost of underwater labor to remove them. Underwater forms may require the use of special non-toxic form oils and special cleaning of previously used forms to not cause pollution. Forms are often hand placed underwater, using stakes or sand bags to secure the bottom edge. The forms should be of a size and weight that can be handled by the diver and not so large that currents make them difficult to control.

Rigid forms

Rigid forms have many advantages. All contractors are experienced in their use above water, they can be reused; and they produce clean lines and good surfaces, an important consideration where the work area extends above water. Their disadvantages include being cumbersome for divers to control underwater, especially in strong currents; being difficult to fit to irregular surfaces caused by extensive deterioration; and not conforming to uneven channel bottoms.

Flexible forms

Flexible forms are easier to work with underwater, and can be used in differing configurations. They may be fabric pile jackets, fabric bags of small or large size, or plastic membranes. Fabric forms are generally lightweight, prefabricated and sewn, field fitted, and fastened with zippers, ties or clips.

These flexible forms pose their own special problems. They tend to billow in currents and can be difficult to place. Some flexible materials may be degraded by ultraviolet light and become ragged. Because they are flexible, they can be difficult to secure away from the substrate to maintain the design thickness and proper cover for reinforcing steel. This is especially a problem for horizontal members and battered members.

Pile Jackets

Pile jacketing systems are commonly used to repair and protect piles. Most systems have three components, a form, a reinforcing system and a grout. The procedures to install an encasement system are virtually the same whether the substrate is timber, steel or concrete. Jacketing systems can alter the characteristics of the piles. They become larger, which may increase scour at the mudline, they add dead load and may increase lateral loads from currents and wave action. A further concern, particularly for thick systems, is their effect on seismic response of the structure.

Pile jacketing systems are widely used, and unfortunately, they are badly used by many. Pile jacketing systems need to be designed for the specific use and installed with the same care that would be given to any structural systems. Pile jacket design considerations include fabric strength, fastener strength, top and bottom closure methods, special provisions for inclined members and leakage. Fabric and zippers must be selected considering the head of the grout and ease of installation. The top and bottom closures must be designed to prevent grout loss and to provide a durable surface. On both vertical and inclined members standoffs, generally plastic, must be used to maintain the proper cover and annular space around the pile. The fabric must be porous enough to allow water passage out of the jacket, but fine enough to prevent loss of cement fines which could weaken the repair and possibly constitute an environmental concern.

In the installation of the fabric form jackets, care must be used to properly prepare the substrate, install the form and any reinforcement, and place the grout to obtain a dense, durable jacket. Substrate preparation is the foundation to good jacket construction. Normally, all deteriorated pile material and marine growth should be removed, although some engineers allow tight fitting marine growth to remain in a large, reinforced concrete jacket. Care must be taken in removing deteriorated pile material so as to not remove too much so that the structure capacity is weakened. Repairs must be planned to provide, where necessary, temporary supports. Where corrosion is active in the pile, reinforcing steel should be cleaned, and additional bars installed to makeup for losses of steel section. As a minimum, steel mesh should be installed in the jacket to provide added integrity and resistance to cracking. Generally plastic spacers are used to hold the reinforcing steel in place away from the existing pile and to hold the fabric form away from the steel.

The fabric form is generally zipped around the pile and the top and bottom of the pile jacket are secured to the pile with metal bands. High slump, small aggregate concrete is then placed within the form through a small tremi pipe inserted at the top or the concrete is pumped upward. Pumping from the bottom is generally preferred because it tends to scour the substrate surface and improve bond.

Care must be taken so that the grout is equally distributed around the pile to prevent the form from being stretched out of shape. Vibration is rarely used, but a diver can probe the outside of the fabric form if consolidation is a problem.

This type of repair can be used for limited sections of the pile or carried all the way into the mud line. For repairs at and above the waterline, it is good practice to carry the jackets to the underside of the pile cap or deck. Otherwise, cracking often develops in the short unjacketed section between the top of the jacket and the underside of the pile cap.

Rigid pile jacket forms used underwater are either removable or stay-in-place. Procedures for removable forms are similar to those for above water use of forms. Rigid forms for underwater use are generally of the stay-in-place variety due to removal costs and lack of appearance concerns.

Probably the most common type of rigid form is a circular fiberglass form, filled with concrete or cement grout. Because the form is loaded in tension, it can be quite light. The form can be sized to permit the installation of additional reinforcing steel if necessary. When these fiberglass forms are filled with a cement grout, the form does not act with the grout but merely contains it, so bond between the grout and form is not critical.

In systems that use a fiberglass form with epoxy grout, the stay-in-place form and grout act as a system, and the bond between the grout and substrate, and the grout and the form is vital to the success of the system. Because of the cost of epoxy, the epoxy-filled space is normally kept to a minimum. The reduced volume of the epoxy grout also reduces the heat generated by the curing reaction.

Several factors affect bond strength. The substrate must be clean, and in warm waters, marine growth can begin adhering to a cleaned pile in a few hours. The forms themselves must be clean, not only of marine growth and silt, but also form wax, dirt, or airborne pollution such as exhaust fumes from vehicles and construction machinery. Again, the bond strength is enhanced by placement techniques which cause a shearing action along the form and substrate, such as pumping or tremie placement upward from the bottom of the jacket.

Some jacketing sysrems include cathodic protection. The State of Florida, for example is using a system for repairing conventionally reinforced and prestressed concrete piles which includes a fiberglass jacket, portland cement grout, and a sacrificial zinc mesh anode embedded in the jacket. In this system, shown in Figure 1, the sacrificial anode is connected to the reinforcing steel or prestressing strands of the pile.

USING DIVERS EFFECTIVELY

In deciding when to use divers and perform underwater construction, and in the actual employment of the divers, it must be remembered that a diver's time is at a premium and rather expensive. With the exception of offshore work, most diving construction operations are in relatively shallow waters. Both surface supplied air and scuba (self-contained underwater breathing apparatus) could generally be used at the depths encountered.

Underwater Concrete Repair 263

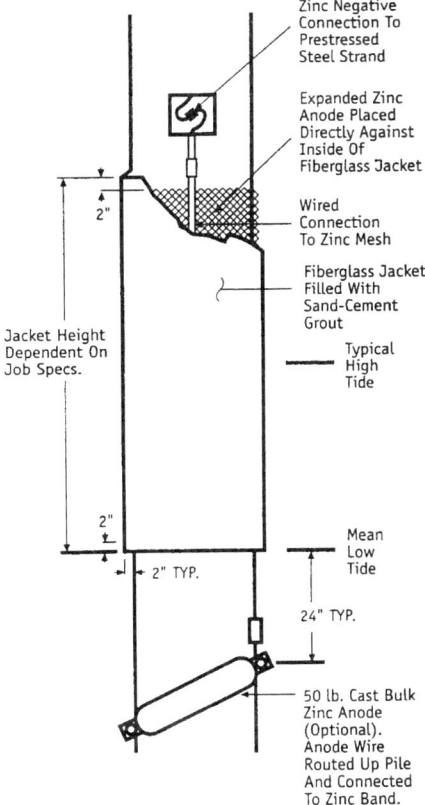

Figure 1 Rigid form pile jacket with cathodic protection

In the United States, safety regulations permit diving with surface supplied air to a depth of 220 feet and with scuba equipment to a depth of 130 feet. Most inshore construction diving operations are well within these limits. As a matter of practice, however, surface supplied air equipment is generally used because of safety considerations and better communications. Even in good diving conditions, divers have many adverse physical forces acting on them and the underwater environment reduces their ability to function efficiently. Work planned for divers, therefore, should be as simple as possible, not requiring precision operations or adjustments. Where it is feasible, as many components as possible should be assembled above water, allowing the diver to perform only installation procedures.

Even in shallow water, a diver's actual production time will be reduced from that obtained from workers above water. Besides time lost to preparing and maintaining diving equipment, placing and removing the diver from the work site, the diver's bottom time is limited.

The deeper a diver works, the less time he can spend on the bottom. The deeper a diver goes and the longer time he spends underwater, the more gases, such as nitrogen, are absorbed by the tissues of the body. When the diver ascends, those gases tend to come out of the tissues and enter the blood stream. The diver must ascend slowly enough so that the gases return to the blood stream slowly. If the diver ascends to quickly, these gases can form bubbles which generally collect near the body joints, causing severe pain. This is known as decompression sickness, or "the bends". To prevent this, the diver may have to make in-water stops as he ascends, or immediately after reaching the surface be placed in a recompression chamber which imposes the same pressure he was subjected to underwater and slowly reduces that pressure to air pressure at the surface. Depending on the diver's working depth, this may take many hours. Table 1 shows the estimated actual production time and time lost to preparation time and decompression for work at various depths.

Table 1 Diver productivity at various depths

BOTTOM DEPTH	ESTIMATED PRODUCTIVE BOTTOM TIME PER DAY
3.0m (10 ft)	6 hours
12.2m (40 ft)	6 hours
18.2m (60 ft)	4 hours
24.4m (80 ft)	3 hours
30.5 (100 ft)	2 hours

Diving safety procedures, whether employed as a matter of good practice or imposed by governing agencies, further reduce the cost-effectiveness of diving operations. Generally, for even shallow operations, to place one diver in the water will require at least two support persons on the surface. As the depth and complexity of the dives increase, additional personnel will be required.

Weather and water conditions can further increase the size and complexity of diving operations. In cold, and inclement conditions, the diver may not be affected, but the above water personnel can suffer and reduce the safety of the operation. One of the greatest impediments to underwater work is strong current. This is the hardest and most costly obstacle to overcome. In currents, divers must be heavily weighted, have something to hold on to or be secured to, or must be protected to remain at the work site. Generally, the fastest current a diver can comfortably work in is about two to three knots, but this is dependent upon what the diver has to hold on to. Ropes may also be used to secure the diver near the work. In very fast moving currents, a diving shield of temporary driven sheet piling or a movable shield of some kind can be installed, usually at significant cost. Figure 2 shows one such shield that has been used in flood stage on the Mississippi River.

Figure 2 Diving shield suspended from barge mounted crane

UNDERWATER CONTRACTING

Most agencies have well defined procedures for contracting for above water construction projects, but are often unfamiliar with the special requirements and challenges of underwater construction. In underwater work, it is generally desirable to allow the contractor greater leeway in achieving the project goals, but it is important to also place special restraints on some parts of the work.

Although many agencies have standard specification for above water work, often they do not include provisions applicable to work underwater. Special provisions have to be written for this work. They should allow the contractor to select the methods and mean of construction, but require approval of those methods and means to ensure the desired product is obtained. One contractor, for example, may wish to use rigid forms while another prefers fabric forms; either one could be acceptable in many situations. One contractor might want to use a tremie to place underwater concrete, while a second contractor might propose letting the concrete free fall through the water; the latter method not being acceptable. One of the most important functions of the engineer in underwater work is to ensure that there is adequate inspection and quality control.

Measurement of quantities underwater for payment and the basis for payment are two areas that can be contentious. The method of compensating the contractor for the basic work or additional work must be selected to protect both the owner and the contractor. The problem might be illustrated by Figure 3 which shows a detail for the repair of a void in the side of bridge pier underwater.

In this case, the contractor was required to remove the loose material with a light underwater jackhammer, install reinforcing steel, install the form work, place the concrete and remove the forms. The work was to be paid for on basis of cubic yards of concrete placed.

In this structure, the existing concrete was rather weak, and it would have been relatively easy for the contractor to remove much more of the existing structure than was intended and place a much larger volume of concrete that originally anticipated.

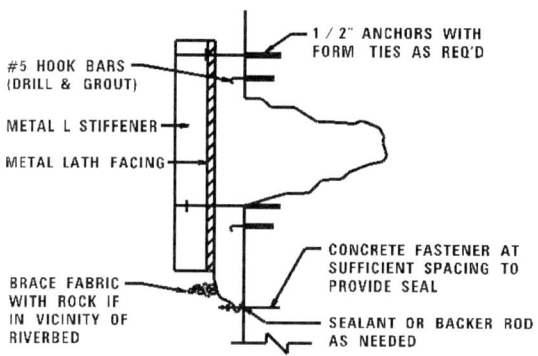

Figure 3 Pier void repair detail

Since the cost of the work included special equipment and a large above water support staff, the unit price for the concrete was quite high and an increase in the volume of concrete placed would result in a much higher payment to the contractor with a relatively little increase in cost to him.

The example above points out the importance of good underwater inspection. Without proper inspection, the contractor could increase the owner's cost significantly. Proper inspection is also necessary to ensure proper preparation of the work at all the stages of the construction. Inspecting a jacketing operation after the work is complete, for example, does not ensure that the preparatory work, such as cleaning, and installation of reinforcing steel has been completed properly.

In above water construction, the work is typically overseen by a resident engineer who is assisted by additional inspectors as required by the job size and complexity. A similar approach can be used for below water work. The use of a full time underwater inspection team is obviously desirable, but the cost can be prohibitive. The use of part time underwater inspectors can be tailored to the project's complexity and cost.

Because of the need for diving capabilities to conduct underwater inspections, some agencies have allowed the contractor to inspect his own work. Underwater photography and video may be used to document the inspection, but this method may not provide sufficient detail to assure conformity with the contract documents.

The data provided to the topside engineer is dependent on the inspector-diver. If the diver does not direct the camera toward significant defects, either purposely or innocently, such defects will not be known to the above water engineer.

Likewise, having the contractor measure the quantity of work performed, for payment purposes, could lead to inflated quantities or at least an appearance of impropriety.

The ideal solution is to have the underwater inspections and measurements for payment be made by an engineer-diver employed by the owner. Both the inspection and the measurements can then be unbiased. In addition, the engineer-diver, because of his technical knowledge and experience, is better able to assess the importance of defects and other conditions he encounters. Even if the contractor's superintendent is conscientious and trying to do a good job, he must depend on the diver who is working below water and out of sight. Especially in underwater work, "You don't always get what you pay for, but you get what you inspect for."

INSURANCE AND SAFETY

The statutory insurance requirements will vary depending upon where the work is performed. Often there are confusing and overlapping requirements. In the United States, for example, the divers and employees in the marine environment are covered by various forms of "no-fault" insurance. Workers Compensation Insurance is a state-mandated coverage protecting employees outside marine areas and in non-navigable waters. U.S. Longshoremen and Harborworkers' Insurance is a federally required insurance for work on and adjacent to navigable waters. Jones Act Maritime Insurance is also a federally mandated insurance, but it covers the members of crews of vessels. In the U.S., determining which insurance coverage is applicable can lead to litigation.

The high risks associated with marine construction activities also lead to high insurance premiums to protect the owner, the contractor and third parties.

Diving operations are closely controlled, for safety, in most countries. In the U.S., generally, diving operation are subject the Federal Occupational Safety and Health Administration Standards for Commercial Diving or a similar state regulation. The regulations include standards for training, equipment, and operations. While the safety of the operation is the responsibility of the diving contractor, engineers employing divers should be at least somewhat familiar with the regulations because they can affect the production rates for construction activities.

ROLE AND RESPONSIBILITY OF THE ENGINEER-DIVER IN THE ASSESSMENT OF UNDERWATER CONCRETE

T M Browne

Collins Engineers Inc.

USA

ABSTRACT. Like land-based structures, waterfront and offshore concrete structures require engineering inspections during the initial construction of a project, as well as periodically during the use of the structure. These structures require engineering inspections to ensure contract drawings and specifications are adhered to during initial construction or rehabilitation, and that continued structural capacity is not compromised by the onset of deterioration. Engineering decisions need to be made by personnel possessing the education and expertise in concrete engineering, as well as being familiar with the existing conditions of the underwater project site. Therefore, personnel professionally chartered or licensed in engineering, and qualified in commercial inspection diving techniques are most appropriately utilized for leading a team to assess and test underwater concrete. This paper will discuss the role and responsibilities of the engineer-diver and their inspection team surrounding the assessment of underwater concrete structures. In addition to a discussion on the duties and obligations of the engineer-diver, an overview of inspection techniques and methods will be included in this paper, as well as other underwater engineering considerations.

Keywords: Engineer-diver, Assessment, Underwater, Concrete, Inspection, Diving, NDT

Terence M Browne, P.E. is a Project Manager with Collins Engineers, Inc., 211 West Wacker Drive, 8th Floor, Chicago, Illinois 60606-1217 U.S.A., Mr. Browne has inspected numerous underwater structures worldwide for private corporations and governmental agencies utilizing commercial scuba, surface-supplied-air, and ROV operations. He is also active in concrete design and research, and is a member of the American Concrete Institute - Technical Committee 305, the Society of Naval Architects and the Marine Engineers, the Permanent International Association of Navigation Congresses, the National Society of Professional Engineers, the American Society of Civil Engineers, as well as a number of other professional organizations.

INTRODUCTION

Although many engineers, contractors, and owners realize the importance of conducting inspections on concrete structures, frequently the testing and assessment of concrete structures located in the most aggressive and constructability-challenging environments are ignored due solely to their difficulty to access, not assess. While providing qualified personnel and a means of getting effective equipment to all portions of a submerged concrete structure is more involved than a similar above water inspection, the underwater assessment and testing procedures are as important, if not more important, and should not be ignored. Numerous principles of concrete engineering, as well as the underwater environment and diving physiology, which should be well known to those in the underwater engineering field, will be briefly discussed in this paper for those readers outside the underwater assessment field but whose work requires knowledge of these fundamentals.

Types of Structures

There are many types of concrete structures located in offshore and inland waters, as well as submerged on land as part of raw or potable water facilities. These concrete structures are vital to local and global commerce, transportation, and recreation. Some types of concrete structures which warrant underwater assessments include: offshore towers and platforms; pipelines; harbor facilities (piers, dockwalls, marinas, vessel moorings, and dry docks); shoreline protection elements (breakwaters, revetments, seawalls, and scour countermeasures); bridge substructures; floating structures; building foundations; raw and potable water intakes, conveyance, and storage facilities; and hydraulic structures (locks, dams, spillways, and outfalls). These structures often utilize either plain, reinforced, or prestressed concrete as their primary structural material.

Types of Inspection Methods

A variety of inspection methods are available to inspect concrete located below water. Some common inspection methods include the utilization of: commercial engineer-divers and qualified technician-divers; commercial construction divers reporting to a topside engineer; engineers operating remote operated vehicles, referred to as ROV's; and engineers as passengers in relatively small manned submarines, referred to as submersibles [1]. Although much controversy is often associated with determining the most beneficial, feasible, and cost-effective inspection method, the use of engineer-divers is frequently employed to maximize the usefulness of the inspection data while minimizing the cost of the inspection. An engineer personally conducting the diving inspection allows a rapid and relatively inexpensive inspection to be performed [2]. The engineer-diver's knowledge of what to look for, and ability to recognize the structural importance of unexpected conditions, adds significantly to the value of an underwater structural inspection. Therefore, this paper will focus on the role and responsibility of the engineer-diver in the assessment of underwater concrete since the majority of concrete structures are located in water less than 65 meters deep. Beyond 65 meters, access involvement in addition to the required inspection tasks are often too time consuming to be effectively and feasibly conducted without the use of ROV's or submersibles.

The cost savings associated with utilizing ROV's and submersibles beyond 65 meters typically limits the engineer-diver's utilization for only the out-of-water evaluation and/or rehabilitation design rather than the physical underwater inspection portion of an assessment project.

ROLE OF THE ENGINEER-DIVER

The role of the engineer-diver includes inspection, evaluation, assessment, analysis, and design, as well as assistance with research, of a structure located underwater. Underwater engineering consultants and various governmental agencies employ engineer-divers to fulfill these roles, resulting in lower overall costs and the elimination of the problems associated with multiple entity projects.

Inspection

The engineer-diver's primary purpose is to detect and diagnose structural defects in a structure. There are very few defects that are limited exclusively to the underwater environment. Therefore, the methods of assessing and documenting these defects are essentially the same below water as they are above water.

Although each type of deterioration is different, they often have much in common and more than one type can be present at the same time. The eight primary concrete deterioration mechanisms consist of: scaling, freeze-thaw damage, osmotic pressure damage, impact damage, spalling, cracking, chemical attack, and biological attack, in addition to quality control issues (poor consolidation, delamination, laitance, segregation, and washout) [3].

The engineer-diver must frequently determine whether underwater concrete elements are flawed due to deterioration, or poor construction quality control standards. Since many existing structures never received a quality assurance construction inspection, underwater concrete elements frequently are constructed differently than shown on the design plans. Although varying batter slopes on concrete piles or widely exposed cold joints in a concrete with severe areas of poor consolidation may be indications of damage or settlement problems, the engineer-diver may or may not be able to discern the true significance of these observations. It is imperative for the engineer-diver to have the necessary experience to distinguish non-structural from structurally significant observations, since an inspection assessment and repair recommendations originate solely from the engineer-diver's perceived views. Therefore, a project's quality assurance policy should always mandate that a more experienced individual review a less experienced person's assessment of an abnormality's significance, especially if there is the slightest doubt regarding a defect. This quality assurance procedure is similar to the review for above water assessments, or engineering calculations performed in a design office. Without quality control/quality assurance (QA/QC) reviews, erroneous assessments may led to unnecessary repairs, or lack of needed repairs.

There are five primary types of inspections performed on concrete structures below water: construction quality assurance, inventory, routine, emergency, and forensic. The following paragraphs describe these inspection types so that the engineer-diver's role may be understood.

Construction inspections are required as part of the QA/QC involvement during the construction of a concrete structure underwater. Inspection tasks may vary from checking the quantity and quality of concrete placed underwater to verifying that proper joints have been sealed in precast concrete pipeline sections. Depending on the construction type and schedule, one final dive or various intermediate dives may be needed to assure conformance with all aspects of the construction specifications [4].

Inventory inspections are required when a previous inspection has never been performed on a structure, or when this documentation has been lost by the owner. Inventory inspections consist of the collection of any available data, establishment of as-built conditions, location of all defects or potential problems, and basic verification of the structural adequacy for the existing conditions.

Routine inspections are required on a periodic basis to assess the extent of any deterioration on a structure since the previous inspection. The maximum recommended interval between routine inspection varies from region to region, although a typical range between four and six years is generally accepted in the concrete engineering profession. In the United States, the Federal Highway Administration (FHWA) specifies a five year maximum interval, while the United States Navy (USN) specifies a six year maximum underwater inspection interval for facilities under their jurisdiction. Internationally, some certifying and insurance agents specify an annual inspection interval depending on the importance of the structure [5].

Emergency and forensic inspections are required for particular reasons such as the occurrence of an event that may have, or already has, damaged a structure. Generally, emergency inspections are performed after significant natural events (such as floods, earthquakes, and hurricanes) or abnormal occurrences (such as vessel impacts or unintentional over-loadings). Forensic inspections are required after an unusual occurrence is detected at a structure. Forensic inspections, or investigations, are usually conducted to detect the cause and extent of a recently developed problem or defect. These inspections typically only require the inspection of specific areas, although it is usually most cost-effective for the owner to authorize a complete inspection, depending on the remaining months until the next routine inspection is required.

The primary purpose of all underwater inspections is to determine the condition of the structural components in the water at the time of the inspection from the waterline to the channel bottom. However, it is the engineer-diver's role to also inspect areas which could be submerged during periods of high water. A form of written documentation should always follow an inspection, even if narrated videography was utilized. The written documentation, whether a formal report or simple rating form, should include a description of the structure, the method of investigation, findings of the investigation, and an evaluation and recommendations based on the findings. Prior to the date of the inspection, the engineer-diver should obtain and review copies of previous inspection reports, design drawings, and any previous maintenance work orders to effectively fulfill the expected role.

The means of access utilized by the engineer-diver during an inspection is commonly either surface-supplied-air (SSA) equipment, or self-contained underwater breathing apparatus (SCUBA) equipment. SSA equipment is usually used when the diver needs to be underwater for extended periods of time, or situations that prohibit a safe free-ascent to the water surface in the case of an emergency. Commercial SCUBA is most beneficially and safely used when the diver will only be underwater for relatively short periods of time and a safe free-ascent to the water surface is possible. Depending on the situation, one access method is usually more feasible and safer than the other. The utilization of SSA equipment is typically safer for long duration, deep, or penetration dives, while SCUBA is typically safer for short duration dives that may involve underwater obstacles or debris.

During an inspection, the engineer-diver is usually expected to perform a visual and tactile examination of all accessible concrete surfaces from the waterline to the channel bottom with particular attention being given to any observed areas of deterioration or apparent distress. However, where concrete surfaces are enormous and an owner's budget is limited, the engineer-diver may seek the owner's approval to focus particularly on only key inspection locations. In addition to the concrete condition, the type of channel bottom material, the location of any scour holes, the presence or absence of riprap, the location of any foundation undermining, and the presence of debris should always be noted during an inspection. Furthermore, the water surface elevation at the time of the inspection should always be surveyed and referenced to a known elevation on the structure, as well as to local datums. This information, along with the visibility, wave height, and current, is pertinent as it describes the particular underwater environment for other interested parties. Videography and photographs frequently are requested to further document the underwater environment and any deterioration. When necessary, a measuring scale for reference along with adequate lighting, appropriate lenses, and a clearwater box should be utilized to obtain the highest quality photographs. Visual documentation using underwater videography and/or photography should be included in a report with the quantity and quality adequate for proper documentation of the findings that will most appropriately represent the condition of the facility [6].

To effectively communicate the extent of the underwater inspection performed on a structure, the following common terminology and methodology have been generally accepted worldwide. Each type of inspection generally requires a combination of examination intensities, designated as Level I, Level II, and Level III, to optimize the funding and inspection time available. These standard levels of examination reflect the required effort exerted during an inspection. The scope of work for a particular inspection is categorized into these levels, and specifies the amount of work required in each level. The minimal procedure prescribed for any inspection is usually a combination of at least two of these levels of examination.

A Level I Examination, also referred to as a general examination, is essentially a cursory observation to detect any obvious gross or major damage which does not involve cleaning of any structural elements, and can therefore be conducted much more rapidly than the other examination levels. The Level I Examination confirms as-built structural plans and detects obvious major damage or deterioration due to overstress (ship impact, ice, etc.), severe steel reinforcement corrosion, or extensive chemical / biological attack. The engineer-diver must rely primarily on visual and/or tactile observations (depending on water clarity) to make condition assessments during a Level I Examination. These observations are normally made over the total exterior surface area of the underwater structure.

A Level II Examination, also referred to as a detailed investigation, is directed toward detecting and identifying damaged or deteriorated areas which biofouling organisms or surface deterioration may hide. Level II Examinations will often require cleaning of structural elements. Since cleaning is time-consuming and labor intensive, it is generally restricted to areas that are critical or which may typify the entire structure itself. The amount and thoroughness of the cleaning are governed by what is deemed necessary to discern the general condition of the overall facility. At this level, some observations of the concrete quality can usually be made. This data should be sufficient to permit calculations of gross estimates of the facility's load capacity. Simple instruments such as calipers, measuring scales, and inspection hammers are commonly used to take approximate physical measurements. However, a small percentage of more accurate measurements may also be taken with more sophisticated instruments for several reasons. The use of more sophisticated instruments will allow validation for a large number of simple measurements, and in some hard-to-measure areas, may be easier and faster to obtain. Where the visual scrutiny, cleaning, or simple measurements reveal extensive deterioration, a small sampling of detailed measurements will also enable gross estimates to be made of the structure's integrity. For example, concrete soundness can be evaluated with an inspection hammer on a small percentage of elements receiving Non-Destructive Testing (NDT) to determine relative results. The results determined by these NDT "spot checks" are used to determine the overall facility load capability.

A Level III Examination, also referred to as a highly detailed examination, always requires the use of NDT techniques, but may also require the use of partially destructive techniques. While these methods are similar to above water testing, it should be noted that working below water usually requires more time to assure adequate samples. The purpose of this type of examination is to detect hidden or interior damage, loss in cross-sectional areas, and material homogeneity below the surface of the water. A Level III examination usually requires prior cleaning. The use of these techniques is generally limited to key structural areas, areas that may be suspect, or to structural members that may be representative of the underwater structure. Similar to the dilemma of having construction divers perform engineering assessments, caution should be taken that the engineer-diver does not venture outside their area of expertise when utilizing construction tools or NDT testing equipment. A construction diver may be needed to physically obtain samples, while a NDT specialist may be needed to assist, retrieve, and/or interpret the results of NDT data.

Typical underwater NDT techniques consist of using ultrasonic testing equipment, rebound hammers, R-meters, ground penetrating radar equipment, impact echo testing equipment, and crack movement meters. These techniques utilize calibrated tools which can more accurately gather data than the engineer-diver's inspection hammer and intuition.

Cathodic Protection Evaluation for Concrete Structures

Although a cathodic survey can typically be performed by corrosion engineers above water, frequently engineer-divers are requested to perform visual and tactile assessments of the anodes and wiring on passive and active cathodic protection systems. Engineer-divers may also be asked to obtain cathodic potential measurements, although the engineer-diver is again cautioned about venturing outside their area of expertise. Complex corrosion studies should be performed only by certified corrosion engineers.

Structural Assessment and Analysis

Frequently a structural assessment must be performed in general terms at the project location. As needed, the engineer-diver should be able to perform analysis computations to quantify the inspection and to inform the owner of the any required load restrictions.

Whether immediately analyzed on location with hand calculations or quickly analyzed with modern computer technology at the office, all computations should be checked similar to original design calculation procedures.

Rehabilitation Design

Often rehabilitation designs are performed by the engineer-diver who is most familiar with the actual project site and deteriorated conditions. This assures the owner that a developed repair or maintenance scheme is the most cost-effective and feasible. The engineer-diver's unique role as an underwater inspector, as well as a designer, affords the opportunity for the engineer-diver to witness and remember the best and worst rehabilitation procedures.

QUALIFICATIONS OF THE ENGINEER-DIVER

The individual qualifications of engineer-divers vary greatly amongst the underwater engineering community since there are generally no uniformly mandated specific training requirements. The underwater inspector must possess both commercial diving (access) skills and engineering inspection (assessment) skills. While the diving industry and the engineering profession agree that both skills are essential, they differ in the opinions to which skill is more important than the other. The truth of the matter is that each skill is a vital requirement. Lack of competence in either skill can result in loss of human lives in the worst case, or a tremendous amount of wasted time and money from inaccurate and misleading data in the best case.

In many countries, various agencies and organizations have developed separate engineering and diving training criteria, although their recommendations are not always specified by the owner of a concrete structure in the interest of selecting the lowest-priced service. However, numerous owners do select their engineer-divers on a quality-based-selection method similar to engineering design and above water inspection projects, instead of the lowest bid-based-selection method often used for construction projects [7]. The use of quality-based selection means that the owner expects the various levels of expertise that are not as pronounced for the services of construction trades such as placing concrete.

Governmental agencies around the world, including but not limited to the United States Navy (USN) and the United States Coast Guard (USCG), have written policies outlining the minimum requirements for inspection personnel conducting underwater structural assessments. However, the USN and USCG only enforce these written policies at their own facilities since training requirement legislation has never officially been passed in the United States. While the USN and USCG mandate that the underwater inspection dive team leaders must be an experienced professionally licensed structural engineer for their projects, the U.S. Department of Transportation - FHWA utilizes the National Bridge Inspection Standards (NBIS) which require underwater dive team leaders to be either professionally licensed structural engineers; Level III or IV NICET (National Institute for Certification of Engineering Technologies under the National Society of Professional Engineers) certified inspectors; or individuals having completed a 120-hour comprehensive inspection course with five years of inspection experience. The worldwide trend is for owners to use only professionally licensed, chartered, or certified personnel to conduct underwater inspections [8].

In the United Kingdom, the Certification Scheme for Weldment Inspection Personnel (CSWIP) is generally accepted as the principal certification body. In 1980, CSWIP first introduced Phase 7 3.1D Certification with a specialized endorsement for the underwater assessment of concrete by inspection divers [9].

Many facility owners usually mandate additional stringent requirements on inspectors and team leaders. Several individual transportation departments, as well as many port authorities, in the United States allow only licensed professional engineer-divers to perform inspections. This requirement mandates that an engineer-diver or technician-diver may assist the professional engineer-diver only while both individuals are below water. In addition to having a professional engineer-diver as a team leader, the USN also mandates that the professional engineer-diver concurrently inspect at least 50 percent of a structure, even if several experienced engineer-divers and technician-divers are also utilized. In the United States, underwater inspections are conducted by at least a three-person team typically consisting of a professional engineer-diver, an engineer-diver, and a certified engineering technician-diver. Unfortunately, a less qualified team is often hired by the owner of a structure for convenience or economic reasons.

The National Council of Examiners for Engineering and Surveying mandates the minimum requirements for a professionally licensed engineer in the United States. Typically, an engineer with a bachelors degree from a EAC/ABET (Engineering Accreditation Commission of the Accreditation Board of Engineering and Technology) accredited curriculum and four years of practicing work experience needs to pass an eight-hour Fundamentals of Engineering Examination and an eight-hour Professional Engineering Examination, in addition to completion of an Ethics and Regulations Examination to obtain a professional engineer license.

Commercial diving operations in the waters of the United States, including associated lands, are governed by the Occupation Safety and Health Administration's (OSHA) Commercial Diving Standards. Furthermore, the regulations of the United States Coast Guard (USCG) also have jurisdiction covering the commercial diving operations in waters of the United States, as well as all vessels or offshore platform structures that require a certificate of inspection issued by the USCG. Although these governmental agencies regulate commercial diving operations, commercial dive training requirements are self-regulated by the diving industry. The OSHA regulations require each dive team member to have the experience or training necessary to perform assigned tasks in a safe and healthful matter.

Since the engineer-diver's tasks are to observe and gather data similar to a scientific diver, the engineer-diver does not have to possess training in construction operations such as underwater cutting, welding, and explosive techniques as outlined in the Minimum Consensus Standards of the Association of Diving Contractors (ADC), the curriculum of the Association of Commercial Diving Educators, and Standard ANSI/ACDE-01-1993 published by the American National Standards Institute (ANSI), utilized for construction divers. Commercial dive training for engineer-divers typically includes professional classroom and field instruction on general requirements, diving physics, medical aspects of diving, and diving equipment utilization with extensive continual in-house training on additional techniques and equipment. Similar to construction divers, engineer-divers undertake a rigorous program of on-the-job training specific to their assigned tasks in order to successfully achieve higher levels of qualification. For example, it is not uncommon for an owner to specify that an engineer-diver have over 500 hours of logged dive time prior to becoming eligible to perform an underwater inspection.

RESPONSIBILITIES OF THE ENGINEER-DIVER

In accordance with the Code of Ethics published by the National Society of Professional Engineers, engineer-divers must perform under a standard of professional behavior which requires adherence to the highest principles of ethical conduct on behalf of the public, clients, employers, and the profession. Some of these ethical responsibilities include following all governing regulations; pro-actively participating in conformance assurance by others; continuing education in the diving and engineering fields with respect to newly developed technology and recently published research findings, and utilizing the most appropriate methods and means available to accomplish an accurate structural assessment of a facility.

It is the responsibility of all dive team members to conduct inspection operations safely. Ultimately, the team leader has the duty of assuring safe operations, and therefore, should always have the ability to stop any operation in which safety of the team is jeopardized. While it is the inspection team's responsibility to immediately inform the owner of any structurally significant abnormalities on a structure, the inspection team should always have the ability to first stop the public's use of a facility until the owner is notified.

Liability Issues

Although the individual engineer-diver may not be fully aware of all the legal and accounting issues associated with the liability performing underwater inspections of structures, it is important that all engineer-divers, as well as the owner's representatives, be cognize that these issues exist.

Due to an inspection environment with typically heavy marine growth, high water turbidity, and low ambient light, the underwater assessment and recommendation decisions often are based on less information than is typically available from an above water inspection. Similar to above water inspections and engineering designs, professional liability insurance should cover all individuals involved in a project for the protection of the engineer, the engineering company, and the owner. In the past, underwater structural inspections were often conducted using construction divers reporting to an engineer on the surface.

The construction divers on those projects were not able to obtain professional liability insurance for the inspection's accuracy, and the engineer's insurance liability typically only covered the recommendations based on the given data. Problems arose from utilizing construction divers who either did not fully realize the significance of their observations, or lacked the technical vocabulary to accurately describe the underwater structural conditions to an engineer on the surface. Even with the use of underwater video cameras, an engineer on the surface is forced to rely on information transmitted to the surface in an environment where tactile observation is often necessary. Having an engineer on the surface to perform an underwater structural inspection from a video monitor usually results in added costs and liability to the construction diver, the engineer, and the owner.

It is recommended that engineers exercise caution when asked to make assessment and recommendation decisions on projects without personal hands-on involvement in the physical inspection. This caution is expressed for instances where other individuals or ROV's are utilized to obtain data essential for engineering decisions.

Along with liability comes the issue of personal insurance requirements. In the United States, engineer-divers must have at least the minimum coverages on Workers' Compensation Insurance; Jones' Act and Longshoremen's Insurance; General Liability and Property Damage Insurance; and Professional Liability Insurance.

Workers' Compensation Insurance covers any personal injuries directly related to, or which occur as a result of the work performed. Although qualified and safety conscious underwater inspection teams generally have few or no history of injuries, the diving community as a whole accounts for several injuries per year. These injuries range from decompression illness to fatalities. These injuries primarily occur to construction divers and recreational divers. The National Underwater Association of Diving Contractors (NUADC) utilizes seven different types of commercial work categories, in addition to five occupational work categories, in documenting the actions being performed when such injuries occurs. Engineer-divers do not have a specific category since accidents occurred by trained and experienced engineer-divers while inspecting structures are uncommon. However, all underwater activities by an engineer-diver should require sufficient safety precautions and adequate insurance.

CONCLUSIONS

Engineer-divers perform a vital role in the successful long-term utilization of concrete. Monitoring construction and the continual condition of concrete structures assures a high level of public safety, serviceability, and investment return to the owner. Furthermore, these inspections can lead to the lowest maintenance and repair costs for a structure's life span as long as the proper maintenance schedule is adhered to. Implementing a variety of inspection procedures and specialized equipment techniques, the engineer-diver has the ability to accurately, quickly, and cost-effectively assess structures. Although an international consensus still does not exist for engineering and diving minimum experience and training requirements in order to perform the underwater assessment of a structure, prudent owners and facility operators recognize the need and benefits of underwater concrete assessments performed by engineering professionals at routine intervals not to exceed a range generally between four to six years.

REFERENCES

1. ALLEN, W. "Underwater Inspections", Concrete International Design and Construction, 1989, Vol. 2, No. 9, Sept., pp. 37-40.

2. CHABOT, J. "Do You Need an Underwater Inspection? Call an Engineer!", Underwater Magazine, 1995, Vol. 7, Nov. 4, Fall, pp. 7-9.

3. PUBLIC WORKS CANADA - DESIGN AND CONSTRUCTION BRANCH, Guidelines for Inspection and Maintenance of Marine Facilities, Circa 1985.

4. COMMITTEE ON MARINE STRUCTURES - MARINE BOARD NATIONAL RESEARCH COUNCIL, The Role of Design, Inspection, and Redundancy in Marine Structural Reliability, Proceedings of an International Symposium, Nov., 1983.

5. U.S. DEPARTMENT OF COMMERCE, Underwater Inspection/Testing/Monitoring of Offshore Structures, USDC Contract No. 7-35336, Feb., 1978.

6. COLLINS, T. "Underwater Inspection: Documentation and Special Testing", Public Works, 1988, Vol. 119, No. 1, Jan., pp. 53-55.

7. GANAS, M. AND BOSWELL, S. "Bridge Diving Inspection and the Competitive Bid System: Problems & Pitfalls", Public Works, 1993, Vol. 124, No. 13, Dec., pp. 49-52.

8. MILLS, G. "Underwater Inspection - The Requirement and the Methods", Proceedings of the 20th Annual British Conference on NDT, Scotland, Sept., 1985, pp. 227-243.

9. MILLS, G. "Underwater Inspection Personnel Certification", Underwater Technology, 1986, Vol. 12, No. 3, Autumn, pp. 10-16.

UNDERWATER CONCRETE CONSTRUCTION

L McLennan
Sika Limited
United Kingdom

ABSTRACT. The paper examines different methods of producing, placing and testing of Underwater Concrete in particular describing the benefits obtained by the use of appropriate admixtures during mix design.

Keywords: Anti-washout concrete, Washout resistance, Tremie.

Mr L McLennan is the Sales and Marketing Manager of the New Construction section at Sika Limited. He has been at Sika for 10 years and is primarily involved with the promotion of admixtures in standard and specialist concrete.

INTRODUCTION

Underwater concrete construction presents special challenges both ecologically and structurally. A fundamental problem often encountered in the placing of underwater concrete is the washout of the cementitous proportion from the concrete mass. This washout leads to a reduction in structural integrity and has detrimental long-term durability effects. One solution to this problem is to erect a structure around the construction area then de-water this prior to placement of any concrete. This procedure, whilst effective, is also time consuming and costly. Alternative construction methods such as tremie or placement by pump within the mass of concrete underwater can provide the engineer with a fast-track construction method. However special care and attention to mix design detail must first be undertaken normally by a series of trial mixes both at laboratory and then at site prior to concrete placement. In order to control this washout we must first design the concrete mix with the most suitable constituents available including where appropriate the use of anti-washout admixtures.

MATERIALS AND SELECTION

Wide ranges of materials are available for use in underwater concrete, some of which can enhance both plastic and hardened properties of concrete. The selection of aggregates is often dependent on local resources but as a general rule rounded gravel and a good quality zone M sand provide the best overall results. The other critical factor in aggregate selection is the ability of the coarse and fine aggregate to produce a well-proportioned combined grading curve with no gaps in the grading. A typical example is shown in Figure 1.

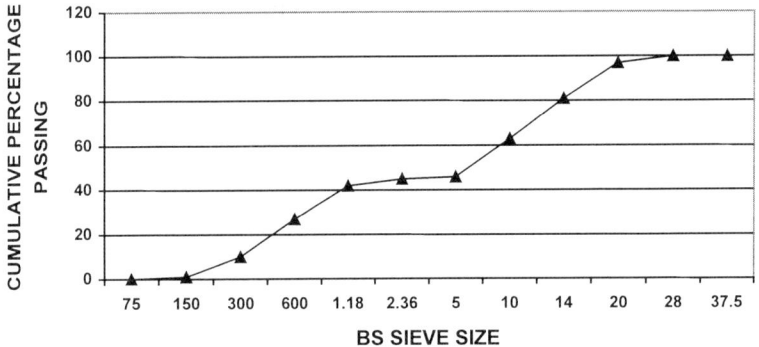

Figure 1 Typical desired grading curve of coarse and fine aggregate

Cements and replacement materials often used in underwater concrete include the following: Portland Cement, PFA, GGBFS and Silica Fume. The selection of these materials is often dictated by local availability but best results are achieved with the use of the pozzolanic replacements such as PFA or SF. Underwater concrete of the highest quality can also be produced with PC only or slag blends. The critical factor in establishing which material to include can be dependent on commercial considerations but in every case a minimum cementitous content of 400kgs should be specified in underwater structural concretes.

The choice of materials can also effect the anti washout properties as well as the self-compactibility of the concrete. During the trial stages these properties need to be assessed to provide the highest quality concrete requiring the least amount of compaction or handling.

ADMIXTURES

Good quality concrete should be manufactured with a low water cement ratio and should be cohesive yet easily placed. These properties are a prerequisite of underwater concrete and for these reasons it is often necessary to use two admixtures in combination, these will normally be a superplasticiser and an anti washout agent.

There are many superplasticisers on the market today and different generic types will impart varying properties to concrete. In the experience of the writer melamine, naphthalene or lignins are unsuitable as they can revert quickly causing a loss of workability particularly detrimental when divers are involved in placing. In addition these types of admixtures generally have dosage ranges of 1-3% and at the higher dosage bands they can make the concrete too cohesive which can effect the flowability of the mix.

Acrylic Vinyl Copolymers have been used since their development in the 80s with great success. These admixtures are dosed between 0.5%-1.5% they also retain workability for a period of up to six hours at the higher dosage and can still produce concrete with a w/c ratio <0.4 but workability in the region of 550mm on the flow table.

Stabilizers or anti washout agents are based on cellulose derivatives and are dosed depending on the degree of thixotropy required. This property and the subsequent dosage can depend on the environment in which the concrete is to be placed. Examples of differing washout properties can be seen in a fast flowing river or in a sheltered lagoon with no tidal forces in evidence. Clearly these different environments require a different design approach to the concrete.

Typical dosages of stabilizers are between 0.4% and 1% of the cementitous proportion. Whilst these products impart anti washout and anti bleed properties to concrete they also increase workability and on a negative note can, at high dosages, create air entrapment in concrete due to their cohesive effect.

MIX DESIGN

Traditional mix design methods are equally suitable to produce underwater concrete as long as the designer incorporates the need for high workability, self compaction and cohesive properties. A good starting point is to begin with an highly workable but pumpable concrete mix.

A typical mix design for underwater concrete is given in Table 1 and properties of the mix are given in Table 2.

Table 1 Typical mix design for underwater concrete

CONSTITUENT	PROPORTION	
Cement	280	kg/m^3
Pfa	120	kg/m^3
COARSE AGG 20-5mm	1050	kg/m^3
Fine Agg	700	kg/m^3
Sikament 10 Superplasticiser	3.5	litres/m^3
Stabliser UCS01	2.0	kg/m^3
Water/Cement Ratio	0.42	
Workability Flow	600	mm

Table 2 Properties of underwater concrete mix

DENSITY kg/m^3	WASHOUT %	COMPRESSIVE STRENGTH N/mm^2	
		7 Days	28 Days
2280	2	29.0	42.0

MIXING

Underwater concrete should be mixed thoroughly, a specific recommendation is difficult due to the wide range of mixing vessels currently available and their different efficiency rates. Tumble drums or dry mixing plants generally need two or three times longer than conventional wet plants incorporating a forced action mixer. Prior to using the mixer for alternative concrete mixes it should be cleaned to remove any residue from the mixer lining. Figures 2, Figure 3 and Figure 4 show the improvement in thixotropy during the addition and mixing of a stabiliser.

Figure 2 Superplasticised concrete

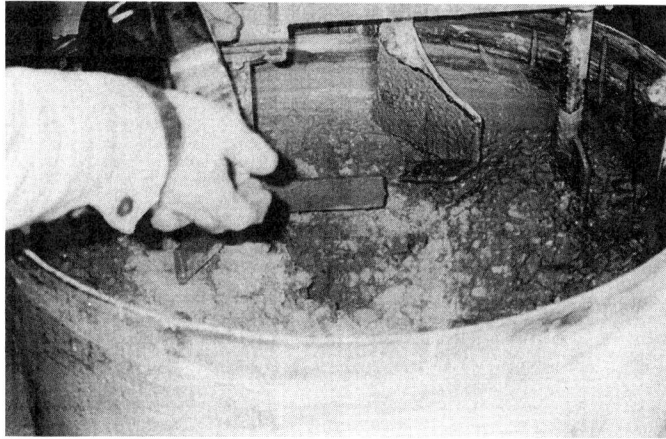

Figure 3 Concrete after addition of stabliser

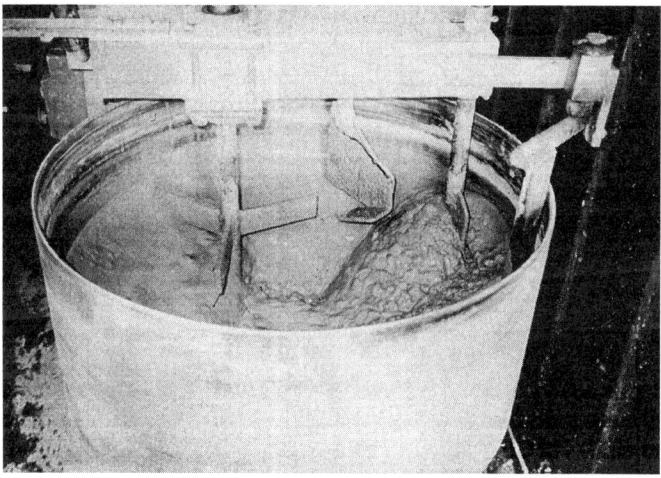

Figure 4 Underwater concrete with improved thixotropy

PLACING AND COMPACTION

Normal placing and compacting techniques do not apply in concrete placed underwater. The concrete must be self-compacting and in some cases where limited access is available it must also be self-placing, Figure 5 Often the environment the concrete is to be placed in has limited visibility and access is only available to divers, for these reasons it is necessary to extend the workable life of the concrete and to ensure that it has the ability to flow in and around the form in which it is to be placed. The use of a pump or tremie is the best placing method. Care should be taken to prevent the end of the pump or tremie from withdrawing from the mass of concrete as free falling concrete increases segregation and latency build up on the surface of the concrete.

Figure 5 Self compacting underwater concrete

When this laitence build up occurs it is necessary to remove it usually by water or air jetting the top surface at an early age as soon as is possible after setting has occurred. The amount of laitence will vary with the free fall distance and mix design constituents.

OTHER METHODS OF PLACING

Pre-placed Aggregate Concrete

This method is also known as grouted aggregate concrete. Large aggregate is poured into the form and vibrated or rodded until it is well compacted. A colloidal grout is then injected through pre-placed pipes which go to the bottom of the section to be cast. The injected grout then displaces water upwards replacing the water with a high quality grout with no segregation and low settlement. This technique is regarded as specialist and, as such, is normally undertaken by contractors with suitable expertise.

TESTING

Anti washout test methods vary across industry the main test method currently adopted by industry is the US Corp of Engineers Test CRD_C61. This involves passing a known volume of concrete through water x no of times within a cylindrical vessel, the mass of concrete is reweighed and the difference mass 1 mass 2 is expressed as a % and called the washout value. Limits for washout can vary from 1% up to 10% figures of less than 6% are generally accepted as good washout resistance using this test method. The general setup of the test is shown in Figure 6.

A comparison of anti and non anti washout concrete tested is shown in Figure 7.

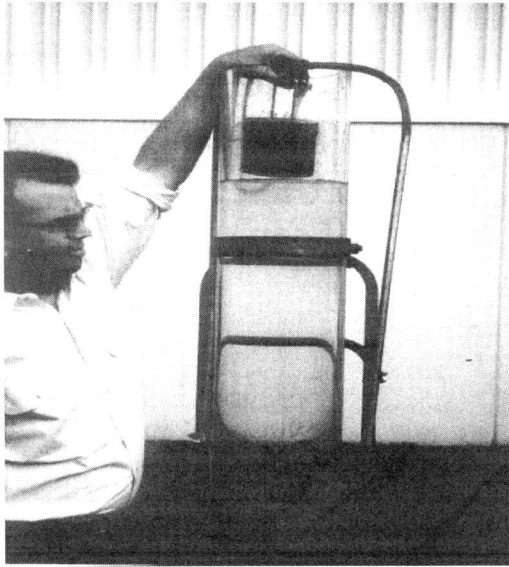

Figure 6 General arrangment of anti washout test

Figure 7 Typical results from concrete without
anti washout (left) and with anti washout (right)

Additional testing should include test cubes to monitor the characteristic strengths and consistency prior to placement. Quality control is paramount as cores taken from underwater concrete can become expensive. Nowadays it is also fairly common to monitor placement with a video connection to a land based monitoring station.

Flow

BS 1881 Part 105 [1] Method of determination of flow is recognised as the best method of determining workability. Flow values are normally in the range of 450-600mm. This represents a more meaningful test for control than the more commonly used slump test which is less accurate at high workability. A typical example is given in Figure 8.

Figure 8 Typical flow measurement of underwater concrete

SUMMARY

1. Grading of fine and coarse aggregate must be well proportioned with no gaps in grading

2. Cements and replacement materials can be used however minimum cementitious content of $400kg/m^3$ should be specified in underwater structural concrete.

3. Superplasticisers should be used in conjunction with anti washout admixtures. Stablising admixture dosages should be carefully monitored so as not to cause air entrapment.

4. Underwater concrete should be workable enough to be self compacting and often self placing.

5. An underwater concrete mix should be tested thoroughly before use by means of the anti washout test and flow test.

REFERENCES

1. BRITISH STANDARDS INSTITUTION. BS 1881 Part 105 1984 Testing Concrete, Method for Determination of Flow, 8pp

THEME FIVE:
FLOOR TOPPINGS AND OVERLAYS

Keynote Paper

THICKNESS DESIGN METHODS FOR SLABS ON GROUND

D Beckett
University College London
United Kingdom

ABSTRACT. Concrete literally provides the shop floor for the vast majority of industrial enterprises and in the UK alone, involves the use of some six million cubic metres per year in the construction of ground supported slabs. Since the 1970's, there have been a number of significant developments in design and construction methods, necessitated by more onerous requirements for economy, loading, surface regularity, crack control and reduced wear. The primary purpose of this paper is to provide an overview of thickness design methods and includes alternatives to the classical work of Westergaard (1925) and the use of fibres (steel and polypropylene) as a replacement/supplement to fabric reinforcement. Comprehensive test data is included.

Keywords: Thickness design, Westergaard et al, Ultimate load, Theories, Comparison of design methods, Test data.

Derrick Beckett was a member of the working party for the second edition of Concrete Society Technical Report No. 34 – Concrete Industrial Ground Floors – A guide to their design and construction and is a member of the Society's Floors Committee. Over the past ten years, he has undertaken numerous tests on large-scale concrete ground slabs and is the author of a number of papers on thickness design. He is currently a Visiting Professor in the Department of Civil & Environmental Engineering at University College London.

INTRODUCTION

Between 1989 and 1995, the author undertook a large number of comparative tests on 3.0 x 3.0 m concrete ground slabs at Thames Polytechnic (now University of Greenwich), Dartford, Kent. These tests were carried out in a purpose-made rig capable of applying a concentrated load of up to 600 kN at any point on the slab. The results of these tests (refer to Appendix C), undertaken on plain, polypropylene, steel fibre and fabric reinforced slabs, indicated that the conventional Westergaard thickness design approach was conservative and that an ultimate load approach (e.g. Meyerhof 1967) may be more appropriate.

In January 1995, the United Kingdom Concrete Society launched a second edition of Technical Report 34 [1] – Concrete Industrial Ground Floors (a guide to their design and construction). This document included an alternative to the classical Westergaard et al thickness design method and was based on the work of Meyerhof [2]. The two approaches were compared by the author in *Concrete* July/August 1995 [3]. It was subsequently brought to the author's attention [4] that a number of ultimate load thickness design procedures had been developed. Thus it was considered necessary to undertake a desk study to review current design procedures which, when supplemented by large scale testing, could lead to the publication of a definitive design manual. As an aid to the desk study, students in the Department of Civil & Environmental Engineering at UCL compared four design approaches as part of a third year special project. The design approaches considered were – Westergaard (1925), Meyerhof (1962), Rao & Singh (1986) and Shentu, Jiang & Hsu (1997). The project report [5] compared the results obtained in relation to slab thickness, load position, radius of relative stiffness and modulus of subgrade reaction applied to plain, microsilica and steel fibre reinforced concrete. The report also correlated theoretical work with load test results and included laboratory test data on modulus of subgrade reaction. The purpose of this paper is to outline the design procedures listed above and to discuss their implications when applied to a specific load position, contact area and concrete grade, but varying the slab thickness and modulus of subgrade reaction.

Westergaard [6]

Following the pioneering work of E Winkler (1867), H Zimmermann (1883) and A Foppl (1922), H M Westergaard (1925) presented a mathematical analysis to compute stresses and deflections in concrete roads making the following assumptions:

(a) The slab acts as a homogeneous, isotropic, elastic solid in equilibrium.

(b) The reactions from the subgrade are vertical only and are proportional to the deflections of the slab.

(c) The subgrade is an elastic medium and its elasticity can be characterised by the force which, distributed over unit area, will give a deflection equal to unity. This is known as the modulus of subgrade reaction k – load per unit area per unit deflection, e.g. N/mm^3.

Westergaard introduced another parameter which is a measure of the stiffness of the slab relative to the subgrade. It is in the nature of a linear dimension and is referred to as radius of relative stiffness l defined by

$$l = [E_c h^3 / 12(1-\mu^2)k]^{0.25}$$

where E_c is the modulus of elasticity of the concrete, μ is Poisson's ratio (0.15) and h is the slab thickness. Westergaard considered three cases for computing stresses and deflections – corner, edge and internal loading, the subgrade acting as if it were made up of rows of closely spaced but independent elastic springs. The modulus of subgrade reaction is equivalent to the spring constant and is thus a measure of the stiffness of the subgrade. Subsequent research led to modifications of the original equations and in the UCL study, the equations given in C&CA Technical Report 550 [7] were adopted. Further discussion of the Westergaard et al approach is given in reference [3].

Meyerhof [2]

In 1962 Meyerhof proposed a method to estimate the ultimate bearing capacity of concrete ground slabs under internal, external and corner loads and this forms the basis of the alternative thickness design method given in Appendix F of the second edition of TR34. The Meyerhof formulae, together with a comparative example, are given in reference [3].

Rao & Singh [8]

In 1986, Rao & Singh presented a slab design method in which collapse loads are predicted by using rigid plastic behaviour and square yield criterion of failure for concrete. The Meyerhof and Rao & Singh approaches are similar with one important difference in that boundary shear equilibrium was ignored by Meyerhof. Further, Rao & Singh consider two modes of collapse:

(a) Semi-rigid: the loading is flexible, like that of a wheel on a pavement and a plastic hinge is formed at the base of the slab and inside the loaded area.

(b) Rigid: the loading is rigid, like a column cast monolithically with the slab and a plastic hinge is formed at the base of the slab around the column

Shentu, Jiang & Hsu [9]

In 1997, Shentu et al used a finite element model assuming a Winkler subgrade to develop a simple formula to determine the load carrying capacity of a plain concrete slab on grade subjected to an interior concentrated load. The load capacity P can be expressed as:

$$P = 1.72 \times [(k \cdot r/E_c) \times 10^4 + 3.6] f'_t \cdot h^2$$

where k = modulus of subgrade reaction
        r = radius of loaded area
        E_c = modulus of elasticity of concrete
        f'_t = uniaxial tensile strength of concrete
        h = depth of slab

As confirmed by test data [(refer to Appendix (2)], failure occurs via the development of radial and circumferential cracks. This would appear to be unlikely for a plain concrete slab. Consider a slab of depth h = 200 mm, k = 0.05 N/mm², E_c = 30 kN/mm², r = 50 mm and f^1_t = 2.4 N/mm², then

$$P = 1.72 \times [(0.05 \times 50/30 \times 10^3) \times 10^4 + 3.6] \times 2.4 \times 200^2 \times 10^{-3}$$

$$= 732.0 \text{ kN}$$

This is an order of magnitude greater than the Westergaard load and the reason for such a large increase is the presence of in-plane compressive forces referred to as compressive membrane action and sometimes known as "arching action". The presence of arching action in suspended slabs (buildings and bridge decks) has been known for at least 40 years and the current Ontario Highway Bridge Design Code (OHBDC) prescribes the use of an empirical design method for bridge slab systems [10]. The author is not aware of its use in ground slab design and this is clearly a matter for further research.

OTHER METHODS

Two methods, included as a supplement to the UCL desk study, are those of A. Losberg [11] and R.A. Baumann & F.E. Weisgerber [12]. In 1978, Losberg proposed a yield line analysis for slabs on ground and advocated the use of structurally active reinforcement rather than the so called "crack reinforcement" which he argued is mainly too weak to prevent the formation of cracks or to control crack widths. The structurally active reinforcement is placed in the bottom of the slab (typical percentage 0.25 to 0.35) and this steel is used to estimate the sagging (positive) moment capacity. The hogging (negative) moment capacity is taken as the flexural strength of plain concrete. Allowance is also made for the influence of shrinkage and temperature variation.

In 1983, Baumann & Weisgerber developed a yield line method to determine collapse loads of slabs on ground. Expressions were derived for the collapse load of a slab with an interior, free edge and free corner load. As with Losberg's approach, reinforcing steel was provided for positive moment only. Comparisons were made with previous analyses by Losberg and Meyerhof and there was reasonable correlation, the Baumann and Weisgerber approach being somewhat conservative. Appendix B (supplement to UCL report) compares the results of Meyerhof, Rao & Singh, Baumann & Weisgerber & Losberg for a given set of slab parameters.

PARAMETRIC STUDY

Space limitations permit the inclusion of only a small part of the parametric study included in the UCL report. The extract relates to a plain concrete slab with a centrally applied load, using the following data:

Modulus of elasticity of concrete	= 30 x 10³ N/mm²
Compressive strength of concrete	= 47 N/mm²
Modulus of rupture of concrete	= 5.1 N/mm²
Direct tensile strength of concrete	= 2.04 N/mm²

Poisson's ratio = 0.15
Contact radius of loaded area = 50 mm

The values of the concentrated load P (kN) were obtained for a range of slab thicknesses (h = 100 to 450 mm) and moduli of subgrade reaction k (N/mm^2). Extremely small values of k were deliberately chosen to demonstrate the influence of k on load and deflection. Sample results are given in Table 1 for a slab with thickness h = 150 mm.

Table 1 Relationship between k (N/mm^2) and centrally applied load P (kN)
(taken from UCL Report [5] - the report also considers corner and edge loading)

k, N/mm^3	Westergaard deflection, mm	CENTRALLY APPLIED LOAD P, kN				
		Westergaard	Meyerhof $M_n = 0$	Rao & Singh $M_n = 0$	Rao & Singh $M_n = M_p$	Shentu
0.3	0.205	83.3	143.2	166.2	332.5	682.2
0.1	0.320	75.1	136.5	155.9	311.8	417.8
0.01	0.838	62.3	127.2	140.9	281.9	298.8
0.001	2.263	53.2	121.9	132.2	264.4	286.9
0.0001	6.244	46.4	119.0	127.2	254.4	285.7
(1)	(2)	(3)	(4)	(5)	(6)	(7)

It can be seen from Table 1 that deflection is more sensitive to changes in the value of the modulus of subgrade reaction k than the collapse load. For a 150 mm slab with k = 0.01 with a load at a free edge and free corner, the Westergaard deflections increase from 0.838 (central load) to 1.85 mm and 4.19 mm respectively. The collapse loads obtained from the Shentu et al equation, column (7), are influenced by the presence of "arching action" as discussed previously, which apparently becomes more significant as the value of k increases. There is reasonable correlation between columns (4) and (5) Meyerhof and Rao & Singh with the negative (circumferential) moment M_n taken as zero. An attempt will now be made to correlate the analytical results with typical test data.

CORRELATION WITH TEST DATA

Figure 1 shows a typical load/deflection relationship for a steel fibre reinforced slab (taken from the University of Greenwich tests, test no. 14 in Appendix C) for which the following data applies, the notation is given in Appendix B. Slab dimensions 3.0 x 3.0 x 0.15 m, concentrated load P applied at centre via a steel plate 100 mm x 100 mm, A = 10^4 mm^2, r = 56.43 mm, b = 64.87 mm, k = 0.035 N/mm^3, h = 150 mm, f_{cu} = 60 N/mm^2, R_{e3} = 60, E_c = 36 kN/mm^2, μ = 0.15, l = 737.57 mm and f_{ct} = 6.15 N/mm^2 (from beam tests to BS 1881).

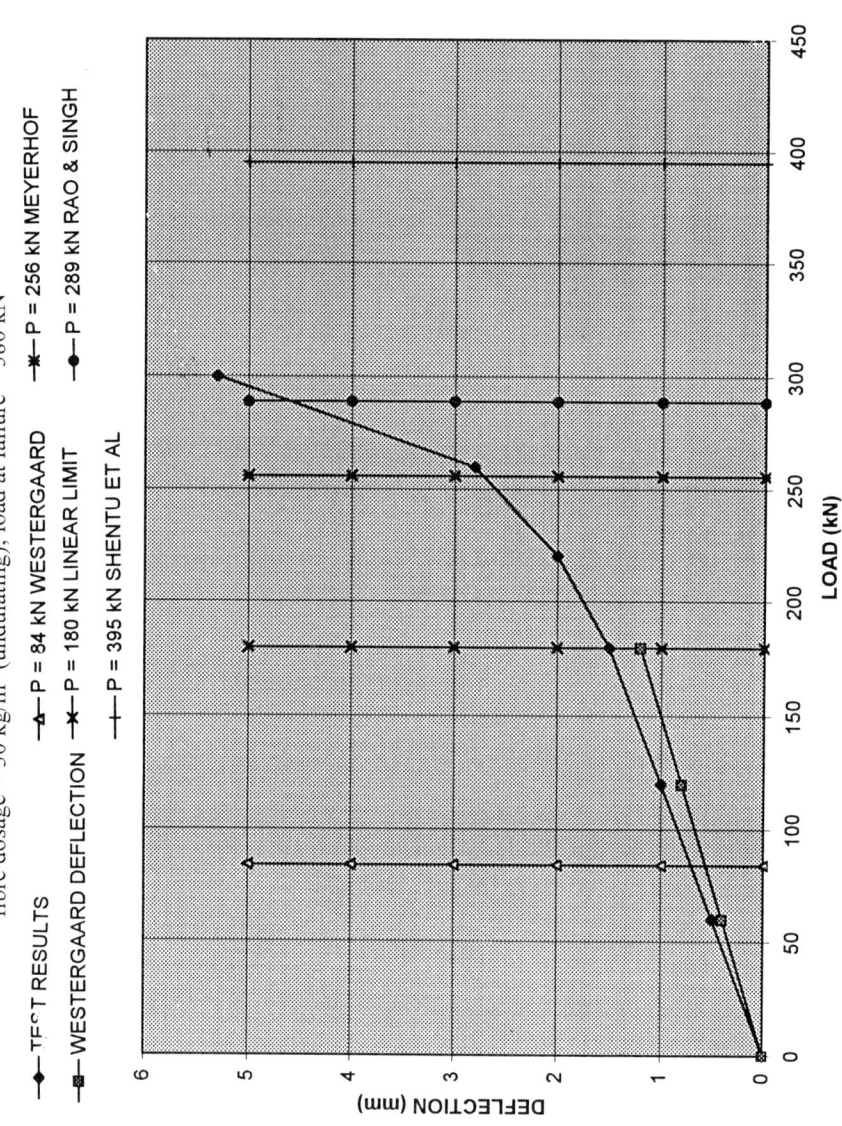

Figure 1 Typical load/deflection plot for steel fibre reinforced test slab

The values of P (kN) are as follows:

P (Westergaard) = 84.21 (1.0)
P (Meyerhof) = 255.8 (3.04)
P (Rao & Singh) = 288.5 (3.43)
P (Shentu et al) = 394.96 (4.69)
P (test) = 380 (4.51)

The numbers in brackets give the ratios related to the Westergaard value of 84.21 kN. The following points should be noted.

(i) The test load/deflection relationship is linear up to 180 kN, that is, 2.1 times the Westergaard load.

(ii) The test load at first visible cracks on the vertical faces of the slab is 260 kN.

(iii) The Westergaard deflection [see Figure (1)] is less than the test value. Using the Westergaard equation, the deflection is 0.066 mm per 10 kN increment of load, whereas for the test the value is increased to 0.0833 mm per 10 kN increment of load. A similar result was obtained by Ioannides et al [13] who concluded from a finite analysis that the Westergaard deflections are unconservative if the ratio of the slab size (L) to the radius of relative stiffness (l) is less than about 8.0 (for internal loading). For the test slab L/l = 4.07 for which Ioannides gives an increase of about 20 percent above the Westergaard value. The test value is about 25 percent.

(iv) The loads from the four analytical methods are shown on Figure 1 and it is clear that the Westergaard approach is conservative with regard to load capacity. Assuming a permissible service load deflection of say 1.0 mm, this corresponds to a service load of 120 kN. This would give a load factor in excess of 2.0 for all of the remaining three analytical methods. However, there are a number of "grey" areas. The Meyerhof and Rao & Singh loads of 255.8 and 288.5 kN respectively assume the development of radial and circumferential yield lines taking account of the flexural toughness of the steel fibre reinforced slab with R_{e3} = 60. The Shentu et al approach also assumes the development of radial and circumferential yield lines together with arching action, but the influence of the steel fibres is ignored and the direct tensile strength of the concrete was taken as 0.4 times the flexural strength (modulus of rupture). Due to the relatively small slab dimensions (3.0 m x 3.0 m) the circumferential yield lines were not developed in the test. In a real design situation with slabs of say 6.0 m x 6.0 m, it is anticipated that circumferential yield lines would develop.

SUMMARY

It is the author's view that with an extension of the desk study supplemented by additional tests on slabs with emphasis on deflection and load transfer at joints, a comprehensive design method can be developed. This should be compiled in a limit state format in a similar manner to EC2/BS 8110 with emphasis on both strength and serviceability (crack control and deflection).

Additional test data is required for load transfer across sawn joints, corner and edge loading. To simulate double leg rack loading, twin loading plates should be adopted. This work has commenced and the results are reported in *Concrete* March 1999.

ACKNOWLEDGEMENTS

The writer is indebted to the industry of William Chow and Yap Chee Keong, students at University College London, in producing the desk study which formed their third year special project and to Dr Richard Bassett, Reader in Geotechnical Engineering for advice on soil/structure interaction.

REFERENCES

1. CONCRETE SOCIETY. Concrete industrial ground floors, a guide to their design and construction. Technical Report 34, 2nd edition 1994.

2. MEYERHOF, G G. Load carrying capacity of concrete pavements, Journal of the Soil Mechanics and Foundations Division, Proceedings ASCE, June 1962.

3. BECKETT, D. Thickness design of concrete industrial ground floors, Concrete, July/August 1995.

4. PAUL SPRIGG. Private communication. Sprigg Little Partnership.

5. CHOW, W AND YAP, C K. Comparative study of analytical procedures for concrete ground slabs, Third Year Special Project 1997/98, Department of Civil & Environmental Engineering, University College London.

6. WESTERGAARD, H M. Stresses in concrete pavements computed by theoretical analysis, Public Roads, Vol. 7, 1926, p. 25.

7. CHANDLER, J W E. Design of floors on ground, Technical Report 550, Slough, Cement & Concrete Association, June 1982.

8. RAO, K S S AND SINGH, S. Concentrated load carrying capacity of concrete slabs on ground, J. Struct. Engrg., ASCE, 112 (12), 1986, pp 2628-2645.

9. SHENTU, L, JIANG, D AND HSU, C T T. Load carrying capacity for slabs on grade, ASCE, Journal of Structural Engineering, January 1997.

10. BATCHELOR, B de V. Membrane enhancement in top slabs of concrete bridges, Chap. 6 Concrete Bridge Engineering : Performance & Advances, ed. R.J. Cope, Elsevier Applied Science 1987.

11. LOSBERG, A. Pavements and slabs on grade with structurally active reinforcement, ACI, 75 (12), 1978, pp 647-657.

12. BAUMANN, R A AND WEISGERBER, F E. Yield line analysis of slabs-on-grade, J. Struct. Engrg., ASCE, 109 (7), 1983, pp 1553-1568.

13. IOANNIDES, A M, THOMPSON, M R AND BARENBERG, E.J. Westergaard solutions reconsidered, Transportation Research Record 1043, 1985.

NOTATION

l	=	radius of relative stiffness
k	=	modulus of subgrade reaction
μ	=	Poisson's ratio
h	=	slab depth
r	=	radius of loaded area
f_t	=	modulus of rupture of concrete
f^1_t	=	direct tensile strength of concrete
R_{e3}	=	toughness factor for steel fibre reinforced slabs
M_n	=	negative (hogging) resistance moment
M_p	=	positive (sagging) resistance moment
P	=	ultimate (collapse) load

APPENDIX A – DESIGN FORMULAE

For the Westergaard et al and Meyerhof formulae, reference [3] should be consulted.

Reference [8] gives the Rao & Singh formulae, reference [11] the Losberg and reference [12] the Baumann & Weisgerber formulae. Note that Losberg presents his results in graphical form plotting $(M_n + M_p)/P$ against a/L and Figure 2 is adapted from Losberg's graphs for a single internal load.

Shentu et al [9]

For internal loading on a plain concrete slab, the collapse load P is given by

$$P = 1.72 \times [(k \cdot r/E_c) \times 10^4 + 3.6] f^1_t \cdot h^2$$

In a test reported in reference (9) with a concentrated load P applied at the centre of a circular slab of radius 2.5 m, contact radius 100 mm, slab depth 160 mm, cylinder strength 25.5 N/mm², uniaxial tensile strength 2.17 N/mm², modulus of elasticity of concrete 2.6×10^4 N/mm² and modulus of subgrade reaction 0.02 N/mm³, cracks developed as follows

P = 240 kN   a circumferential crack was found at the top surface of the slab at a radius of 1.25 m.

P = 310 kN   the circumferential crack extended downwards.

P = 380 kN the bottom radial cracks intersected with the circumferential crack extending down from the top resulting in significant settlement at the centre of the crack.

P = 410 kN slab collapsed.

Applying the above data to the collapse load formula, that is

$$P = 1.72 \times [(0.02 \times 100/2.6 \times 10^4) \times 10^4 + 3.6]\, 2.17 \cdot 160^2 \cdot 10^{-3}$$

$$= 417.5 \text{ kN}$$

A good correlation with test result.

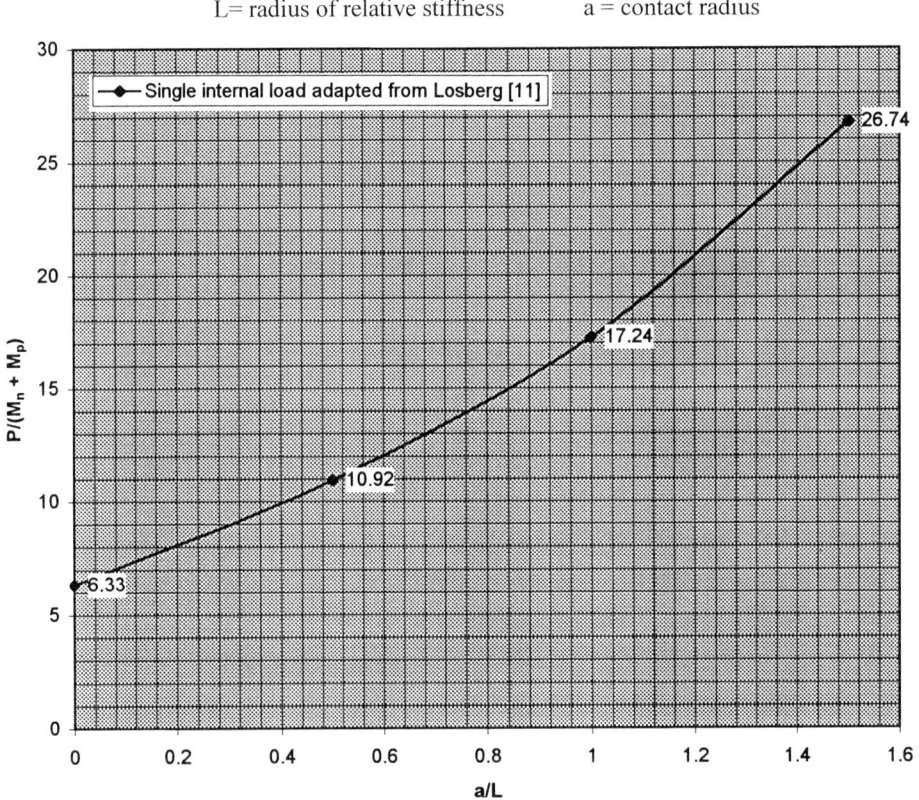

Figure 2 Single internal load: Plot of P/(Mn + Mp) against a/L (adapted from Losberg [11])

APPENDIX B – SUPPLEMENT TO UCL REPORT [5]

A comparison of four ultimate load design methods was made using the following data:

Slab depth h = 150 mm; contact radius a = 56.4 mm (100 x 100 mm base plate); modulus of subgrade reaction k = 0.05 N/mm^3; radius of relative stiffness L = 645 mm; a/L = 0.09; M_n = ultimate negative (hogging) moment; M_p = ultimate positive (sagging) moment; P = applied ultimate load (internal).

The results are given in Table 2:

Table 2 Comparison of ultimate load design methods for a single internal load

	$P/(M_n + M_p)$	COMMENTS
Meyerhof [2]	7.00	simplified formula
Rao & Singh [8]	8.01	radius of collapse zone is (0.6 + 2.3 a/L) L
Baumann & Weisgerber [12]	8.70	radius of collapse zone is 2L
Losberg [11]	7.00	refer to graph, Fig. (2)
Yield line theory [1]	6.28	fan type failure mode with a = 0

APPENDIX C

Table 3 Summary of comparative tests on 3.0 m x 3.0 m concrete ground slabs undertaken at the University of Greenwich

TEST NO. & SLAB DEPTH	SLAB REINFORCEMENT	LOAD AT FIRST OBSERVED CRACK kN	LOAD AT FAILURE kN	COMMENTS
1 C (150 mm)	No reinforcement	180	200	Cracks through full depth of slab and punching
2 C (150 mm)	FABRIC A142 (200 x 200 x 6) top-cover 25 mm	200	> 320	Maximum load attained, no cracks on top surface
3 C (150 mm)	2 LAYERS FABRIC A193 (200 x 200 x 7) 25 mm top & bottom cover	280	380	Cracks through full depth of slab and punching
4 C (150 mm)	HOOK ENDED (60/100) @ 30 kg/m^3	290	> 345	Maximum load attained, no cracks on top surface
5 C (150 mm)	HOOK ENDED (60/100) @ 20 kg/m^3	220	350	Cracks through full depth of slab and punching
6 C (150 mm)	HOOK ENDED (60/100) @ 20 kg/m^3	200	300	Cracks through full depth of slab and punching
7 C (150 mm)	HOOK ENDED (60/80) @ 20 kg/m^3	260	390	Cracks through full depth of slab and punching
8 C (150 mm)	HOOK ENDED (60/100) @ 30 kg/m^3	240	340	Cracks through full depth of slab and punching
9 C (150 mm)	HOOK ENDED (60/80) @ 30 kg/m^3	320	360	Punching, no cracks on top surface
10 C (150 mm)	POLYPROPYLENE @ 0.9 kg/m^3	170	230	Cracks through full depth of slab and punching

C denotes central loading,

Table 3 (cont'd) Summary of comparative tests on 3.0 m x 3.0 m concrete ground slabs undertaken at the University of Greenwich

TEST NO. & SLAB DEPTH	SLAB REINFORCEMENT	LOAD AT FIRST OBSERVED CRACK kN	LOAD AT FAILURE kN	COMMENTS
11 C (150 mm)	NO REINFORCEMENT	200	200	Cracks through full depth of slab and punching
12 C (150 mm)	UNDULATING @ 30 kg/m^3	260	370	Cracks through full depth of slab but no visible punching
13 C (150 mm)	UNDULATING @ 20 kg/m^3	220	260	Cracks through full depth of slab and punching
14 C (150 mm)	UNDULATING @ 30 kg/m^3	260	380	Cracks through full depth of slab and punching
15 C (150 mm)	HOOK ENDED (60/100) @ 20 kg/m^3	290	350	Slab failed in punching
16 C (150 mm)	STRAIGHT FIBRES (19 mm) @ 22.5 kg/m^3	220	240	Cracks through full depth of slab and punching
17 C (130 mm)	HOOK ENDED (60/100) @ 20 kg/m^3	240	320	Cracks through full depth of slab and punching
18 C (175 mm)	A142 MESH IN TOP (cover 35 mm)	250	> 470	Load cell removed at 470 kN
19 * E (150 mm)	HOOK ENDED (60/100) @ 30 kg/m^3	120	290	First circumferential cracks in top of slab at 210 kN

C denotes central loading,
E denotes edge loading, slab depth shown in brackets (eg 150 mm), loading plate 100 mm x 100 mm
* For test no. 19, the centre line of loading plate was 150 mm from slab edge.

DELAMINATIONS OF BONDED CONCRETE OVERLAYS

S J Sopko
Ryan-Biggs Associates
United States of America

ABSTRACT. This paper will discuss the causes, evaluation, and repair of delaminated concrete overlays, through four case studies. Bonded concrete overlays if not properly designed, detailed and constructed, will fail by horizontal cracking, which will develop in the interface between the bonded overlay and the substrate concrete, or in the top surface of the substrate concrete. The development of the horizontal cracking failure plane can be a result of improper surface preparation of the existing substrate, the improper application of the concrete overlay, or improper specification of methods and materials. In each of the case studies presented, the cause of the cracking failure was different, but yielded the same end result, a horizontal crack which resulted in a delaminated overlay.

Keywords: Bonded concrete overlays, Delaminations, Delamination repair, Overlay construction.

Stephen J Sopko is a professional engineer working as a consulting engineer since 1974. He is an associate in the firm of Ryan-Biggs Associates located in Troy, New York. He is responsible for the structural design of buildings and has completed numerous investigations, including the development of repair details, for concrete structures. Mr. Sopko received a bachelor's degree in civil engineering at Manhattan College in 1973 and a master's degree in civil engineering at Rensselaer Polytechnic Institute in 1975.

INTRODUCTION

Placing bonded concrete overlays is one of the most common ways of restoring deteriorated concrete slabs, leveling existing concrete slab surfaces or protecting existing concrete slabs from chemical and chloride attack. Topping slabs are meant to extend the life of the existing structure in lieu of removing and replacing it. In other instances, bonded concrete slabs may be used on precast concrete to provide a stiffer section through composite action. In all of these cases, the topping slab is meant to be bonded to the substrate concrete. However, bonded concrete overlays, if not properly designed, detailed and constructed, will not perform adequately. The adequate performance of such an overlay is highly dependent on the proper preparation of the existing concrete substrate, using the proper bonding grout, and the design of the concrete topping slab.

Prior to the application of a concrete bonded topping, the substrate concrete must first be prepared by removal of any top surface concrete which is deteriorated or not sound. The top surface of the slab must be adequately cleaned to remove all materials which would affect the bond between topping and existing concrete. Moisture content of the existing slab is a key element in the successful placing of the topping slab. If the moisture content is too low, there is a possibility moisture loss from the bonding material and overlay will result in a poor bond, cracking, and curling of the overlay, resulting in a delaminated topping. Similarly a too high moisture content will also affect the bond between the topping slab and substrate concrete, again resulting in a delaminated topping slab.

The design and construction of the topping slab is also a very important aspect of a successful bonded concrete overlay. The topping slab must be designed to minimize shrinkage which will result in cracks and curling of the topping, affecting the life of the overlay. Strength, durability and compatibility are also key aspects of the topping slab. The bonding grout and topping material must be compatible with the existing substrate concrete to minimize differential stress caused by differential shrinkage of the two materials. The strength of the topping material should be designed for both long-term and early loading. In many cases, high early strength is required when the repair area needs to be put back in use soon after the topping is placed. Another key aspect is durability. A topping is normally placed due to concrete deterioration. The topping slab must be designed to accommodate the durability requirement for chloride, chemical or freeze/thaw requirements.

CASE HISTORY ONE

The first case history presents the failure of a thin 25-mm bonded concrete topping placed on an existing lightweight concrete slab. Over 6,500 square meters of the topping was placed.

Description

Our investigation began in 1990, shortly after placement of a new concrete topping on an existing lightweight concrete slab. The thin topping was used to level the existing slab. The original eight-story building was constructed in the 1930s. The framing system consisted of steel beams and girders encased in concrete. A 100-mm lightweight concrete slab spans between the beams. The building was abandoned in the early 1980s and remained protected from weather but unheated. Renovation of the structure began in late 1989 with application of a concrete topping on the existing lightweight concrete floors. Soon after the topping installation, extensive cracking and delaminations occurred.

Testing Program

An extensive testing program was undertaken to determine the cause of the topping's failure. This program consisted of core samples, bond tests, a visual survey of the crack pattern, and a delamination survey. Cores were taken in both the delaminated and non-delaminated areas. Petrographic examination of the core samples was conducted. Again petrographic examinations were taken of both delaminated and non-delaminated cores.

Discussion

The results of the testing program indicated that the topping delamination and cracking were not related, but a result of two separate causes. From the petrographic analysis, it was learned that delaminations occurred at or just above the contact surface of the bonding primer and concrete substrate. At this location, the petrographic exam indicated a dusty laitance on the surface of the existing concrete slab in the delaminated areas. This dusty laitance was not present where the topping was still bonded.

Bond tests were then conducted on the topping adjacent to the core samples. From this testing it could be determined that adequate bond was present where the topping was not delaminated.

The petrographic analysis also revealed the topping cracking was unrelated to the delamination problems. From the testing it was determined the topping slab lost moisture to the substrate lightweight concrete slab. It was determined that the existing concrete slab was not adequately presaturated prior to the topping installation, resulting in a loss of moisture from the topping, causing drying shrinkage cracks in the topping.

Conclusions

From the testing program it was determined that the causes of the topping delaminations and cracking were unrelated, but both were a result of improper subsurface preparation. The existing lightweight concrete slabs were inadequately clean in the delaminated areas and were not properly saturated in the extensively cracked areas.

Adequate slab preparation is extremely important in the long-term performance of a clean topping.

Our investigation revealed that the concrete topping in the delaminated and cracked areas required removal. The 25-mm topping thickness was not adequate for long-term use to act as an unbonded topping. The results of the bond tests indicated that where the topping was still bonded it, could be left in place and would perform adequately for the longer term.

It was also determined that the drying shrinkage of the topping was completed, so only the areas of cracked topping had to be removed. It was concluded the topping had undergone a year of drying and if uncracked, the likelihood of future cracking was minimal.

CASE HISTORY TWO

The second case history presents the failure of 1000 square meters of a 50-mm 30-mpa concrete topping. The topping was placed over a reinforced structural concrete flat slab constructed in the 1960s. The topping was placed to increase slab drainage.

Description

Our investigation began in 1992, while placement of the concrete topping was approximately 75 percent complete. Cracking and delamination of the recently placed topping was occurring. The delaminations were widespread and were increasing. It appeared the delaminations first started at the topping construction joints and were growing backwards, affecting large sections of the topping. The contractor was preparing the existing slabs as required by the specification. The work included scabbing the top surface of the slab to roughen the surface for bond, and vacuum-cleaning the surface after scabbing. The slab delaminations were observed within 28 days after placement. A latex acrylic was used as the bonding grout.

Testing Program

An extensive testing program was undertaken to determine the cause of the delaminations. The testing program included core samples, petrographic examination of the samples, and delamination testing. The delamination testing program indicated the delaminations started at the topping construction joints and were increasing daily. Core samples were taken in the delaminated areas and bonded areas.

Discussion

When first discovered, the slab delaminations were tested, marked, and the delaminated concrete removed. Before replacing the topping in the repair area, the contractor performed additional delamination tests and discovered additional concrete had delaminated adjacent to the repair area. The contractor requested our firm investigate the cause of the problem. Since the topping had not yet been replaced in all areas, the existing slab preparation was visible.

Observations made of these areas indicated the slabs were adequately roughened and cleaned. In addition, the topping mix design and test reports showed an adequate mix with low slump was being placed by the contractor.

Cores were taken and sent for a petrographic analysis. In addition, the concrete topping in the areas of the delaminations was removed, to determine where the failure plane was occurring. Results of slab removals indicated the failure plane was in general in the top surface of the existing concrete slab. Sections of existing slab were still intact with the topping. The petrographic analysis indicated the top surface of the existing concrete slab was overworked during construction and was carbonized to a 15-mm depth. This cracked a weak plane in the top surface of the concrete slab. Preparation of the existing slab roughened the top 5 mm, thus not removing weak concrete. Drying shrinkage of the topping exceeded the tensile strength of this weakening plane, resulting in delamination in the top surface of the concrete.

Conclusions

Based on core samples it was determined the entire slab surface was affected to a 15-mm depth with weakened concrete. Due to the industrial use of the slab area undergoing renovation, it was decided to remove the entire topping already placed.

The existing top concrete surface was then scarified to a 20-mm depth to remove the weak top surface. The surface was then vacuum-cleaned and the topping replaced.

CASE HISTORY THREE

The third case history presents the failure of 5,000 square meters of 75-mm bonded lightweight concrete topping placed over an existing concrete slab.

Description

Our investigation began in 1996 when cracking began in the lightweight concrete topping placed over existing concrete slabs in a five-story building originally constructed in the 1930s. The topping was placed in areas of cinder concrete used below floor tiles that were being removed. The topping was reinforced with welded wire fabric. An acrylic-polymer latex was used as the bonding agent. The topping was placed in the summer and fall of 1995.

The building was not heated during the winter. The topping cracking was first noticed in January, 1996, soon after our investigation began. In addition to cracking, large areas of topping were also delaminated.

Testing Program

A testing program of core samples, petrographic analysis of the concrete topping and bonding grout, bond tests, and delamination tests were undertaken.

Discussion

In many cases, the core samples were recovered, showing the delamination occurred at the bottom of the topping at the bond line.

This was confirmed by the petrographic analysis. It appears the bonding compound had set up prior to the placement of the topping, thus creating a failure plane. It was also determined that the concrete topping had been placed with a high 125-mm slump and had gone through the winter unheated.

Bond tests were performed in several areas where the topping remained bonded. These tests indicated a variable bond strength in these areas, with some remaining questionable.

Conclusions

From the testing and analysis of the topping data, it was determined that both application of the bonding grout and the topping mix design were causes of the problems. Cracking was a result of concrete shrinkage due to the high slump and unheated conditions.

In addition to high slump a pea aggregate was used, increasing the shrinkage potential of the topping concrete. Topping delaminations were a result of inadequate workmanship in placing the bonding grout.

Due to the use of the building (the floor areas were to be carpeted and used as apartment), it was decided to gravity-feed epoxy into the cracked areas and leave the topping in place in all areas. It was determined the 75-mm reinforced topping was adequate as an unbonded topping and did not have to be replaced. The cracks were epoxy-filled with sand after the building was heated, to minimize any future cracking.

CASE HISTORY FOUR

The fourth case history presents repairs to a concrete topping used in repair of an underground parking garage.

Description

Our investigation began in 1995 during structural repair to an underground reinforced concrete parking garage. A concrete topping varying from 50 to 100 mm was being placed to restore a four-level, 1,500-square-meter parking garage. Original reinforced concrete flat slabs had undergone extensive deterioration and were in need of repair.

An epoxy bonding agent was used to bond the repair layer to the existing slab. After placement of the repair layer, several large areas of delamination were found. Core samples were taken to determine the cause of delamination. From these samples it was determined the bonding grout had set before topping placement.

Conclusions

The cause of the topping delaminations was inadequate placement of the bonding grout and topping. Due to the tight construction schedule, requiring the space be reopened in 45 days, removal and replacement of the delaminated areas could not be achieved, as the topping required 28 days of curing before a protection membrane was placed over the slab.

It was decided to epoxy the horizontal delamination to rebond the topping slab to the concrete substrate. This was accomplished by drilling holes at 150-mm spacing and low-pressure-injecting epoxy into the horizontal crack. Delamination tests and cores were taken to confirm the topping was bonded to the original concrete.

OVERALL CONCLUSIONS

The main causes of concrete overlay delamination and failures are poor substrate preparation and inadequate application of the bonding material. Most failures occur at the bond line between the top surface of the existing slab and the bottom surface of the overlay. All contaminants, including dust, efflorescence, laitance, oil, grease, and any other material present on the top surface of the concrete must be removed to adequately prepare the slab top surface. Usually a three-step method must be employed. First, preclean the surface to remove all stains, grease, oil, etc. Second, the top surface of the existing slab must be prepared; this includes roughening the surface to provide a better bond and removing all unsound concrete. Removal of unsound concrete is of extreme importance. In many cases the top surface of an existing slab may be in poor condition due to improper curing and finishing operations at the time of construction, or concrete deterioration from use. If this unsound concrete is not removed, failure will likely occur, when tensile stresses developed during the curing of the topping exceed the bond strength.

The third and final step is a final cleaning. This is to remove any dirt and dust resulting from the surface preparation.

Another major concern is the moisture content of the existing substrate slab. The moisture content must be adequate to insure the bonding material will adhere the topping to the substrate concrete. A too-high moisture content will affect the bonding of the topping as vapor pressures will exceed the bond strength, resulting in delaminations. If the slab is not properly saturated, rapid moisture loss from the topping slab will likely result in cracking.
Once the topping slab has been placed, delamination pull-off tests can be performed to insure adequate bond and long-term performance. If delamination and bond tests indicate questionable bond, removal and replacement or other corrective action may be required.

The bonding material is another important aspect of good performance of a concrete overlay. A hard-to-use material will create problems. Epoxy and acrylic bonding materials require workers who are familiar with the products.

These materials will perform more than adequately if proper precautions are undertaken. If the topping is not placed in the proper time frame, the topping will not adhere to the bonding agent, as the bonding grout may dry. The material that performs the best will be compatible both with the existing concrete substrate and the concrete overlay. In many cases a sand cement slurry is the best material to use, as it is compatible with the original concrete substrate and is easy to work and mix. The existing concrete surface should be saturated, with no standing water present. In dry conditions it may be warranted to keep the slab wet for a period of time before placing the topping slab to prevent rapid moisture loss and cracking in the overlay.

The bonding grout should be scrubbed into existing concrete substrate, and topping placed before it drys. Once the topping is placed it should be cured, preferably water-cured for several days to minimize shrinkage. The topping should be designed and reinforced to minimize shrinkage. The largest possible aggregate size should be used for the topping thickness. Construction joints should be placed in the topping over construction joints in the original framing. The concrete topping mix should be designed with durability for the expected use, including freeze/thaw cycles and chemical attack.

RECOMMENDATIONS

To achieve a well-bonded, good performing bonded concrete topping, the following recommendations should be considered.

1. Analyze the existing structure to determine the cause of the concrete distress and deterioration.

2. Take core samples and perform petrographic analysis to determine the quality of concrete at the top surface of the existing structure.

3. Preclean the slab top surface, removing all dirt, debris, oil, grease, etc.

4. Remove all unsound concrete and poor quality concrete from the top surface which will affect bonding the new topping.

5. Roughen the top surface of the existing concrete.

6. Final-clean the top surface of the concrete, removing all dirt and dust accumulation.

7. Pre-saturate the existing slab to a wet condition with no standing water.

8. Apply an appropriate bond coat.

9. Apply the concrete topping at lowest slump possible. The topping should be designed for:

    a. Early strength where applicable as well as long-term service;

    b. Durability for freeze/thaw and chemical attack;

    c. low shrinkage to minimize differential drying stresses.

10. Wet-cure the concrete topping if possible for a minimum of 7 days.

USE OF CONSTRUCTION REMAINDER SOIL IN PAVEMENT CONSTRUCTION

M Ohno
Nagoya Municipal Industrial Research Institute

K Fukai
UNITEX Co Ltd
Japan

ABSTRACT. Pavement construction works on road surfaces such as sidewalks and factories' sites were executed, for the purpose of recycling construction remainder soil produced by construction works etc. Used materials are construction remainder soil, sand, gravel, cement and solidification agent. The most suitable mixing conditions in which more than $10N/mm^2$ of uniaxial compressive strength (7days) were gained were selected from the results of spare test. No observed cracks on road surfaces were recognized, namely the satisfactory results were obtained on executing construction works.

Keywords: Construction remainder soil, Solidification agent, Ordinary portland cement, Soil cement concrete, Road surface pavement.

Mr M Ohno is director of the concrete technology. Nagoya Municipal Industrial Research Institute carries out the technological supports of research developments for minor enterprises on relative technology of mechanical, metallurgical, chemical and electronic industries.

Mr K Fukai is develops solidification agents for civil engineering using original company technology and manages the developed manufactured goods. Development of relative technology using soil cement concrete is systematically pushed forward.

INTRODUCTION

Construction remainder soil is soil and mud generated as by-products in correlation with civil engineering works and such constructions. The total annual amount generated from construction sites nations bank reaches approximately 437 million cubic meters (fiscal 1998).

On the other hand, the recycling rate of construction remainder soil from public civil engineering projects, reclamation works and such (hereafter, remainder soil) totals no more than about 30%. The amount of generated soil increases annually in correlation with the city redevelopment and the effective use of underground space.

However, it is difficult to secure the disposal sites for the remainder soil that isn't recycled, for the reason that appropriate sites decrease due to advancing urbanization, reducing sea reclamation projects and other factors. Consequently, the transport distance of remainder soil is far away and then the illegal disposal of remainder soil increases. These circumstances obstruct the smooth progression of construction works.

In consideration of these conditions, the ministry of construction has issued the "Construction Remainder Soil Use Technology Manual" in 1994 which pursued improved recycling rates of remainder soil.

The principal current soil stabilization methods conducted to recycle remainder soil are as follows. First method is the "physical and mechanical stabilization method" conducted through the replacing of inadequate soil, uniformalizing of grain size, and moisture content, to improve rolling compaction and other physical property values. Second method is the "chemical stabilization method" which adds, mixes or injects cement , lime ,solidification agents and such. The main objective of these stabilization methods is to heighten "soil compaction" performance by creating appropriate soil for use through the improvement of remainder soil quality.

This research seeks directly "solidified soil" using a newly developed "special civil engineering solidification agents (here after, solidification agents)". A highly reinforced soil cement concrete is developed in line with this objective and road surface tests are executed. A report concerning the compaction properties of stabilized soil using solidification agents will also be provided.

SUMMARY OF EXPERIMENTS

Materials Used

The materials used consisted were of ordinary portland cement and remainder soil, sand, ballast and gravel were used as the aggregate. The solidification agent is an inorganic compound and is a reddish purple liquid (product name :Unimaster). The solidification agent was weighed up to the necessary quantity, then mixed into water and finally agitated. The generated soil consisted of soil samples collected from construction work sites and generated soil from construction sites. The soil was used in the state of its natural moisture content.

Indoor Experimental Method for Soil Cement Concrete

The generated soil is classified in the following manner according to the type of soil and conditions of generation.

Sandy soil : Soil with a large sandy (gravelly) quality like pit sand
(some fine grain content).

Sandy soil + silt : Sandy(gravely) soil with silt which has a fine grain content of 25 - 30% or more silt is highly viscous.

Silt : A viscous soil with high fine grain content. For example, diluvial viscous soil like Kanto loam, excavated soil from a alluvial viscous soil sub soil and dredged soil. It is viscous soil that highly contains fine grain soil.

When the generated soil contains wood pieces, metallic pieces, concrete clods, coarse gravel and other impurities that are large enough to obstruct; such must be removed using a screen or other tool as a preparatory process. Further, the collected soil is used at its natural moisture content.

Tables 1 - 4 indicate mixing conditions and strength test results for the soil cement concrete used in the experiments. A forced mixing type mixer is used to mix the soil cement concrete. The size of the sample was $\Phi 5 \times 10 cm$, the sample is stripped after being aged for one day and then cured indoors until the test date. The strength test is measured in accordance with the "Soil Unconfined Compression Test Method (JIS.T.511 - 1990)" standard of the Japanese Society of Soil Mechanics and Foundation Engineering. An Amsler type 5ton universal capacity tester is used for the measurements.

COMPACTION PROPERTIES OF STABILIZED SOIL USING SOLIDIFICATION AGENTS

Road Paving Method

In regard to sandy soil compaction tests were conducted on soil that was stabilized by a cement-solidification agent. The testing method observed JIS.A. 1210 "Soil Compaction Testing Method using Tamping." The "E" tamping method, "b" sample preparation method, drying method and non-repeating method were executed. Table 5 indicates the stabilization mixing conditions, Table 6 shows the strength testing results. From the consideration of the unconfined compression test results, dry density moisture content curve and other information introduced the confirmation that the compaction properties of soil stabilized with cement solidification agents were improved.

METHODS FOR PAVING ROAD SURFACES WITH SOIL CEMENT CONCRETE

A construction test was conducted in which soil cement concrete was used for paving the top layer of a road surface. The objectives of the experiment were to determine construction engineering problems and to establish the construction system.

Table 1 Mixing Condition Specification for Sandy Soil

No	W/C (%)	S/C	Cement (kg)	Water (kg)	Sandy-Soil (kg)	Solidification Agent (kg)	Compressive Strength(7d) (N/mm²)
1	175	6.2	200	350	1.240	0.0	5.2
2	175	6.2	200	345	1.240	4.0	5.5
3	173	6.2	201	341	1.243	6.0	5.7
4	172	6.2	201	337	1.247	8.0	5.8
5	165	6.2	204	326	1.264	10.2	6.1
6	164	6.2	203	324	1.265	16.2	5.7
7	162	6.2	203	319	1.257	20.3	7.4
8	160	5.0	234	375	1.162	0.0	5.2
9	154	5.0	238	361	1.179	4.8	6.1
10	151	5.0	239	354	1.187	7.2	6.8
11	144	5.0	243	341	1.207	9.7	7.8
12	135	5.0	249	324	1.234	12.4	8.1
13	128	5.0	253	304	1.256	20.3	9.7
14	120	5.0	259	284	1.282	25.9	10.1

Table 2 Mixing Condition Specification for Silt

No	W/C (%)	S/C	Cement (kg)	Water (kg)	Silt (kg)	Sand (kg)	Compressive Strength(7d) (N/mm²)
1	251	1.5	300	753	450	0.0	0.4
2	194	1.5	361	699	434	108	1.1
3	160	1.5	409	655	368	245	2.1
4	127	1.5	472	598	319	389	4.1
5	108	1.5	517	556	233	543	6.1
6	165	1.0	434	716	434	0.0	1.1
7	130	1.0	510	663	408	102	3.0
8	118	1.0	541	639	325	216	5.0
9	92	1.0	602	554	298	364	7.8
10	79	1.0	679	539	204	476	9.5

Table 3 Mixing Condition Specification for Sandy soil + Silt

No	W/C (%)	S/C	MEASURED VALUE (kg/m³)						Compressive Strength(28d) (N/mm²)
			Cement (kg)	Water (kg)	Solidification Agent (kg)	Sandy-Soil Silt (kg)	Sand (kg)	Balas (kg)	
1	66	4.8	340	187	38.0	1648	-	-	21.4
2	85	8.1	218	153	31.6	865	728	182	17.9
3	85	8.1	218	153	31.6	865	182	728	22.6

Table 4 Mixing Condition Specification for Sandy Soil

No	W/C (%)	S/C	MEASURED VALUE (kg/m³)						Compressive Strength(28d) (N/mm²)
			Cement (kg)	Water (kg)	Solidification Agent (kg)	Sandy-Soil (kg)	Sand (kg)	Balas (kg)	
A	74.5	8.3	213	135	23.6	1067	355	356	27.9

Table 5 Mixing Condition Specifications for Stabilized Soil

W/C (%)	MEASURED VALUE (kg/m³)			
	Cement (kg)	Water (kg)	Solidification Agent (kg)	Sandy-Soil (kg)
74.5	213	135	23.6	1067

Table 6 Unconfined Compression Test Results for Stabilized Soil

AGE AT TEST	COMPRESSIVE STRENGTH (N/mm²)
3 days	3.3
7 days	5.0
28 days	6.1

The construction site was the unpaved section within the yard of A company which was located in Aichi Prefecture. The construction procedure involved the mechanical excavation of soil at the construction site, screening of generated soil to remove impurities and then the use of the soil as a paving material. The site soil used was the soil with a silt-clayey content of around 60% and a sandy content of around 40%. Table 7 indicates the mixing conditions for the soil cement concrete that was used.

The material were mixed according to the above mixing conditions. The slump of the mixed materials was thirteen to fourteen centimeters. The thickness of the constructed road surface ranged from ten to fifteen centimeters in accordance with the locations to be laid. The total laid area of the soil cement concrete was 320 m². The finished construction surface has shown no generation of cracks or such and it was confirmed that the good condition has been maintained.

Table 7 Mixing Specifications for Soil Concrete

W/C (%)	MEASURED VALUE (kg/m³)			
	Cement (kg)	Water (kg)	Solidification Agent (kg)	Generated Soil (kg)
178	120	200	13.0	1722

CONCLUSIONS

At this time when the recycling of construction generated soil is demanded, effective use of generated soil as a road base course, low cost pavement, top layer pavement or other material would have significant meaning. In the scope of these experiments, the authors are able to confirm that the strength of soil cement concrete is equivalent to the traditional asphalt materials as physical properties. The authors will continue to conduct construction and other experiments and strive to accumulate physical property values and other data.

REFERENCES

1. HONDA and YAMADA. Recycling of Construction By-products and Waste, Energy Conservation Center, 1994.

2. KAWAMURA ET AL. Soil Cement Concrete Mixing and Strength, Summaries of Academic Lectures of the Architectural Institute of Japan's Annual Convention, 1993, pp.1231-1232.

3. NAGAISHI ET AL. Research of Farm Roads on Slopes (NO.2), Ministry of Agriculture, Forestry and Fisheries Shikoku Farming Test Center Report, 1996, pp.29-34.

4. MACHIDA ET AL. Sludge Stabilization Technology for Environmental Improvement, Cement. Concrete Journal, 1989, No.511, Sept, pp.94-103.

5. KAZUHIRO ISITANI. Special Cement for Soil Mud and Sludge, Cement Concrete Journal, 1991, No.535, Sept, pp.94-103.

TECHNIQUES FOR THE EARLY-LIFE IN-SITU MONITORING OF CONCRETE INDUSTRIAL GROUND FLOOR SLABS

S A Austin

P J Robins

J W Bishop

Loughborough University

United Kingdom

ABSTRACT. This paper describes the instrumentation of a concrete industrial ground floor slab, to further the understanding of its early life behaviour. The slab internal strains were monitored using vibrating wire embedment gauges, which were carefully positioned, both at joints and within bays, to ensure data was obtained which could be used in the development of a mathematical model of the early age behaviour. Other instruments were used to monitor the ambient temperature, relative humidity and wind speed.

Keywords: Ground floor slabs, Early-life, Site instrumentation, In-situ strain measurement, Drying shrinkage.

Dr Simon A Austin and **Dr Peter J Robins** are Senior Lecturers in Structural Engineering in the Department of Civil and Building Engineering, Loughborough University.

Jonathan W Bishop is a Research Assistant in the Department of Civil and Building Engineering, Loughborough University.

INTRODUCTION

There have been major advances in the field of concrete industrial ground floors in the last decade, with the adoption of machine laying techniques for large area pours and associated improvements in standards. However, modern applications are placing increased demands on floor performance in terms of flatness, durability and abrasion resistance. As the boundaries of slab performance are being extended, the design guidance available to engineers is becoming out-dated because, for the most part, it is based on empirical rules, rather than on any detailed understanding of the processes involved in the early life of the slab. This lack of understanding of the slab's behaviour, coupled in some cases with an inappropriate application of the empirical rules, can lead to inappropriate designs - some conservative (in terms of reinforcement and/or thickness), some inadequate, resulting in cracking and failure.

This paper shows research aimed at advancing the knowledge of how concrete floors behave through in-situ monitoring during and after construction. The method of installing the gauges is presented, along with data on early life movements and temperature changes in the slab.

OVERVIEW OF THE RESEARCH PROJECT

This research is being undertaken as part of a three year project funded by the Engineering and Physical Sciences Research Council (EPSRC) under the Materials for Better Construction Programme: Phase 2. The aim is to advance the understanding of how industrial concrete ground floors behave, through in-situ monitoring during and after construction. In particular it will improve our understanding of how the early age development of the concrete's material properties can be made to interact with the joint arrangement and the climate and curing to form a lightly stressed concrete that can efficiently (i.e. with reduced slab thickness and less reinforcement) carry the subsequent stresses due to the imposed load. The in-situ monitoring of the slabs and their environment will provide information on the timing and magnitude of early movements and the associated climatic conditions. This data will assist us in the development and assessment of a computer model of the early life behaviour of industrial ground floor slabs. The instrumentation will also provide data on the nature of the loading induced stresses and strains as the slab enters service.

For industrial ground floor slabs, design is broken down into two parts, namely: the structural design, which is the process of ensuring the slab can carry the imposed loads; and the detail design, which considers the stresses in the slab during it's early life and tries to prevent cracking. The latter includes drying shrinkage cracking which can also affect the durability and load carrying capacity of the slab. The effectiveness of the detail design can, therefore, have a bearing on the structural design, as residual stresses in the slab will reduce its load carrying capacity in-service and could cause failure. Detail design is further affected by the client's perception of what constitutes a failed slab, a common attitude being that a slab which has cracked has failed, even when the cracks could be classed as cosmetic and would not reduce serviceability or durability.

The main guidance available in the UK for both detail and structural design is in the Concrete Society document Technical Report 34 [1]. It divides industrial floor slabs into two main categories: those cast in strips and those cast with large area pours.

The relative percentages of the different types of floor now being constructed in the UK are shown in Table 1. Types 3 and 4 are expanding markets with 2 and 5 dropping. Long strip construction has seen large decreases in recent years, but is now showing only very slow decline as most high bay racking warehouses require this type of floor in order to meet flatness requirements.

Table 1 Relative proportions of current types of floor construction in the UK [ACIFC]

1	Long strip	10%
2	Nominally reinforced large area pour (sawn joints - A142 mesh)	50%
3	Nominally reinforced large area pour (sawn joints - 10-20kg/m^3 fibres)	20%
4	Fibre reinforced jointless large area pour (35-40kg/m^3 fibres)	5%
5	Heavily reinforced large area pour (sawn joints - A252/A393 mesh)	10%
6	Slabs on piles (specialised and rare at the moment but increasing)	5%

SELECTION OF INSTRUMENTATION

A number of key features have been identified as being of relevance to the study of the early age behaviour of concrete ground floor slabs. These fall into three interdependent categories:

1. Information on the movement of the concrete is of prime interest, particularly that associated with the shrinkage and creep of the slab. For long strip and jointed large area pour construction this will involve instrumenting both the joints and the concrete within bays, whilst jointless large area pours will need additional instrumentation monitoring the slab movement relative to the sub-base.

2. Movements in the concrete are dependent on the concrete's internal relative humidity and temperature, which need to be determined by internal monitoring.

3. The internal state is dependent on the atmospheric environment which controls the rate of moisture loss from the slab. Important factors here are the air speed, ambient temperature and relative humidity.

By choosing appropriate sensors these parameters can be monitored automatically using a computer controlled data logger, which allows the sensors to be sampled at different frequencies to suit their rate of change and the age of the slab. Site access is then only required to download the stored data onto a laptop computer.

No details of previous in-situ instrumentation of concrete ground floor slabs could be found in the literature, although Kim and Lee [2] have carried out laboratory tests on the differential drying shrinkage of concrete slabs using embedment strain gauges. It was therefore necessary to carry out laboratory investigations to identify possible problems with using embedment strain gauges and to establish how early in the life of the slab reliable data could be obtained from the sensors.

Three types of vibrating wire embedment strain gauge were tested for sensitivity, robustness and reliability by casting them in an 800x500x180 mm mould and subjecting it to forced drying. The resulting plot of strain against time is shown in Figure 1.

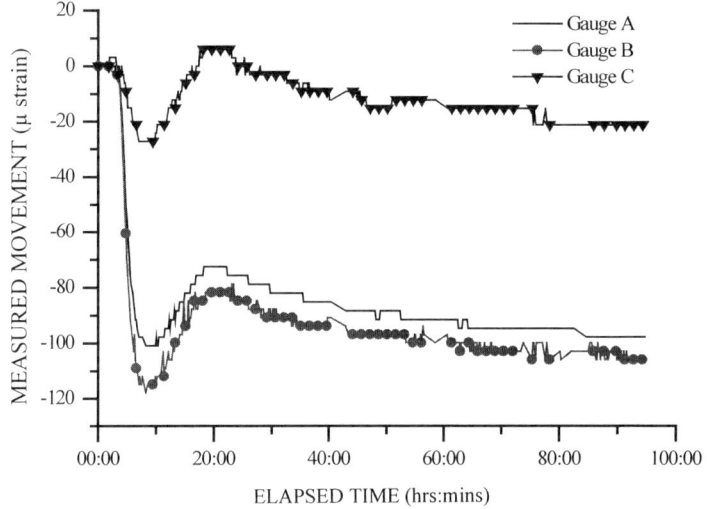

Figure 1   Shrinkage strains in the laboratory slab

The laboratory testing showed that all three gauge types proved robust enough to survive the casting process undamaged. The sturdier construction of Gauge C made it the least sensitive of the three gauges. Such a gauge is better suited to applications on site, where the gauges will be more exposed during the casting process, though the early readings may not be as accurate, as discussed below. Lachemi et al [3] report on long term testing of a suspension bridge using vibrating wire strain gauges. Readings were taken every two hours, but none were made until 24 hours after the concrete had been cast, because it was felt only after this period would the readings be useful. Our laboratory trials produced reliable and repeatable readings from all of the gauges after only two hours, though these early strain readings must be interpreted with care as at these early ages the relative stiffness of the concrete with respect to the gauge is low and, therefore, the effect of the gauge on the measured strains could be large. The effective elastic modulus for the gauges, determined by measuring their stiffness and dividing this by the area of the flange plates, was around 50 N/mm^2 for types A and B and 1kN/mm^2 for type C. The much greater stiffness of type C explains why it is the least sensitive during the early age of the concrete.

SITE INSTRUMENTATION

Close co-operation was required between the consultant, the contractor and the research team in order to co-ordinate the site work. Once a suitable project had been selected the client was approached for permission to instrument the slab; possible locations had to be identified for the strain gauges once the joint locations had been fixed, with nearby fixing points for the data logger and the ambient sensors. The construction sequence was obtained from the contractor to determine which of the possible locations fitted with the construction schedule. The bay chosen was in an infill strip which necessitated extra planning to provide a protective duct for the cables which had to pass under the adjacent strip on their way to the data logger. The intention had been to place the gauges in the middle of the strip, but after consultation with the contractor, it was decided that they would be better placed nearer one edge of the strip (Figure 2) in order to minimise the disruption to the pouring schedule, and to minimise the risk of damage to the strain gauges.

Figure 2   Casting strain gauges into a long strip slab

The laboratory trials allowed the selection of instrumentation based on performance, whilst also giving experience in programming the data logger and in the methods required to connect all of the instruments. The laboratory trials also showed how much wiring would be required to install all the sensors in the slab and as a result an interface box was built, which took all of the leads from the strain gauges and linked them to 25 pin RS232 connectors, to simplify the assembly process on site. All that was then required on site was to plug in these connectors, rather than to fix over 100 wires in very cramped and dusty conditions. This removed the possibility of wiring mistakes being made and simplified the procedure for revisiting the site once the normal monitoring period had ended.

Site instrumentation was carried out with all three types of gauge to allow further testing under site conditions, and to provide a degree of insurance against failure of one type of gauge. The gauges were located as shown in Figure 3, enabling the monitoring of both concrete strains and the movement at a contraction and a tied joint. Six strain gauges were located at section B-B, with one of each of the three types fixed 70mm and another 180mm from the bottom of the slab. All the gauges for measuring concrete strains were mounted on steel wire tensioned between reinforcement chairs, which gave good directional stability whilst minimising the interference to the gauge and the disruption of the concrete, Figure 4.

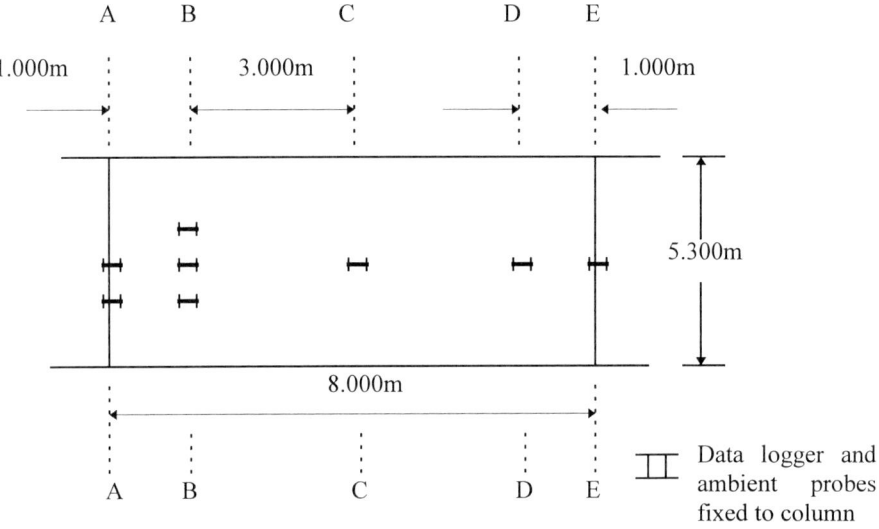

Figure 3 Plan view of the gauge locations in the slab

Section C-C had three gauges of type B in the direction of the slab at 60, 120 and 180mm from the slab bottom. One further gauge was mounted 120mm from the slab bottom perpendicular to the direction of the strip in order to measure any lateral shrinkage. Section D-D was monitored with 2 gauges of type B, which were fixed at the same depths as at B-B. This vertical distribution of gauges was intended to establish a vertical shrinkage profile through the slab as a result of differential drying shrinkage. The lateral distribution of gauges at section B-B also allowed a comparison of shrinkage at different distances from the slab edge, giving an indication of the effect of the edge restraint.

Two type C gauges were mounted across the saw-cut tied joint at section E-E. As little movement was expected at this point these were 140mm (5.5") long, the same length as the concrete movement gauges. The contraction joint at A-A required custom 254mm (10") gauges to be made in order to measure the larger movements expected across this joint, which could be in the order of several mm. These were of the same construction as type C in order that they were robust enough to survive any lateral or vertical movement which might occur at the joint. These gauges were also fixed 70mm and 130mm from the slab bottom as at E-E, however, due to the length of the gauge it was not possible to fix them between reinforcement chairs and they were wired directly to the reinforcement chairs instead (rain gauge fixing technique   Figure 5).

All the gauges were monitored at 10 minute intervals (except during saw cutting when the gauges at the joints and immediately adjacent were monitored at 1 minute intervals) until the slab was 2 days old, when the measurement interval was changed to 30 minutes and then 1 week after the slab had been cast this interval was further increased to 4 hours.

The increasing intervals reflect the decreasing rate of shrinkage expected as the slab ages. As the data logger only has a finite storage space, increasing the frequency of readings allows a greater period between site visits to download data.

Figure 4   Strain gauge fixing technique   Figure 5   Fixing method for 254mm gauges

The concrete internal conditions were monitored by thermistors inside the strain gauges. These were primarily for temperature correction of strain readings, but supplemented by thermocouples they allowed an internal temperature profile to be determined. The concrete temperature readings were taken every time the strains were recorded to allow temperature correction of strains and a comparison between the thermocouple and thermistor measured values. The concrete internal relative humidity was not monitored due to problems interfacing the sensors with the data logger. However, these problems have now been overcome in the laboratory and several chilled mirror probes will be installed in the next site instrumentation. Ambient conditions were monitored hourly by an ambient temperature and relative humidity probe fixed to a column adjacent to the instrumented strip. Two anemometers were also located here to give data on the air speed over the slab (two instruments being installed to overcome the wind shadowing effects of the column).

SLAB MEASUREMENTS AND INTERPRETATION

Concrete Movement

The strain gauges in the slab were successful in monitoring the internal movements. As in the laboratory trial, reliable readings were obtained from all of the gauges after about 2 hours. The plot of movement against time for the strain gauges at section C-C can be seen in Figure 6. The gauges show an early shrinkage followed by a period of expansion and then further shrinkage.

This behaviour is consistent with the various phases of concrete early age behaviour described by Brüll & Komlos [4] and Kasai *et. al.* [5]. In laboratory experiments they found the early age behaviour could be broken down into four phases as follows:

First phase - normally lasts from one to six hours, water forms on the surface of the concrete, no appreciable volume changes occur.

Second phase - primary (plastic) shrinkage, lasts up to 5 hours, the shrinkage is proportional to the cement content whilst the relationship to the water/cement ratio is uncertain.

Third phase - begins once the maximum shrinkage has been achieved, usually lasts for 10 - 12 hours, volume changes are very small (possible slight swelling) as continuing shrinkage is offset by expansion of the cement paste due to hydration.

Fourth phase - secondary (drying) shrinkage, begins after the concrete has set, continues indefinitely, the rate of shrinkage is approximately 20 times slower then the primary shrinkage, and the magnitude is directly proportional to the water/cement ratio.

The maximum third phase expansion recorded in their experiments was about 30% of the second phase shrinkage, which also agrees with the laboratory test carried out at Loughborough, however, the relative magnitude of the third phase expansion measured on site is greater than expected. This may be due to the gauges not recording all of the plastic shrinkage, which occurred in the second phase, as the relative stiffness of the gauges to the concrete was still too high.

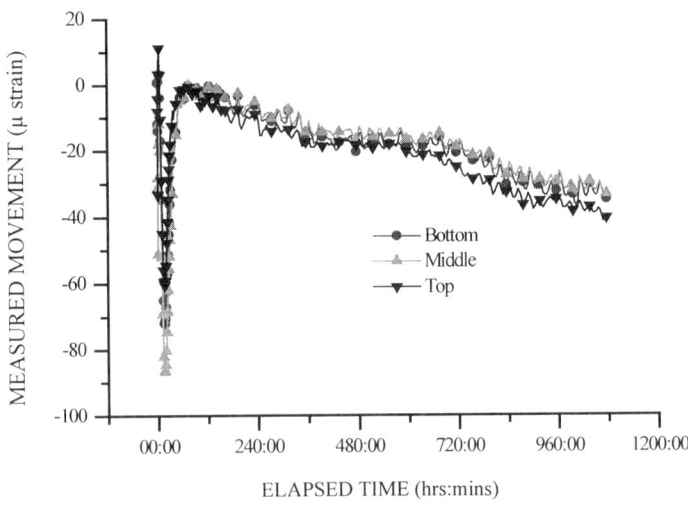

Figure 6   Influence of the strain gauge depth on the measured strain

Similar levels of drying shrinkage were recorded by the bottom and middle gauges, with the top gauge measuring higher levels of shrinkage. The diffusion of moisture through concrete, which has been identified [6] as the main process governing the drying shrinkage of concrete, is a slow one-dimensional process progressing downwards from the surface of the slab. The top gauges are near the surface and will, therefore, be exposed to drying shrinkage sooner than the lower gauges, thus producing a greater amount of drying shrinkage at any given time t.

All of the gauges showed a variation in strain due to temperature effects, which can be seen in the diurnal variation in Figure 8. After 72 hours the temperature in the slab had stabilised and the daily variation was small ~1°C. Applying the temperature corrections of ~10 µstrain, which account for the different thermal coefficient of expansion for the strain gauge wire and the concrete, does not reduce the amplitude of the daily strain variations, thus proving that they are real changes in strain and not just effects of the different thermal coefficients of expansion of the concrete and the gauge components. The effects of the slab thickness can also be seen, whereby the gauge in the bottom of the slab takes longer to respond to the changes in temperature than does the gauge in the top of the slab.

Figure shows the recorded movement at the contraction joint. Although the slab was saw cut when it was 24 hours old, no movement of the contraction joint occurred until it was 37 hours old. This movement coincided with a drop in the slab internal temperature, which followed the ambient temperature change at point T (34 hours) in Figure 8a.

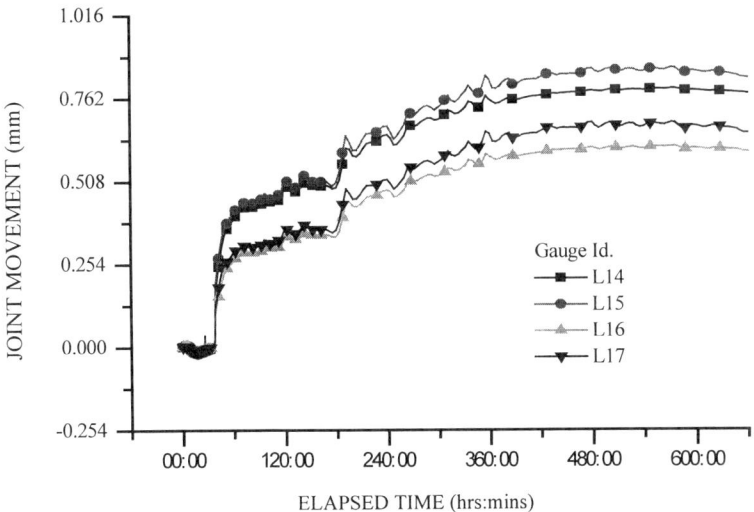

Figure 7   Contraction joint movement

This temperature change will have caused tensile thermal stresses in addition to the drying stresses already present in the slab. The sum of these stresses must have exceeded the tensile strength of the slab causing the joint to fail and open.

Once this initial movement had occurred differences between the gauge readings can be seen. Two trends are apparent: gauges L14 and L15, which were nearer the centre of the slab, recorded the most movement, and the gauges near the top of the slab, L15 and L17, moved more than the gauges in the bottom of the slab. The first trend is indicative of edge restraint caused by the dowel bars linking the strip to it's neighbour, which had been cast several days earlier. The second trend is believed to be caused by the differential drying of the slab and the frictional restraint acting on the bottom surface, but this cannot yet be confirmed as no data on moisture movement was obtained for this slab.

The maximum tensile force per unit width of slab, F_t, set up in a slab due to frictional restraint of shrinkage according to Losberg [7] and TR34 [1] is:

$$F_t = \mu \Delta t \frac{L}{2} \qquad (1)$$

where μ is a dimensionless friction coefficient,
Δ is the unit weight of the concrete (kN/m^3),
L is the contraction joint spacing (m),
t is the slab thickness (m).

This method assumes that the only restraint to shrinkage comes from the subgrade friction, which results in a triangular stress distribution between joints. The friction coefficient given in TR34 for a polythene sheet on a sand blinded sub-base is 1.5. Experimental work by Timms [8] on the effectiveness of friction reducing materials found that the friction coefficient for first movement on a polyethylene slip membrane was about 0.9 and 0.5 for subsequent movement in the worst case, although there was also a dependence on the thickness of the slab, with thicker slabs having a reduced value of the friction coefficient. The friction coefficients given in TR34 for design contain a factor of safety of about 2, which is intended to account for variation in the surface of the sub-base.

The formula of equation (1) does not take account of any additional restraint provided by joints nor any restraint provided by adjacent strips. As has already been stated, the results shown in Figure 7 suggest that restraint does occur at the joints and that there is a reduction in strain along the edge of the strip caused by the dowel bars tying the strip to its neighbour.

Concrete State

The internal temperature was recorded by 18 thermistors and 4 thermocouples. After casting the temperature can be seen to rise, with a peak value of 24.7°C reached after 14 hours when the ambient temperature was 7.2°C (Figure 8b). This time period is in broad agreement with work by Wang and Dilger [9] who found the peak heat rate for a normal portland cement concrete to occur after 10 hours. The temperature would be expected to continue to rise slightly after this time until the rate of heat evolution had dropped to below the rate of heat loss to the environment. After 75 hours the internal temperature followed the ambient temperature, albeit with a 9 hour delay and a reduced amplitude. Good agreement was found between the thermocouples and the thermistors at all locations.

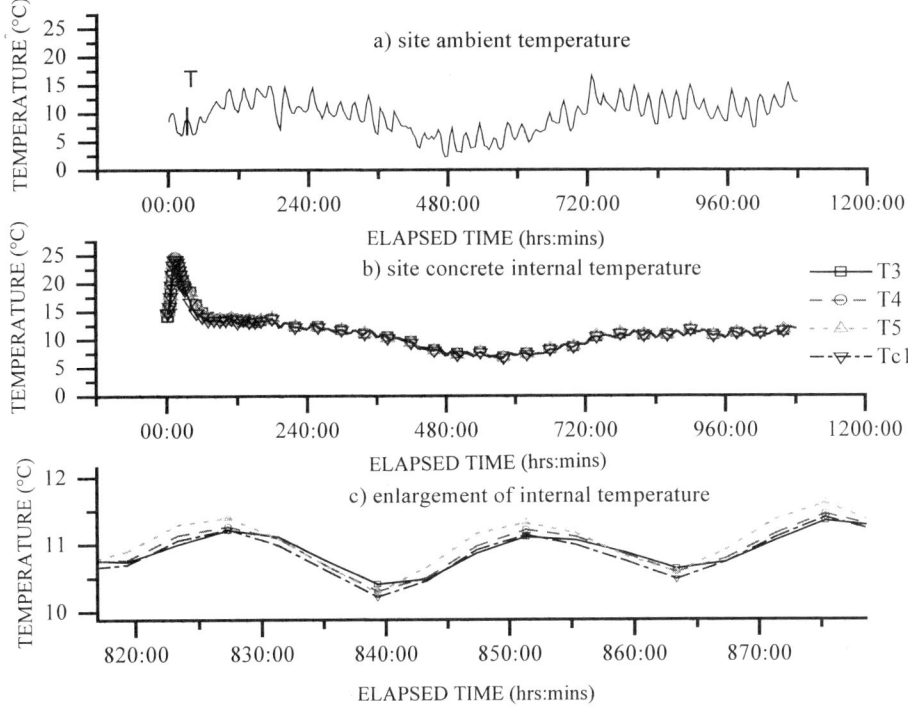

Figure 8   Temperature profiles in the slab

The highest temperature occurred at mid-slab, which would be expected as initially heat was being lost to the ground and to the air both of which were colder than the concrete. The internal and ambient temperature as measured by some of the sensors can be seen in Figure 8. Sensors T3, T4 and T5 were in the bottom, middle and top of the slab respectively. Reference to Figure 8c shows that the temperature variation is greatest in T5, with the other sensors lagging slightly when the ambient temperature changes. This is in agreement with the profile found by Venkatasubramanian [10], where the sensors near the surface of the slab show the greatest variation in temperature, experiencing the daytime highs and the night-time lows, whilst the sensors located deeper in the slab show a reduced amplitude of temperature variation, which lags behind the upper sensors.

Atmospheric Conditions

The ambient temperature varied between 1.2 and 17.8°C as night-time low and daytime high respectively, with a mean and standard deviation during the monitoring period of 9.3 and 3.1°C respectively (Figure 8b). The relative humidity was over 80% during the casting and had a mean of 78% and SD of 10% over the monitoring period.

The wind speed measurements gave a maximum of 1m/s (2.2mph) and a mean value of 0.1m/s. Air movement was very variable with only two periods of fairly constant movement, from 37-84 hours and from 246-322 hours. Sensor 2 generally gave higher readings than sensor 1, which due to the sensor locations indicates that the prevailing air currents were across the building rather than down it's length.

CONCLUSIONS AND DETAILS OF FURTHER WORK

The instrumentation was successful in recording strains in the slab and the ambient conditions until 45 days after casting when the logger had to be removed. None of the gauges were damaged and the method of fixing the gauges in place proved effective and quick, so as not to disrupt the casting. The thermistors integrated in the strain gauges and the thermocouples gave similar temperature readings. For this reason, as there are also significant cost benefits, future instrumentations will not include thermocouples. Having identified a successful instrumentation technique, studies can be carried out on jointed and unjointed large area pours to determine if they behave in similar ways.

The data has shown the phases of early shrinkage as seen in the laboratory trials and reported in the literature, although further work may be needed to interpret these strains fully. A vertical shrinkage profile is apparent after less than 60 hours, with lateral differentials indicative of edge restraint being apparent before this.

The possibility has been identified of thermal shrinkage being the trigger of the initial cracking and movement at the contraction joint. It is hoped that this can be verified by data from further sites or by computer modelling of the internal and ambient conditions leading up to the failure.

Work is currently being carried out on a finite element model of the early life behaviour of concrete slabs. This model will take account of the heat evolution due to the hydrating cement and the one-dimensional diffusion of moisture out of the concrete as well as time dependent material non-linearity and creep. The frictional restraint from the subgrade has already been modelled, but new terms to account for edge stiffening caused by adjacent strips can now be included. Finally, smeared or discrete cracking models will be used to predict the locations and size of cracking zones. The final model should enable the sensitivity of the different environmental and design parameters to be assessed, thus allowing design recommendations to be made for engineers.

ACKNOWLEDGEMENTS

We acknowledge the financial support of the EPSRC (Engineering and Physical Sciences Research Council) and thank the members of the project steering group, the ACIFC (Association of Concrete Industrial Flooring Contractors) and our industrial partners BRC Ltd., Burks Green & Partners, The Concrete Society, Face Consultants Ltd., Fibercon UK Ltd., Fibermesh Europe, Stanford Industrial Concrete Flooring Ltd., Stuarts Industrial Flooring Ltd., The Sprigg Little Partnership, Tarmac Topmix Ltd. and the Quarry Products Association. The research team is particularly grateful to Booker for granting access to their project Bentley site for the instrumentation.

REFERENCES

1. BARNBROOK, G. AND BECKETT, D. Technical Report 34: Concrete Industrial Ground Floors - A guide to their Design and Construction, 2nd Edition, The Concrete Society, Slough, 1994, 145 pages.

2. KIM, J K AND LEE, C S. Prediction of differential drying shrinkage in concrete, Cement and Concrete Research, 1998, Vol. 28, No. 7, pp. 985-994.

3. LACHEMI, M. ET AL. First Year Monitoring of the First Air-Entrained High-Performance Bridge in North America, ACI Structural Journal, 1996, Vol. 93, No. 4 (July-August), pp. 379-386.

3. BRÜLL, L AND KOMLOS, K. Early shrinkage of hardening cement pastes. In: Fundamental Research on Creep and Shrinkage of Concrete. 1st Edition. (Ed: Wittman, F.H.) Martinus Nijhoff Publishers, The Hague, 1982, pp. 239-248.

5. KASAI, Y, ET AL. Volume change of concrete at early ages. In: Proceedings of International Conference on Concrete of Early Ages, Paris, 6-8 April 1982. Edition Anciens ENCP, (Ed: Rilem) Association Amicale des Engénieurs, Paris, 1982, pp. 51-56.

6. GRZYBOWSKI, M AND SHAH, S P. Model to predict cracking in fibre reinforced concrete due to restrained shrinkage, Magazine of Concrete Research, 1989, Vol. 41, No. 148, pp. 125-135.

7. LOSBERG, A. Pavements and slabs on grade with structurally active reinforcement, ACI Journal, 1978, Vol. 75, No. 66, pp. 647-657.

8. TIMMS, A G. Evaluating subgrade friction-reducing mediums for rigid pavements, Highway Research Record, 1963, Vol. 60, pp. 28-38.

9. WANG, C. AND DILGER, W H. Modelling of the development of heat of hydration in high-performance concrete. In: Advances in Concrete Technology - Proceedings of the Second CANMET/ACI International Symposium, Las Vegas, Nevada, USA, 1995. (Ed: Malhotra, V.) American Concrete Institute, Detroit, Michigan, 1995, pp. 73-82.

10. VENKATASUBRAMANIAN, V. Temperature variation in a cement concrete pavement and the underlying subgrade, Highway Research Record, 1963, Vol. 60, pp. 15-27.

MEASURING THE NEAR-SURFACE MOISTURE CONDITION OF CONCRETE SLABS AND GYPSUM SCREEDS

R P West
M L O'Neill
Trinity College Dublin
A D Rynhart
Tramex Limited
Ireland

ABSTRACT. In this paper a comparison is made between different methods for establishing the moisture condition or relative humidity of concrete slabs and gypsum screeds as they dry out. A series of slabs and screeds were monitored over time until an equilibrium had been reached between the moisture retained in the slab and the ambient humidity in the environment. The merits and limitations of the various methods are discussed and a correspondence between their relative quantitative readings is established.

Keywords: Concrete slabs, Drying out, Gypsum screed, Hygrometer, Moisture condition, Relative humidity, Vapour emission test.

Dr Roger P West is Senior Lecturer and Director of the Structural Laboratory at Trinity College Dublin. His research interests in concrete lie in durability, rheology and new materials. He is Secretary of the Irish Concrete Durability Committee and is the current Vice-Chairman of the Council of the Irish Concrete Society. He is also a former Chairman of the Structures and Construction Section of the Institution of Engineers of Ireland, and is a Fellow of that Institution.

Miriam L O'Neill graduated with an honours degree in Civil Engineering from Trinity College Dublin in 1997. Since then she has been a Research Assistant and postgraduate student at that university, researching moisture movement during the drying of concrete slabs.

Alan D Rynhart is the Managing Director of Tramex Ltd., a company that specialises in the development, manufacture and marketing of moisture detecting instruments for many industries, including the construction industry.

INTRODUCTION

A recurring problem in the construction industry at large is the establishment of the point in time at which a concrete slab or gypsum screed has dried sufficiently to allow an impermeable covering of any kind to be applied. If the covering is applied too early, vapour pressure from the residual moisture in the slab can cause the covering to be debonded, manifesting itself as tiles lifting, timber shrinking, paint or vinyl blistering, etc. If the covering is applied after an unnecessarily lengthy drying period, precious time is lost for the building contractor and, probably, also for the end-user.

Many of the various non-destructive testing techniques[1] used to assess the moisture condition of the slab at any stage in the drying process are well known, particularly the hygrometer and the vapour emission test (the ASTM test, which involves determining whether condensation of moisture forms on the underside of a sealed plastic sheet, is not a quantitative test, and so is not considered here). Other frequently used tests include the Protimeter relative humidity test, Speedy moisture meter (also known as the carbide bomb test) and the oven-drying method, all of which will be investigated in this paper. A relatively new method involving a hand-held meter, the Concrete Moisture Encounter, will also be included in the methods reported because it provides a suitable contrast to many of the more established techniques.

The objective of conducting these tests on carefully controlled specimens is to allow a comparison to be made between them with regard to the accuracy and nature of the information that they provide and their convenience of use.

It is worth noting that the processes involved in the drying out of concrete and gypsum are complex and difficult to predict analytically [2]. However, in essence, the mechanism is one of slow migration of vapour to the surface of the concrete and relatively fast evaporation of moisture from the surface. At some stage in the usually lengthy drying process, an equilibrium is reached between the relative humidity of the ambient environment and the surface of the slab, and between the core and surface of the concrete (the rate of diffusion of the vapour within the slab is determined largely by the internal open pore structure and the moisture gradient). Therefore, when a slab is deemed to have dried sufficiently to apply a covering, this does not imply that the slab is completely "dry", but rather that this equilibrium state has been achieved under the prevailing environmental conditions at that time. What each of the tests investigated in this paper is actually measuring will be carefully considered in this context.

EXPERIMENTAL SET-UP

Specimens

A total of six 750x750x100mm specimens were cast, two in concrete (with a 35N20 mix) and two each of a nominal 30N and 40N gypsum screeds (30 and 50mm thick respectively) cast on 50mm concrete bases, separated by a layer of 1000 gauge visqueen.

Table 1 Concrete slab mix constituents

Normal Portland Cement	360 kg/m^3
20mm crushed aggregate	760 kg/m^3
10mm crushed aggregate	380 kg/m^3
Medium sand	730 kg/m^3
Water	168 l/m^3
"P7" plasticiser	1.8 l/m^3

The concrete mix was as given in Table 1, which yielded a slump of 100mm and an average 28 day cube strength of 49.9 N/mm^2. The gypsum screed was a pre-blended proprietary mix ("Gyproc" from British Cement) and achieved 28 day strengths of 47.5 and 52.0N/mm^2 respectively.

Considerable care was taken to ensure that the initial moisture content of each pair of specimens was identical. Following three days of curing under wet hessian and polythene following casting, the samples were sealed on the vertical faces using a polyvinyl acetate to prevent moisture loss. The upper and lower horizontal surfaces (750mm square) of each specimen were allowed to breath freely by supporting them at the corners only.

Test Methods

Method statements were written for each test method [3]. It would be inappropriate to describe these in detail here, but a brief description of what parameters are being measured in each test is important if the results are to be sensibly interpreted.

Hygrometer Test

This test measures the relative humidity of an entrapped pocket of air immediately above the concrete surface. It can take up to eight days (though more usually four) for equilibrium between air and surface to be established and, clearly, this is undesirable. A threshold of 75% humidity is usually specified for the safe laying of floor coverings.

Protimeter Relative Humidity Test

This probe is inserted into a sealed, pre-drilled hole of depth 45mm, and measures the average near-surface relative humidity. Equilibium is usually reached within two hours, as determined by consecutive equal readings, taken at 10 minute intervals.

One expects the results of this test to be higher than an equivalent hygrometer reading if there is a moisture gradient in a drying slab.

Vapour Emission Test (VET)

This test (using the "Vaprecision" brand of VET) measures the gain in weight of an anhydrous calcium chloride sample within a plastic tent which is sealed onto the concrete surface. It takes between 60 and 72 hours to complete and results are expressed in pounds per $1000ft^2$ per 24 hours, where a value of 3 lbs/($1000ft^2$x24 hrs.) is the often-quoted safe threshold for laying a covering[4].

Oven-Drying Test

This test measures the quantity of evaporable moisture (by weight) of a sample extracted by drilling a hole between 30 and 45 mm into the specimen. The result is determined by drying the sample in an oven at 105°C for 3 days.

Speedy Moisture Meter Test

A sample drilled from the surface to a depth of 50mm is tested in a calibrated container which yields a moisture content in percentage weight of concrete. While the results are obtained in a matter of minutes, due to inconsistencies in the readings (as a consequence of moisture loss on rapid heating of the drill bit, despite attempts to reduce its temperature), this test was abandoned and will not be considered further.

Concrete Moisture Encounter (CME)

This device [5] passes an impulse between electrodes on the surface of the concrete and uses the impedance offered by the near-surface concrete and gypsum to give an indication of the moisture content (%). It has been calibrated for both concrete and gypsum separately and gives an instantaneous reading.

Test Sequencing

Given that each pair of slabs had initially the same water content and curing conditions, the most important procedure in the testing program was to ensure that the entire surface areas of the pairs of slabs should be in the same moisture condition at the point of testing so that the test results could be compared on an equal basis. Hygrometer and VET tests were conducted separately and simultaneously on each pair and every slab/screed was tested using the Protimeter, Over-drying and CME at either end of the periods over which the more lengthy tests were conducted (for example, in some cases, eight days apart for the longest Hygrometer tests). The need to seal the slabs locally during the lengthy hygrometer and VET tests meant that a plastic sheeting template had to be made to cover the remaining exposed portions of the slab/screed to guarantee that equal drying conditions over the entire test areas were maintained over the life of each slab.

Note that the ambient relative humidity was also recorded and varied between 60 and 80% over the 140 days of the test program.

TEST RESULTS

The variations in the various readings taken with respect to time are presented in Figure 1a to 1e.

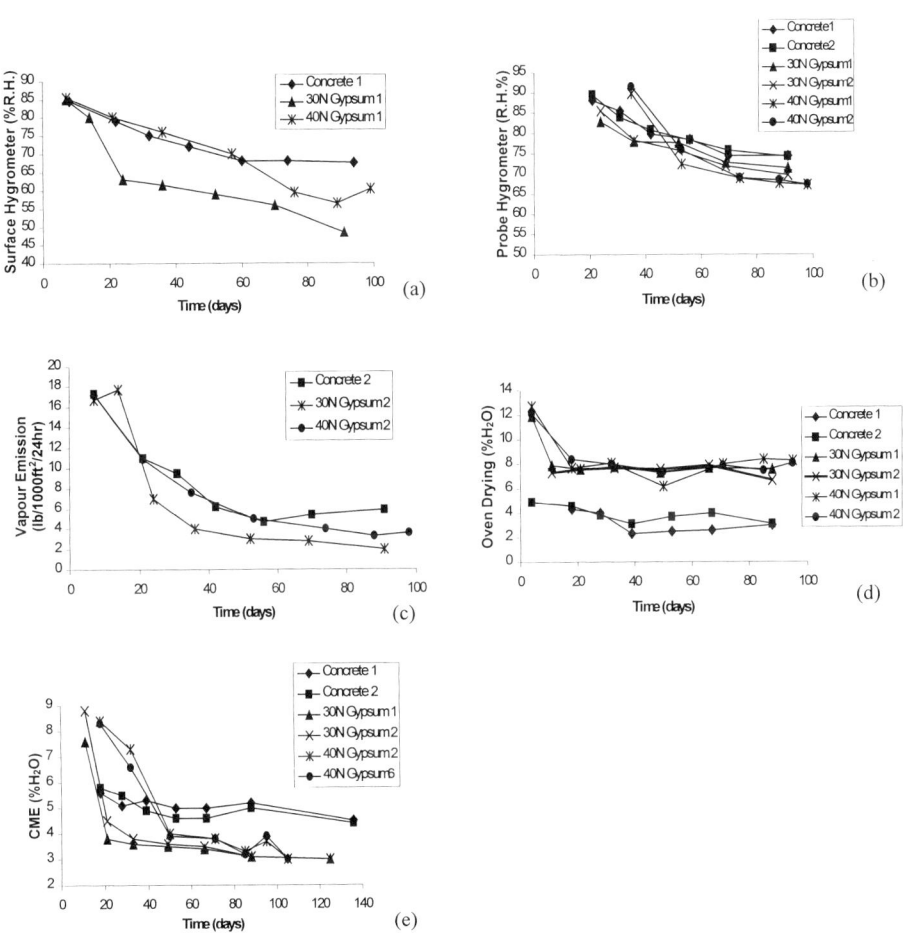

Figure 1 Variation in relative humidity and moisture content readings with time for a) hygrometer b) Protimeter c) Vapour emission test d) oven-drying e) concrete moisture encounter.

Hygrometer

It may be observed that the concrete slab reached equilibrium with the environment after about 60 days at a relative humidity of 68%. Following this, minor fluctuations will occur as the ambient relative humidity varies and, , some moisture may be re-absorbed into the surface.

The 30N gypsum relative humidity appears to drop suddenly in comparison with the 40N screed, but it should be recalled that the latter was only 30mm thick (the 40N screed is 50mm thick). The "final" values are, of course, lower than the concrete because the mix constituents, chemistry and pore structures are different.

Protimeter

It is encouraging that the results for each pair are reasonably clustered together (Figure 1b) and that the average values are all higher than the hygrometer equivalent. This is because the measurement is taken below the surface where the moisture content is higher as the slab is drying out. At 90 days the concrete relative humidity is about 75% and is only just reaching equilibrium, which is in contrast with the hygrometer results. Again, the 30N gypsum reacts faster and all gypsum screeds are approaching lower equilibrium humidities.

VET

Figure 1c confirms the hygrometer's indications that equilibrium for concrete occurs at about 60 days and can fluctuate beyond that point. The lowest reading is 4.2 lbs/($1000ft^2$x24hrs.) which is just above the recommended threshold for floor covering. The gypsum samples are still gradually drying after 90 days, as observed above.

It is notable that the readings for all samples indicate rapidly diminishing moisture evaporation with time over early stages of drying and so are particularly sensitive (and, therefore, useful). Note that the test itself does not appear to be influenced by the overall ambient conditions in that the test measures the emissions from the surface, which depends on whether equilibrium has been reached, not on the absolute value of the ambient relative humidity.

Oven-drying

The variation in the moisture content, as determined by oven-drying (Figure 1d), is small and may be influenced by some moisture being lost while drilling, despite regularly changing the drill bit to keep the temperature down. The gypsum, being a softer material, seems to be less prone to this problem and exhibits a sharp fall over the first 20 days, as seen with the hygrometer. The significant differences between the results for concrete and gypsum can be explained by their different composition, porosity and the fact that the chemically combined water is known to be less tightly bound for gypsum. While the results here are the least satisfactory for determining the state of drying out, further tests (involving chipping off larger samples and pulverising prior to drying) are being undertaken.

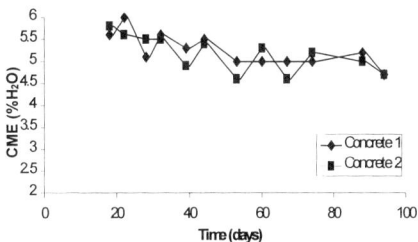

Figure 2 Detailed CME readings with time for the concrete slab

CME

The results in Figure 1e are closely paired in their groups and a wide range of results are observed for the gypsum. The results for each case concur well with the most indicative of those above, namely that the thicker 40N gypsum screed takes longer to dry at the early stages as compared with the 30N gypsum, the concrete reaches equilibrium at about 60 days (at 4.8% moisture content) and can fluctuate thereafter.

It is worth noting that the equilibrium reading for concrete is influenced by several factors including drying conditions (relative humidity and temperature) and, to a lesser extent, mix constituents. For example, in an intense drying environment, the equilibrium moisture condition for the CME would be below 3%. Hence, this test result does not drop to an absolute threshold equilibrium (as the VET does) and if the circumstances are not known beforehand, the test need to be undertaken at regular intervals to establish when the drying curve bottoms out.

The CME is a sensitive test in that when specimens were covered during the hygrometer and VET tests, the CME readings before and after these tests (that is, over a 5-8 day period) increased considerably.

Figure 2 illustrates that, as the moisture migrated towards the surface but could not evaporate (as the surface was covered during the hygrometer/VET tests), the CME readings always rose noticeably, reflecting the revised moisture condition in the near-surface concrete (interestingly, similar trends have been reported for the VET [6]). The values given in Figure 1e are the lower bound envelopes of these fluctuations.

COMPARISONS

It is instructive to compare some of these test results to see if a strong correlation exists between them. It is expected that if the raw data (that is, test against time) shows a strong trend for both tests being compared, then the comparison will also have a discernible trend. These comparisons are shown in Figure 3a to 3f.

Hygrometer vs. VET

These tests (Figure 3a) show a significant trend, where the early changes in the VET test give rise to a relatively insensitive correlation for high VET values.

There is some discontinuity in the later results (at low relative humidity) as the slab absorbs moisture from the atmosphere. Note that, in this case, the 75% threshold is not closely correlated with the supposedly equivalent 3lbs/(1000ft^{2}x24hrs.) for the VET.

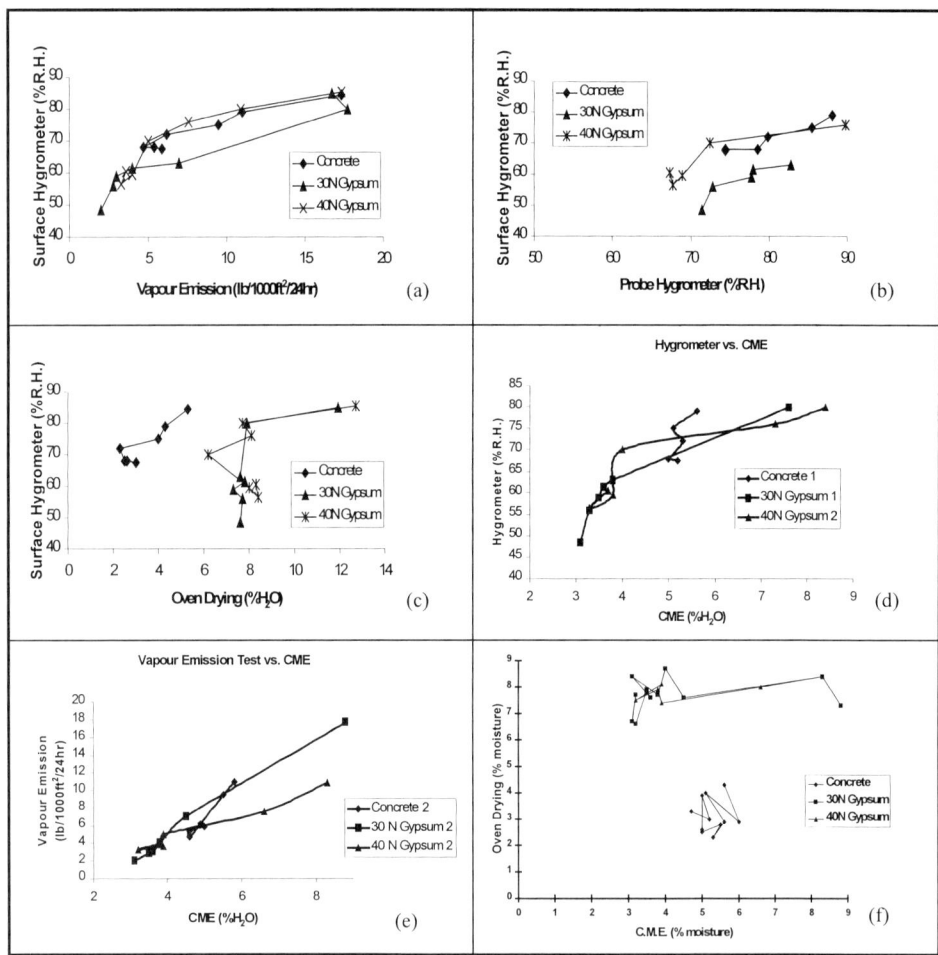

Figure 3 Comparative test diagrams for a) Hygrometer against VET b) Hygrometer against Protimeter c) Hygrometer against oven-drying d) Hygrometer against CME e) VET against CME and f) Oven-drying against CME.

Hygrometer vs. Protimeter

Figure 3b illustrates the consistently higher value for the Protimeter humidity as compared with the hygrometer, although the trends correlate well for concrete and gypsum.

Hygrometer vs. Oven-Drying

Although some trend is evident in Figure 3c, the steepness of the curves over a range of humidities reflects the poor quality of the raw data for the oven-drying test, as explained previously.

Hygrometer vs. CME

While the gypsum results in Figure 3d show a strong trend over a wide range of CME readings, the concrete results are less encouraging. Also the range of CME readings is small (5.8 down to 4.8%) for the equilibrium of these particular concrete slabs to be attained.

VET vs. CME

The trends here (Figure 3e) are stronger as the underlying raw data are good. In each case a low value of VET corresponds to those portions of the CME results that have clearly reached equilibrium, even though these tests measure different phenomenon in the investigation of the moisture condition.

Oven-Drying vs. CME

These two tests measure the moisture content directly and Figure 3f illustrates that the correlation between them is weak. In particular the gypsum CME readings seem to be sensitive and cover a wide range of values, whereas the corresponding oven-drying results have, again, reflected difficulties in the sampling technique employed.

A complete set of correlations between all five sets of data can be found in West and O'Neill [3]. In general, if the raw data against time is consistent, the correlations are strong and, therefore, these tests may be of considerable benefit in practice, depending on circumstances.

CONCLUSIONS

This paper has compared five techniques used for establishing how dry a concrete slab or gypsum screed is at any point in time. While it is true that most of the tests measure slightly different aspects of the same basic process, they also give an indication as to when an equilibrium state between slab and environment has been reached. It is recognised that this state will depend on location and mix, amongst other things.

As a practical cessation in the evaporation of vapour from the surface is the important criterion for laying a floor covering, the Protimeter is not as useful as some of the others, for although it is a relatively fast test, it measures moisture below the surface and is partially non-destructive.

In this respect, the oven-drying test, while often used, requires considerable care in sampling and one could not be certain in these tests that the results were reliable or that the moisture condition registered represents the equilibrium state, unless other tests were conducted simultaneously.

The hygrometer is also frequently used but can require a long period for a stable reading to be achieved during any one test. However, safe thresholds have been established from experience.

The VET has the advantages that it is clear from the test when the slab has reached equilibrium if testing is done at intervals and, again, safe thresholds are known from experience. It takes several days to register a single reading.

While the CME gives an instantaneous reading, these readings can be influenced by ambient conditions. Therefore, it is necessary to establish what is a safe moisture condition before the slab can be considered to be adequately "dried out". However, in the absence of this knowledge, the taking of regular readings allows the monitoring of the drying with time so that the equilibrium state becomes apparent from the emerging trends.

In summary, it may be concluded that if the point at which a covering is to be safely applied in a concrete slab or gypsum screed needs to be established, one convenient and cheap method would involve using the CME at intervals to indicate when moisture equilibrium with the environment had been reached. Then, if necessary, a more conventional test (for example, the VET or hygrometer) could be undertaken to confirm the CME indications.

ACKNOWLEDGMENTS

The Authors would like to acknowledge with gratitude the financial support given to this project by Tramex Ltd. They would also like to record their thanks for the kind assistance of Sarah Prichard in the reproduction of the figures.

REFERENCES

1. PARROTT, L, J. A review of methods to determine the moisture conditions in concrete. BCA, 1990, pp 27.

2. PIHLAJAVAARA, S, E. Estimation of drying concrete at different relative humidities and temperatures of ambient air with special discussion about fundamental features of drying and shrinkage. in Creep and Shrinkage, Ed. Z P Bazant and F H Williams, John Wiley, 1982, pp 87-107.

3. WEST, R, P, AND O'NEILL, M, L. Measuring the moisture condition of concrete slabs and gypsum screeds using the Concrete Moisture Encounter. Confidential Tech. Rept., Trinity College Dublin, 1998, pp 76.

4. SUPRENANT, B, A. Moisture movement through concrete slabs. Concr. Constr., Vol 42, No. 11, 1997, pp 879-885.

5. O'NEILL, M, L, WEST, R, P, AND RYNHART, A. The measurement of the moisture condition of concrete and gypsum using the concrete moisture encounter. RILEM Conf. on Non-destructive testing and experimental stress analysis of concrete structures, Kosice, 1998, accepted for publication.

6. SUPRENANT, B, A, AND WARD, R, M. Are your slabs dry enough for floor coverings? Concr. Constr., Vol. 43, No. 8, 1998, pp 671-677.

DESIGN OF STEEL FIBRE REINFORCED FLOORS ON FOUNDATION PILES

H Thooft
Bekaert
Belgium

ABSTRACT. The presented paper gives an insight on the latest developments on the use of steelfibers in structural pile supported ground floors. An extensive research was done at the German university of Braunschweig to investigate the behaviour in serviceability and ultimate limit state of these type of floors. In the end a combined reinforcement, "Dramix RC-80/60-BN + re-bar" proved to be the optimum solution re. cost and performance. With the findings of the test program existing yield line design methods were adapted to this combined reinforcement system. Where applicable the design is in line with the Eurocode 2. In view of the technical and commercial advantages over a period of 1 year more than 100.000 m² of these floors were cast.

Keywords: Pile supported ground floors, Serviceability limit state, Ultimate limit state, Yield line design method, Eurocode 2.

Ing Hendrik Thooft is Dramix product manager at the company Bekaert, based in Belgium, and is accountable for the application flooring world-wide. He is member of the Dutch CUR committee and has recently become ACI member.

INTRODUCTION

After more than 20 years the use of steel fibres in ground supported industrial floors has become generally accepted. With their advantages in terms of price and performance the use of steel fibres for this application continues to grow.

Continuous research has been taken place to increase our knowledge into the behaviour of steel fibre reinforced concrete in ground supported slabs.

The addition of steel fibres into the concrete mix results in a post cracking tensile strength, which is a function of tensile strength of plain concrete as well as the type and amount of steel fibres. In particular, impact and toughness are increased. For the use of steel fibers in piled supported floors this toughness plays a decisive role for the load bearing and deformation behaviour of these slabs.

The equivalent flexural tensile strength and toughness can be tested according to several standards or recommendations such as JSCE-SF4 (Japan), ASTM C 1018 -94b (U.S.A), TR34 (UK), DBV-Merkblätter Faserbeton (Germany), CUR 35 (The Netherlands), NF P 18-409 (France), NBN B 15-238 (Belgium), or similar standards.

Traditional piled floors with pilecaps, in contrast to flat slabs, are additionally influenced by temperature and shrinkage strain and horizontal pile reaction. Therefore pilecaps should be avoided because of the possible restrain due to horizontal soil interaction. Vertical restrain forces occur due to uneven pile settlement and can highly influence the moment redistribution within the slab.

TEST SET-UP

The aim of the test was to investigate the load bearing capacity and deformation of a suspended DRAMIX® reinforced floorslab.

The 5 x 5 m slab was supported by 9 columns of 20 x 20 cm. These columns were 2 m apart. To simulate the behaviour of a suspended floor with larger pile distances the slab thickness was reduced to 140 mm.

The slab was loaded in the centre of each span. The rigid columns were equipped with the load cells and could be lowered to simulate uneven pile settlement.

A diagram of the test set-up is given in Figure 1

Three slabs were tested.

While the first slab only consisted of SFRC without any additional reinforcement (Reference slab; Slab 1), a second slab with additional reinforcement of 10 mm bars in the column strips was tested (Slab 2).

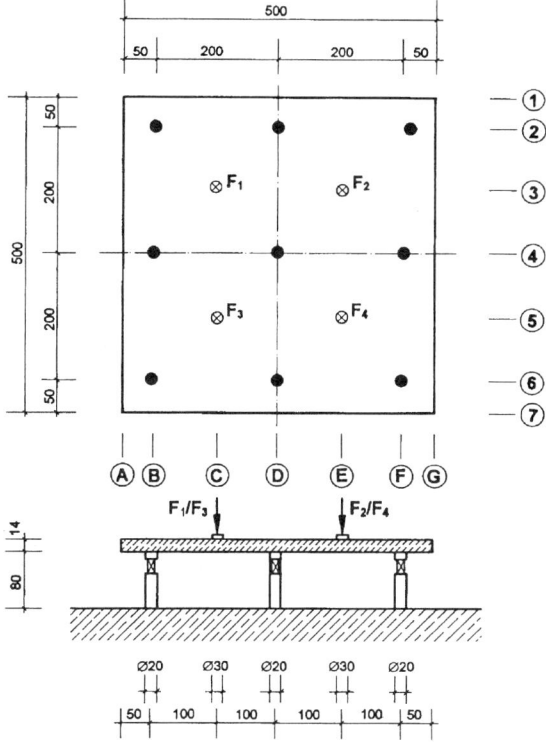

Figure 1 Schematic view of the test set-up

The third was a prestressed slab with internal unbonded 0,6 ''-tendons, placed in the neutral axis, added to the SFRC (Slab 3).

The first and second slab test will be further detailed.

CONCRETE COMPOSITION

DRAMIX® Steel Fibers

A dosage of **40 kg/m3** Dramix® RC-80/60-BN was added to the concrete mix. The Dramix® RC-80/60-BN is a wire drawn fiber with a length of 60 mm and a diameter of 0.75 mm. For easy and homogeneous mixing the fibers are glued into bundles. This fiber with its very high length/diameter ratio of 80 is the best performant DRAMIX® steel fiber used in industrial floors. The fibers were added in the batching plant.

Concrete Mix

The concrete used was of a C 45 quality, mixed at the batching plant, transported by truck mixer to the laboratory and pumped with a normal concrete pump.

Table 1 Concrete mixes of the two slabs tested

	SLAB 1	SLAB 2
Cement	360 kg/m3 CEM I 32,5 R	360 kg/m3 CEM I 32,5 R
Fly ash	100 kg/m3	100 kg/m3
W/C ratio	0,46	0,53
Sand 0/2	703 kg/m3	681 kg/m3
Gravel 2/8	279 kg/m3	280 kg/m3
Gravel /16	766 kg/m3	748 kg/m3
Plasticizer	0,5 % (C)	0,5 % (C)

Additional Reinforcement

While "slab 1" only had 40 kg/m3 Dramix® RC-80/60-BN "slab 2" was additionally reinforced with steelcages (6 Ø 10) spanning from pile to pile in both directions.

TEST RESULTS

General

To enable a thorough analysis of the load bearing capacity and deformation, the test was completely monitored by computer controlled measuring equipment. Loads, deformations and reaction forces in the piles were all measured. The propagation of cracks and the crack widths were measured by a video measuring equipment. The test carried out on slab 1 and 2 was identical.

Loads and Forces

Reaction forces

Between the columns and the slab hydraulic supports were placed to measure the reaction force in 8 out of the 9 supports. The column in the middle was fixed to ensure the stability of the slab. As the hydraulic equipment accurately measured the reaction forces, the force in the ninth column could be calculated.

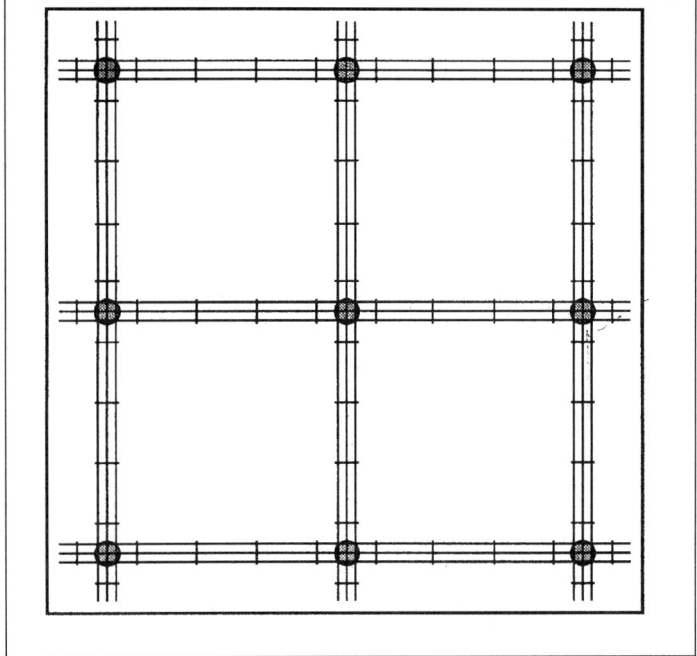

Figure 2 Lay-out of the re-bar reinforcement

Serviceability load

The slab was loaded in three phases.

In a first phase the slab was loaded with a theoretical calculated serviceability load of 50 kN. The loading criteria for the test is in Figure 3. During the first loading cycle the load on the first couple F1-F2, was progressively increased by intervals of 10 kN while the second load couple was simultaneously decreased. Subsequently the slab was completely loaded ($F_1=F_2=F_3=F_4=50$ kN).

Starting from the serviceability load all four area's were loaded 10.000 times with the serviceability load (Top load 50 kN / Bottom load 20 kN) followed by another 10.000 cycles at a 20 % increase on the serviceability load. (60 kN / 25 kN). The frequency of the loading cycle was 0,2 Hz.

Measurements of deformations, cracks and crack widths were taken at 10, 100, 1000, 3000 and 10.000 cycles.

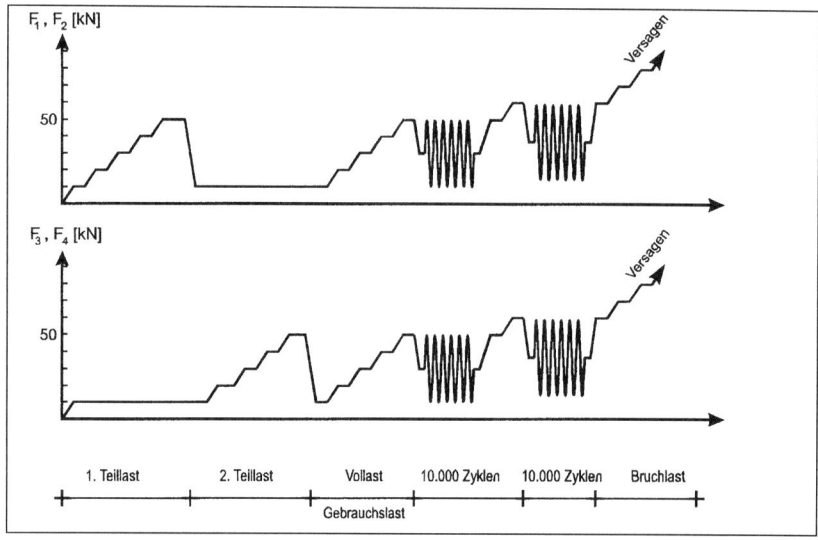

Figure 3 Loading cycle

Ultimate load

Starting from the increased serviceability load (60 kN) the load on the slab was increased in intervals of 5 kN.

After each interval the deformation and crack propagation was measured. Maximum load was achieved at a deflection of 3 mm for "slab 1" and 40 mm for "slab 2".

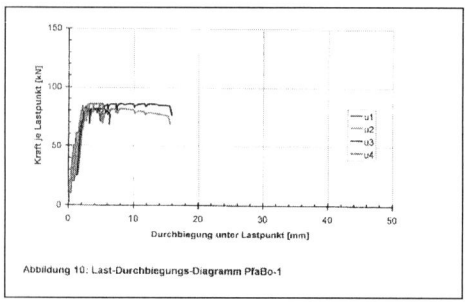 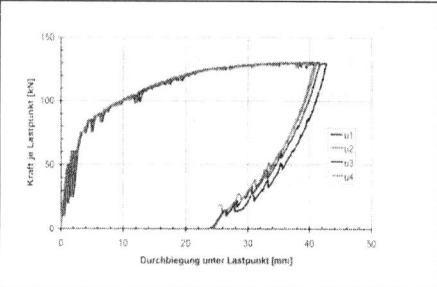

Figure 4 Load deflection curve of the four pointloads
Left - slab 1 : Only steelfibers
Right - slab 2 : Steelfiber + re-bar

Deformations

To learn more about the deformation of the slab 16 displacement gauges were placed under the floor, where the yield lines (= Cracks) were supposed to occur. At the perimeter of the slab 4 gauges were placed to measure the rotation of the edges.

The horizontal displacement of the slab was also measured.

Cracks and crack widths

Table 2 Crack width in function of the load

LOAD (kN)	CRACK WIDTH (mm)			
	Slab 1		Slab 2	
	Top	Bottom	Top	Bottom
50/20	0,06	-	0,05	-
50/40		0,04		-
50/50	0,07	0,05	0,07	0,04
60/26		0,16		0,17
83		0,2		
130				0,44

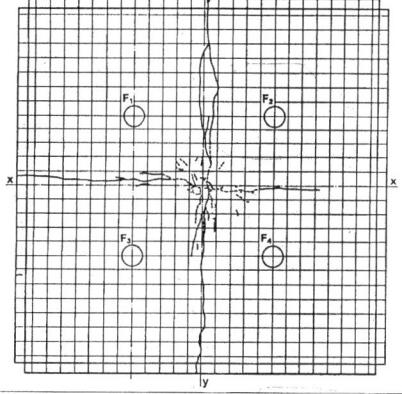 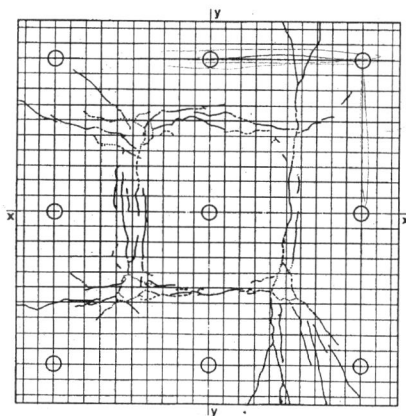

Figure 5 Cracks at the top (Left) and bottom (Right) of "slab 1"

DESIGN PRINCIPLES

Bending

The bending moments are calculated using the yield line model of *Antoni Sawczuk* and *Thomas Jaegher* modified by the University of Braunschweig.

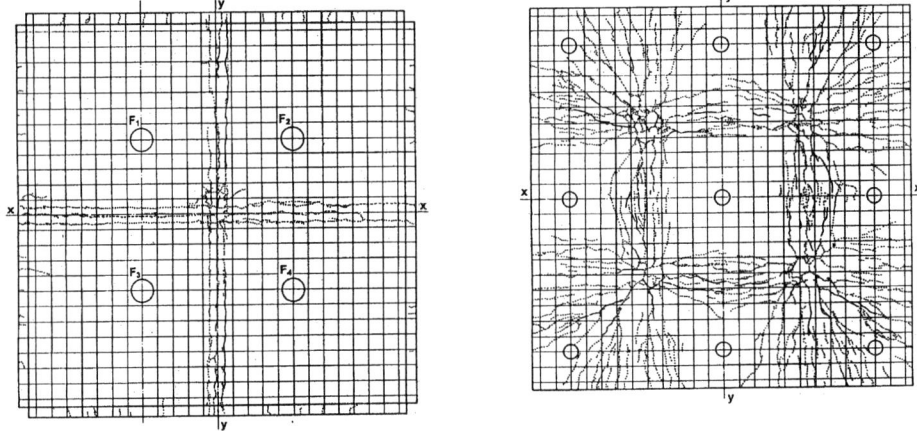

Figure 6 Cracks at the top (Left) and bottom (Right) of "slab 2"

Yield line theory is best way of designing a SF-floor in the Ultimate Limit State. The safety factors given in the different standards (British standard and Eurocode) are all based on ULS design.

3 different locations on the floor are calculated
- Centre fields
- Edge Field
- Corner field

For each location the combination Re-bars and steelfibers is calculated (Global yield lines) and also the area between the steelcages with only steelfibers (Local yield lines).

Global yield lines

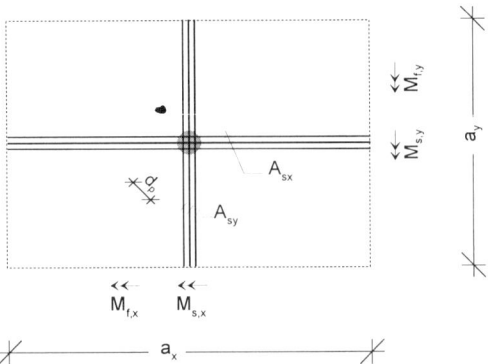

Local yield lines

Possible yield lines forming in the area between the steelcages

Punching Shear

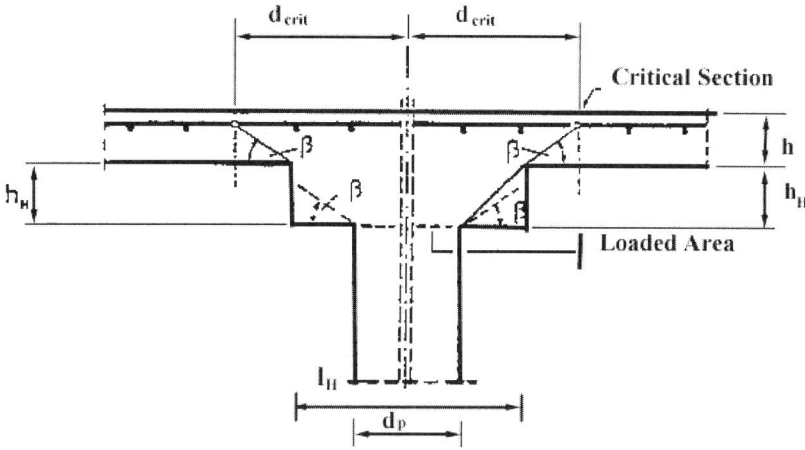

Figure 7 Example ; Slab with column heads $l_H < 1,5h_H$

The design of punching shear is completely in line with the Eurocode 2.

1. For a loaded area the critical diameter and perimeter is calculated.

$$d_{crit} = 0,5 * \max(d_p, d_p^*) + 1,5h$$
$$u = 2\pi d_{crit}$$

2. In function of the load and the location of the pile the shear stress is calculated

$$v_{Sd} = \frac{V_{Sd} * \beta}{u}$$

3. The shear resistance of a section is calculated with the following formulae

$$v_{Rd1} = \tau_{Rd} * k * (1,2 + 40\rho_1) * d$$

* For further details we refer to the EC 2.
* τ_{RD} is composed of a part concrete and a part SF concrete in line with the Dramix ®guidelines.

4. The admissible shear resistance must be higher than the occurring shear stress

352 Thooft

Material Characteristics

SF Concrete

Bending

The stress-strain diagram of the steel fibre concrete is as shown below. This graph is in line with the DRAMIX® guidelines.

The equivalent flexural stress of the SF concrete has to be assessed by beam test as described in TR34 of the concrete society or similar standard/recommendations.

This characteristic equivalent flexural stress is function of the dosage and the quality of the steelfiber. Due to the high L/D ratio the RC-80/60-BN offers the highest equivalent flexural stress for a given dosage.

In line with the codes only characteristic values are used !!!

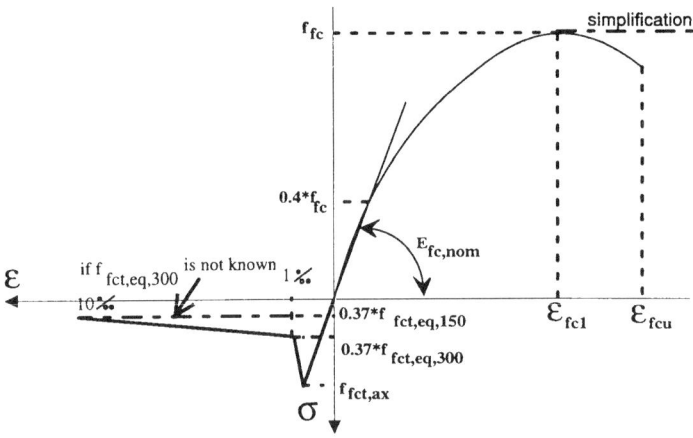

Figure 8 Stress strain diagram of steelfiber concrete

The combination of the steel bars and the DRAMIX steelfibers are calculated using the following deformation diagram. (Figure 9)

Shear Resistance

See Dramix® product data sheets

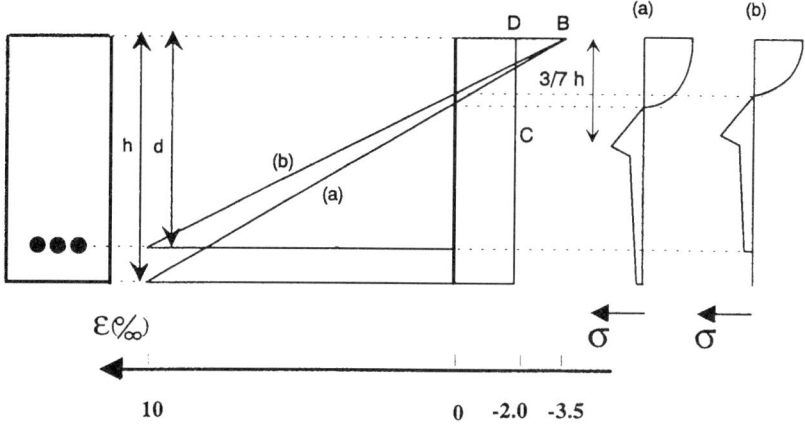

Figure 9 Ultimate limit state for steel fibre and re-bar concrete

Steel

For the steel a steel quality of BE 500 is been used in the design

Safety factors

The safety factors are in line with the Eurocode 2

Material factor
Steel $\gamma_s = 1,15$
Concrete $\gamma_c = 1,5$
Steelfiber concrete $\gamma_f = 1,5$ (DRAMIX Guidelines)

Load factor
Dead load $\gamma_G = 1,35$
Variable loads $\gamma_Q = 1,5$

Creep factor
SF Concrete $\alpha = 0,85$

Remark

1. The real safety between slab test done at the University of Braunschweig and this design method is between a factor of 5 to 7.

CONCLUSIONS

1. The time and location of the cracks appearing in the slab confirm with yield line theory.

2. The theoretical load calculated using yield line theory for suspended slabs corresponds with the values found in the test.

	Slab 1	Slab 2
Theoretical ultimate load	72,8 kN	128 kN
Measured ultimate load	81,6 kN	129,9 kN

3. The impact on dynamic loading was also simulated in the test.

4. N.V. BEKAERT S.A. has filed an international patent application PCT/EP98/00719 for a reinforcement combining steel bars with steel fibres for a floor on piles. Executing piled floors according the principle of "slab 2" can only be done after official authorisation of nv. Bekaert sa.

REFERENCES

1. DRAMIX GUIDELINES : Design of concrete structures, Steel wire fiber reinforced concrete structures with or without ordinary reinforcement.

2. SAWCZUK, A.JAEGER : Grenztragfähigkeits-Theorie der Platten, Springer Verslag Berlin /Göttingen / Heidelberg.

3. GOSSLA, U. FALKNER, H. : Pfahlgestützte Bodenplatten. In Braunschweiger Bauseminar 1997 -   Innovatives Bauen, Heft 136 des iBMB der TU Braunschweig. ISBN 3-89288-115-4

4. DEUTSHER AUSSCHUβ FÜR STAHLBETON : Hilfsmittel zur Berechnung der Schnittgröβen und Formänderungen von Stahlbetontragwerken, Ausgabe 1992, Heft 240

5. EUROCODE 2 : ENV 1992-1-1 : Design of concrete structures Part 1 : general rules for buildings

FLOORS – THE WORKING SURFACE OF INDUSTRY

A Dennis

Sika Limited

United Kingdom

ABSTRACT. This paper will consider the technology behind a variety of surface treatments that may be applied to concrete and highlight the potential pit falls and benefits of such treatments. Importantly, this paper will show the typical problems that, unfortunately, far too many clients experience. The vital keys to ensuring success are considered whereby, a systematic approach is used to select the right finish for a given situation, such that problem are avoided and designed out from day one. Getting things right first time means that the essential areas such as preparation of concrete and detailing are also covered.

Keywords: Epoxy, Polyethylene terephthalate (PET), Upcycling, Osmosis, Damp proof membrane (DPM), Detailing, Substrate preparation.

Alex Dennis is Market Manager for the Manufacturing Industry at Sika Limited, UK. He has been with the company for 5 years and has a wealth of experience in the use of construction chemicals for the maintenance and upgrade of Industrial Facilities, specialising in high performance flooring systems for the clean-room related industries.

INTRODUCTION

The last few years have seen a major increase in technological innovations that have developed within the construction materials industry. One such an innovation has been the enhancement of man-made resin technology, bringing synthetic resin materials to the fore as a method of improving the surface qualities of floors within all areas of industry.

The floor in any part of an industrial facility - be it a factory, a warehouse, a kitchen or even a plant room - is one of the most important parts of a buildings fabric. Concrete in many cases does not satisfy the demands that are now made upon it due to the increase in technology within all types of industrial production processes.

Industry nowadays demands a floor system which is able to meet the needs of all types of working environment.

Sika Limited is recognised internationally as a symbol of quality and innovation, and has responded to the needs created with modern industry - the need for high performance, durable flooring solutions. Today's discussion is based upon the correct approach required in finding that successful solutions and addressing the long term problems and pitfalls associated with defective floors, allowing your resin floor to be tailored according to your needs.

What is an Industrial Floor?

A floor is significantly more than a flat surface enabling relatively easy movement of goods and traffic from one location to another. It is an industrial road or pavement, a store, a drain, a discharger and container of industrial effluent and corrosive fluids and a protection against both chemical and mechanical attack.

The previously mentioned technological innovations within industry have introduced greater emphasis on health and safety at work, efficiency, environmental acceptance and the need for a more conclusive and tolerable working environment

A floor is one part of the building that cannot be avoided, you must work on it, sit on it, truck, build or support on it, store, spill on it, lift from it, drop or throw on it. Quite simply, an industrial floor is the working surface of industry. This may be nothing new but how many people appreciate it?

What are the Consequences of Industrial Floor Finishes

a) Major Reconstruction

b) Loss of Production

c) Access Problems

d) Breakdown of Machinery

e) Contamination of products

f) Hygiene Problems

g) Product Damage

h) Safety Hazards

i) Premature failure in Service

j) Environmental Contamination

k) Poor Working Environment

We have all had or known of experiences of floor failure and the consequential costs in both human and financial terms that can result.

Despite the importance of floors in industrial installations, specifications in floor finishes generally reflect short term budgets rather than long term durability, hence inviting failure.

THE SYSTEMATIC APPROACH

As with any refurbishment project the principles of philosophy remain the same. There are some fundamental questions that we must ask of ourselves:

What has gone wrong?

What are the symptoms? - What can we see?

Why has it gone wrong?

What is the cause of the problem? - All to often we may look to treat the symptoms without first establishing the true causative factors of the problems we face.

How can we put it right?

A specification needs to be detailed, technically correct and practical. It must fully reflect what the client is wanting to achieve.

How can we protect and prevent it from happening or happening again?

In essence, we must carefully assess ways of protecting against aggressive influences and hence prevent the reoccurrence of problems.

Who should be involved?

Careful consideration must be given to who should be involved on any project. Usually the team would consist of an engineer, an architect, a surveyor, a materials manufacturer and a specialist contractor though this is not inflexible.

The same approach would apply also to a new build situation, as follows:-

a) What can go wrong?

b) Why does it go wrong?

c) How can we put it right?

d) How can we protect/prevent it happening?

e) Who should be involved?

THE PROBLEMS

To successfully evaluate the required performance of a floor we need to be aware of the defects that can occur.

In some instances adequately finished concrete with the correct properties can provide the required finish.

If insufficient care is taken in its production, it is neither suitable as a finish itself, nor suitable for any subsequent finish to be applied to.

Let us consider some typical flooring problems to help answer some of these questions:

Cracking

We must always establish the cause of any cracking before treating the symptoms. How we deal with this problem will depend upon the nature and cause of the crack. Why is it there?

Absence of Waterproof Membrane

Blistering is a typical symptom caused by the absence of a waterproof membrane or indeed failure of an existing membrane.

There is now revolutionary and proven technology to overcome such problems, which can also further serve fastrack construction, by enabling the application of finishes to "green" concrete.

Existing Coating Failure

A loose and flaking coating is often indicative of poor or inadequate preparation. Equally, an inadequate specification in heavily trafficked areas may result in wear through and early coating failure.

Lack of detailing

This is one of the common causes of floor failure. Poor detailing in areas where arises are subject to heavy trafficking, poor detailing at upstands, poor detailing at gulleys and drains and poor detailing of joints will all be sites for premature failure of the floor.

Bad housekeeping

This is one that many of us overlook. Once a floor system is installed, good housekeeping policy must ensure the floor is cleaned and maintained on a regular basis.

VITAL KEYS TO SUCCESSFUL PERFORMANCE

Design

Design is a critical process, it can be the basis for problems or the reason for their absence. All too frequently, insufficient attention is paid to the selection of floor finishes. They are for instance, often classified in specifications for new works as "internal finishes", and are largely ignored until the building and almost certainly, the structural slab is constructed.

In refurbishment projects and maintenance programmes, floors are often reviewed only after the commissioning or even installation of plant and machinery.

"Cost effective design is meeting pre-determined criteria"

Detailing

Before we consider selection of any floor system we must pay attention to the details in the floor.

What is floor detailing? Why is floor detailing so important?

It is a fact that these small but critical areas of the floor cause problems totally out of proportion to their value. They cause the biggest headache to the consultant/specifier and if incorrectly designed the biggest headache to the client once in service.

Floor details include such areas, as gulleys, pipes, drainage channels, joints, coves to upstands etc. All of which are possible sites for premature failure in every floor system if not detailed correctly.

Substrate Preparation

Preparation is vital and again fundamental to the success of any floor system. All too often systems fail due to inadequate preparation.

"Failure to prepare is to guarantee failure"

Good Housekeeping

a) Regular cleaning.

b) Essential maintenance; damage will always occur.

c) Correct selection of type of mechanical traffic, such as tyres and rubber wheels.

d) Protect areas subject to very severe point loading by possibly upgrading the specification locally.

e) Avoid discharge of aggressive fluids and ensure regular removal of spillages.

f) Selection of correct cleaning regime.

DESIGN CRITERIA

Let us explore this point further.

There are a number of design criteria which require detailed consideration on any project. The following summary, with brief comments, should form the basis for the evaluation and selection of most floors - both in new and refurbishment works.

Design Life

This is possibly the most fundamental criterion and is certainly the first question to ask when selecting a floor finish. What is the design life of the plant - 2, 5, 10 or 20 years? Is regular maintenance feasible or desirable? Whatever the answers, the floor finish specification must meet the design life expectancy for the intended maintenance-free period.

Structural Loading

Both the static and dynamic loadings imposed during construction, production, refitting and maintenance should be considered. While a floor topping must be capable of withstanding these demands, it should be remembered that it can only function as well as the substrate to which it is applied, i.e. the structural concrete slab or screed.

Traffic and Mechanical Wear

Physical requirements for resistance to impact or abrasion must be given consideration. Often the greatest wear or exposure occurs in localised areas - trucking aisles or sections around specialised plant, for instance, which may require different treatment to the general floor area.

Chemical Resistance

In addition to assessing the effects on the floor of individual chemicals present in the plant, the consequences of potential chemical reactions must be considered. What happens if any of the chemicals become mixed on the floor? What is their concentration or dilution, both at the time of spillage and after evaporation? Higher temperatures usually increase the aggressive nature of chemicals - particularly, strong acids and alkalis - so temperatures should be assessed in all areas of potential spillage, including production, warehousing and distribution.

Note: A simple automatic washdown system or sparge pipe incorporated into areas of highly aggressive chemical spillage can extend the life of the floor finish - even in extremely onerous conditions.

Slip Resistance

Pedestrian traffic areas require varying degrees of slip resistance, depending on whether the environment is wet or dry. This is principally a question of reconciling surface finish demands with spillage risk. The greater the profile, the greater the slip resistance; however, ease of cleaning and hygiene requirements become more difficult to maintain as the surface profile increases, so balance must be achieved. It is advisable for differing degrees of slip resistance to be designed into the floor to suit the specific requirements.

Note: It may be sensible to involve Health and Safety Officers and trade unions when making decisions regarding slip resistance. An interesting European development, now to be seen in some UK factories, is the use of smooth, easily cleaned production floors in conjunction with the provision of slip resistant footwear for the staff.

Hygiene

Many modern industries, e.g. pharmaceutical, cosmetic, food, beverage, chemical and electronics, now have similar, very demanding hygiene requirements. These progressive industries need "clean room" environments, i.e. floors must be totally dust free, without cracks or angled corners, and easily cleanable, yet still satisfying other individual requirements, such as specific levels of chemical and mechanical resistance. In addition Sikafloors meet the requirements of BS 5295.

Crack-bridging Ability

This relates to structural loading, particularly dynamic loading. What effects do plant vibration and traffic wear have on the floor? How important is it that cracks do not appear? In specific areas of the project, for instance, in mezzanines, in production facilities where aggressive liquids are present, or in "clean room" areas, the floor finish must be dynamically crack-bridging. Alternatively, sufficient stress relief or movement joints need to be incorporated into the substrate during construction to prevent subsequent cracking.

Waterproof Membrane

It is crucial to ensure that there is a waterproof membrane present beneath the slab - if no records exist, and physical inspection is not feasible, this must be determined by testing. Situations where the membrane is defective or absent there are solutions with a new unique epoxy cement technology.

Solvent / Taint Free

There are increasing restrictions on the use of solvented products because of the hazards they present. This is creating a trend towards water based or solvent free products not only in situations where food or products may become tainted, but also, in new construction and usage in enclosed environments where health and safety restrictions may come into play.

Temperature

Thermal shock is a major cause of premature industrial floor failure. It is important to consider not only the temperature of operating machinery and the products in the processes, but also the temperature of adjacent areas. In certain areas, where activities such as autoclaving, cooking, sterilising or blast freezing are carried out close by, extremes of temperature are the norm. These areas may, therefore, require special treatment.

Colour / Aesthetics

An attractive and pleasant environment can assist in increasing productivity and improving industrial relations. Although lighting also plays a part in this, awareness of the psychological significance of colour is growing rapidly.

Colour allows easy and immediate visual identification of different process areas, such as production and packing. It can also be used as an effective aid to safety, defining trucking aisles, wet areas, machinery or chemical exposure risks.

Conductive / Anti-Static

There is an increasing demand by many industries for conductive or anti-static floor coatings. These are used either to prevent electrical interference with sensitive electronic equipment, or to avoid a build-up of static electricity, which may generate sparks creating an explosion risk.

To satisfy these demands it is essential to determine the degree of electrical resistance (conductivity) necessary. In these areas it is normal to specify a conductive floor with an electrical resistance of between 10^{-4} and 10^{-8} ohms dependent on requirements.

Note: The method and mode of earthing (to ring mains, lightning conductors, etc) should be approved, in accordance with relevant local technical and health and safety regulations.

Ease of Cleaning

A flooring system which is installed without prior consultation on "in-service" cleaning procedures can pose problems. Whether machines, high-pressure exposure steam lances or traditional mops and brushes are to be used on the surface, ease of cleaning should always form an integral part of the flooring selection process.

Repair / Maintenance

As part of the normal "wear and tear" at any production plant, some damage to the floor can occur, which necessitates local repair. It is, therefore, essential to ensure that the floor can be repaired simply using the original materials, which must demonstrate good adhesion and compatibility to the surrounding undamaged flooring.

NEW TECHNOLOGIES

P.E.T. Tech®

New paths in formulating synthetic-resin coatings have been established by Sika Limited.

Under the signet P.E.T. tech® the group introduces a complete family of sophisticated polyurethane-coating materials into the market. A significant proportion of the products consists of recycled Polyethylene Terephthalate (P.E.T.) manufactured according to a world-wide patented, Sika owned Upcycling® procedure.

The unique characteristics of this new version of plastics-recycling is the fact that the P.E.T. is chemically converted (Upcycling® technology) into a high-value new raw material. Tailor-made components for different end products can be manufactured by selection of reaction partners and reaction parameters.

The basic material PET itself is available in practically unlimited quantities as beverage bottles (e.g."Coca Cola") are the most popular and well-known constituent part of this versatile material.

For instance, one kilogram of the new industrial flooring material Sikafloor® 325 P.E.T. tech® contains a corresponding P.E.T. portion of approximately one 1,5 ltr thick-walled, Coca Cola-type bottle.

Jordan Grand Prix Limited - a case for PET technology

Jordan Grand Prix selected Sika Limited to assist in assessment of their design criteria and to provide specialist resin flooring systems for the Formula One team's multi million pound manufacturing facility at Silverstone, Northamptonshire. The centre was purpose built for the design, manufacture and building of Formula One racing cars using the latest technologies available.

Performance criteria were analysed in conjunction with Jordan and the new Sikafloor 325 PET tech system was selected for most areas including race shops, machine shops, autoclaves, composites, stores and pattern shops in order to provide a durable "high performance" flooring system capable of withstanding the rigors of industrial use, whilst maintaining a high aesthetic value.

It was recognised by Jordan that the PET system not only offered ecological benefits but also provided technological advantages over the more conventional systems. For example, the system is inherently flexible and less susceptible to scratching than traditional epoxy based flooring systems.

Jordan Grand Prix were the first company in the UK to install the new Sikafloor® PET tech system the most exciting recent breakthrough in resins - 'one company at the cutting edge of technology supporting another'.

Upcycling® technology reveals new components made of a high-quality secondary raw material. Using this technology Sika is contributing its share to the world-wide responsibility of chemical industries. The end products combine the highest of Sika quality with added value for the environment.

SUMMARY

It is apparent that to achieve successful and durable flooring solutions, we need to design and detail floors in accordance with identified performance requirements.

In conjunction with this it is necessary to use quality materials that have a track record of success within the type of industry for which its finishes are intended. These also must be applied by competent applicators.

Follow these guidelines and your resin floor really will be *"Better by Design"*

＃ THEME SIX:
NON-FERROUS REINFORCEMENT

Keynote Paper

NON-FERROUS REINFORCEMENT

R H Scott
University of Durham
United Kingdom

ABSTRACT. The use of fibre reinforced plastic (FRP) reinforcement is a relatively recent development in the design and construction of reinforced and prestressed concrete structures. Its use has been prompted by the corrosion problems which have afflicted structures constructed with conventional steel reinforcement. For this reason, many of its field applications to date have been in bridge decks subjected to hostile winter environments. It is also beginning to be used in repair work where FRP composites are externally bonded to the concrete surface. This paper reviews the different types of FRP reinforcement currently available and illustrates their use by reference to a number of case histories.

Keywords: Fibre reinforced plastic (FRP), Reinforcement, Composites, Reinforced concrete, Prestressed concrete.

Dr R H Scott is a Reader in Engineering at the University of Durham. He spent ten years in industry, designing a wide range of structures in reinforced concrete, structural steelwork, load bearing brickwork and timber before joining the University of Durham in 1978. His research interests are concerned with the behaviour of reinforced concrete structural elements, particularly the measurement of reinforcement strain and bond stress distributions.

INTRODUCTION

Non-ferrous reinforcement, in the form of discontinuous discrete fibres, has been used since ancient times as over 2000 years ago straw fibres or animal hairs were being used to reinforce air-dried loam bricks in order to eliminate crack formation or to preserve load carrying capacity when crack formation did occur. These uses have been perpetuated in modern times since fibres are now added to concrete in order to increase tensile strength, by delaying crack growth, and to increase toughness, by enabling stresses to be transmitted across a cracked section. Desirable properties of fibres are that their tensile strength, elongation at failure and modulus of elasticity should be higher than similar values for the concrete itself. Low creep is desirable and their Poisson's ratio should be similar to that for concrete in order to avoid induced lateral stress [1].

There are many types of non-metal fibres, each with differing characteristics [2]. *Organic fibres* have average tensile strength, low rigidity but high ductile strain. Their colour stability is largely unaffected either by ageing or environmental effects. For these reasons they are used for crack control, sandwich panels, refuse containers and the like. Different types are polypropylene fibres (high alkaline resistance), polyvinyl alcohol fibres (high alkaline resistance, good ageing), polyester fibres (good resistance to acids and alkalis, but low bond strength), polyacrylonitrile fibres (high modulus, good interfacial bonding with cement), polyaramide fibres (good mechanical properties and resistance to chemical corrosion) and plant fibres (very variable properties, but almost unlimited supply). *Glass fibres* are mostly made from glass types E, S and R glass. They suffer from the problem that exposure to alkalis destroys the glass network and thus reduces their strength. Their strength also tends to reduce with ageing, particularly in humid environments. However, research is in hand to improve the problem of alkaline resistance. *Carbon fibres* show a wide range of good mechanical properties. They are very resistant to high temperatures and chemical attack and are particularly well suited to the strengthening of plastics and metals. Currently, their use is restricted due to their relatively high cost. *Ceramic fibres* are usually short fibres used mainly for thermal insulation. *Silicon carbide fibres* are thermally stable and resistant to chemicals and *asbestos fibres* are extremely fire resistant but their use is complicated by their carcinogenic effects.

Organic fibres, particularly polypropylene, are the commonest type of non-metallic fibres used in plain and structural concrete. The quantity, type and length used depend on the specific application. Practical fibre contents range from 0.1 to 1.0% by volume which provide both plastic state and hardened state benefits. Fibres inhibit micro-cracking while the concrete is setting and reduce bleeding and plastic shrinkage cracking. The compressive strength of the hardened concrete is relatively unaffected but tensile properties are improved due to the fibres delaying crack growth. Toughness is significantly increased since fibres allow stresses to be transmitted across a cracked section thus permitting much larger deformations to occur. The use of these fibres is widespread, particularly in ground bearing industrial floor slabs where they are very effective at controlling shrinkage cracking. Other applications, which take advantage of the increased toughness, are in the outer casing of conventionally reinforced driven piles, which are obviously subjected to significant impact loads, and in precast concrete panels, which may suffer accidental handling damage.

Organic fibres do not replace conventional steel bar reinforcement. Their contribution to structural performance is by enhancing the properties of the concrete mix and their properties and use are well researched and documented.

Other suitable fibre families for use in reinforced concrete are glass, carbon or aramid (an abbreviation of aromatic polyamide). These are difficult to handle, however, and so may be combined with resins to form composites. This leads to the interesting possibility of replacing all the main reinforcing steel in both reinforced and prestressed concrete members with reinforcement which is entirely non-ferrous. Considerable advances have recently been made in this field and over fifteen road and pedestrian bridges have been constructed using fibre reinforced plastic reinforcement. This use of fibre reinforced plastics is a new and exciting development in structural design which will now be considered in more detail in this keynote paper.

FIBRE REINFORCED PLASTICS

Corrosion of reinforcing steel is one of the major causes of deterioration in reinforced concrete structures. This is particularly the case with bridge decks where the situation is exacerbated by the frequent use of de-icing salts. Many procedures have been proposed to deal with this problem including the development of new materials. Composite materials, such as fibre reinforced plastic (FRP), are being developed as alternatives for conventional reinforcing steel. This work began over forty years ago and has led to materials which are light, easy to handle and corrosion free. Developments are on-going in Europe, North America and Japan with the first use of glass FRP *rods* being in a prestressed concrete bridge in Germany in 1980. However, the Japanese have probably led the field overall and they were the first to use carbon FRP *strands* in a prestresed concrete bridge structure.

FRP reinforcement is made of carbon, aramid or glass fibres impregnated with a resin and is available in the form of rods, ropes and grids. It has a strength to mass density which is 10 to 15 times that for steel, excellent corrosion resistance, low axial coefficient of thermal expansion and, except for glass FRP, excellent fatigue properties. Its mass is only about one seventh to one fifth of that for steel. It is also non-magnetic which has led to applications in offshore structures and magnetic levitation trains being considered. Disadvantages include high cost (5 to 50 times that for steel), low modulus of elasticity, low failure strain and possible deterioration when exposed to ultra-violet radiation. Aramid fibres can deteriorate due to water absorption and the durability of glass fibres in concrete has still to be resolved. There are also problems anchoring FRP reinforcement when it is used for prestressing, which will be considered later. The different types of FRP reinforcement are described below, using material drawn from the comprehensive reviews by Erki and Rizkalla [3] and Minosaku [4]. The properties and types of FRP reinforcement are summarised in Table 1.

CARBON FIBRE BASED REINFORCEMENT

Carbon fibre reinforced plastic (CFRP) reinforcement is made as rods, tendons and ropes using polyacrylonitrile or pitch-based carbon fibres. It has the highest tensile modulus of elasticity of FRP reinforcement (around 65% of that for steel) but its maximum strain at failure of between 1.2 and 2% is low when uses in reinforced concrete structural elements are being considered. This is illustrated in Figure 1 which gives diagrammatic comparisons of the tensile stress-strain relationship for prestressing wire with those for CFRP and the other FRP's discussed below.

Table 1 FRP Reinforcement
(from Erki and Rizkalla [3])

TYPE		SUPPLIER	ULTIMATE TENSILE STRENGTH (GPa)	ELASTIC MODULUS (GPa)	STRAIN AT FAILURE (%)
Carbon Fibre					
CFCC		Tokyo Rope, Japan	1.8	137	1.6
Leadline		Mitsubishi Kasei, Japan	1.8	147	1.3
Jitec	1-25 mm dia.	Cousin Frère, France	1.5	125	N/A
Bri-Ten Carbon-HS		British Ropes, UK	1.48	136	1.1
Aramid Fibre					
Parafil Ropes	Type A	ICI Linear Composites, UK	0.62	12	5.3
	Type F		1.93	77.7	2.5
	Type G		1.93	126.5	1.5
Phillystran	3/8 in dia.	United Ropeworks, USA	1.90	124	N/A
	1/2 in dia.		1.81	124	N/A
	5/8 in dia.		1.77	110	N/A
Arapree		AKZO and Hollandsche Beton Groep, Holland	2.8	125-130	2.3
N/A		Teijin, Japan	1.9	54	3.7
FiBRA Rod	8 mm dia.	Mitsui, Japan	1.39	64.8	2.1
	12 mm dia.		1.20	64.8	2.1
	16 mm dia.		1.21	64.8	2.1
Glass Fibre					
Polystal E-Glass		Stabag AG and Bayer AG, Germany	1.67	51	3.3
	S-Glass		2.09	63.8	3.3
Isorod		Pultall Inc., Canada	0.7	44.8	1.8
IMCO		IMCO Inc., USA	1.034	41.4	N/A
Jitec	1-25 mm dia.	Cousin Frère, France	1.0-1.6	35-50	4.0
Kodiak	10-25 mm dia.	IGI Int. Grating, USA	0.69	50	N/A
Plalloy		Asahi Glass Matrex, Japan	0.59-0.69	24.5-39.2	N/A

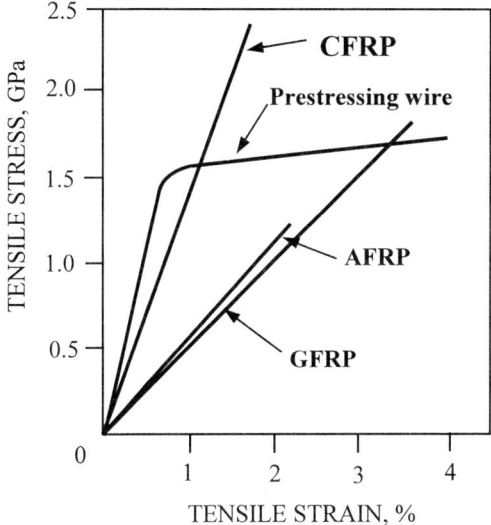

Figure 1 Comparison of Stress-Strain Relationships
(after Erki and Rizkalla [3])

Polyacrylonitrile CFRP

Carbon fibre composite cable (CFCC) is made from wires which consist of polyacrylonitrile fibres impregnated with resin. These are twisted to form a fibre core which is wrapped with synthetic yarns to form a strand. Ropes are made from single, seven, nineteen or thirty-seven wire strands. Diameters vary from 3 mm up to 40 mm and failure loads from 10 kN up to 1100 kN. The resin matrix is hardened by being heat cured. The ropes are sufficiently flexible to be coiled on drums for transport and, prior to heat curing, CFCC can be shaped making it suitable for use as shear links. Its principal uses, however, are as the main tension steel in reinforced concrete members or as prestressing tendons in prestressed concrete members.

Pitch-Based CFRP

Pitch-based round and deformed CFRP bars are made from coal tar pitch-based continuous carbon fibre and epoxy resin. Plain bar diameters range from 1 mm up to 17 mm and deformed bar diameters from 5 mm up to 17 mm. These bars are used in post-tensioned prestressed concrete structures.

Further types of CFRP reinforcement are under development. These include tendons comprising FRP rods in a long helical pitch all contained within a plastic sheath. The FRP may be pultruded continuous carbon fibre, or aramid fibre in a vinyl ester resin matrix. A zero axial coefficient of thermal expansion is claimed for this product.

Applications

CFCC has been used in at least six structures. CFRP strands were first used as tendons in Shinmiya Bridge, a *pre-tensioned* prestressed concrete road bridge constructed in 1988 in Ishikawa Prefecture, Japan. This bridge spans 5.76 m and is 7.0 m wide. The CFRP strands used were 12.5 mm diameter and composed of seven wires. Six strands were provided in the bottom flange and two in the top flange of each of the twenty four main beams.

Epoxy coated reinforcing bars were used for the shear stirrups. The same Prefecture later constructed a pretensioned cycling road bridge in 1991 which used CFRP strands to inhibit corrosion from the nearby ocean.

CFRP strands formed part of the tendons in a *post-tensioned* prestressed concrete road bridge constructed in Germany in 1991. This bridge spans 80 m and is 11.2 m wide. Four large diameter cables, each consisting of 19 CFRP strands 12.5 mm diameter, were used which were anchored by a wedge system.

CFRP rods were first used as tendons in a two-span prestressed concrete road bridge (one span pre-tensioned, the other post-tensioned) constructed in 1989 in Kitakyusyu City, Japan. The rods were used in one of the two *post-tensioned* beams which have a solid rectangular cross-section and span 17.55 m. The tendons consisted of eight cables each made from eight CFRP rods 8 mm diameter and anchored by a wedge system.

ARAMID FIBRE BASED REINFORCEMENT

Aramid fibre reinforced plastic (AFRP) reinforcement is supplied as ropes or rods. It has a higher failure strain than CFRP reinforcement. The three types of fibres used are Kevlar (29 and 49), Twaron and Technora.

Ropes are made from a closely packed core of continuous high strength synthetic fibre filaments encased in a polyethylene sheath. Their weight is only one seventh that of steel wires. They have been used for prestressed fibreglass pedestrian bridges and have been proposed for use in externally strengthening concrete beams. Diameters range from 7 mm up to 140 mm.

Rods are made in the form of pultrusions of aramid fibre and epoxy resin, the latter providing the inter-fibre bonding. Diameters range from 3 mm to 8 mm and the material can also be produced in flat strips with cross-sections varying from 0.5 x 20 mm to 5.6 x 20 mm. An alternative production method is to braid multiple bundles of fibres and then impregnate the bundles with epoxy resins. The resulting rods are available in diameters from 3 mm to 16 mm and they can be wound on to drums to facilitate transport.

The braiding process permits larger rod diameters to be manufactured than is possible with the pultrusion process. Braided rods, which have surface protrusions and depressions, have a better bond performance than pultruded rods which are essentially smooth. Rods of both types can be used for flexural and shear reinforcement and also for both pre-tensioning and post-tensioning.

Applications

The first use of AFRP rods was in a prestressed concrete road bridge built by the Sumitomo Construction Company, Japan, on the site of one their research facilities. One span was a pre-tensioned composite slab spanning 12.5 m and the other a post-tensioned box girder spanning 25.0 m. AFRP rods were used for all the tendons in both spans. AFRP reinforcement was also used for the stirrups and reinforcing bars in the pre-tensioned span and for the external cables in the post-tensioned span. The rods were 6 mm diameter and, when used for post-tensioning, were anchored by bonding with an inorganic mortar in cylinders of metal or FRP.

Other applications in Japan have been in slab bridges, One, a road bridge, has three 12 m spans and used braided AFRP tendons, 14 mm diameter, in three of its 21 pre-tensioned beams. In another case, AFRP bands were used as tendons in a post-tensioned prestressed concrete pedestrian bridge spanning 54.5 m and built in 1990. Each band had a cross-section of 4.86 x 19.5 mm. Eight bands were combined to make a single cable and 16 cables were used. AFRP bands were also used as tensile reinforcement. Cables were anchored by inserting eight AFRP bands into a sleeve and filling it with expansive mortar.

GLASS FIBRE BASED REINFORCEMENT

Glass fibre reinforced plastic (GFRP) reinforcement is the cheapest type of FRP reinforcement. E-glass and S-glass are both used. Most GFRP is unsuitable for use in prestressed concrete members due to its very low transverse shear strength which poses problems when designing suitable anchorages. However, bars comprising E-type glass fibre filaments in an unsaturated polyester resin matrix have been successfully used in prestressed work. The bond behaviour of GFRP reinforcement can be improved by roughening the surface with quartz sand. A similar effect can be achieved by applying external fibre windings.

Applications

GFRP reinforcement has been little used in Japan apart from an experimental trial in lattice-form in a pre-tensioned prestressed concrete footbridge. However, three bridges have been constructed in Germany, the first being in 1980 on the outskirts of Dusseldorf. This bridge, having a 6.5 m span, was essentially experimental. The other two bridges are the Ulenberg-Strasse Bridge, a two-span road bridge, also in Dusseldorf and built in 1986, and the Adolf-Kiepert Footbridge, a two-span pedestrian bridge built in Berlin in 1989. The Ulenberg-Strasse Bridge is post-tensioned with 59 cables each consisting of 19 GFRP rods, 7.5 mm diameter. Similar rods were used in the Adolf-Kiepert Footbridge, but this time as external cables for the partially prestressed members.

ANCHORAGE OF FRP REINFORCEMENT

FRP reinforcement's high ratio of axial tensile strength to lateral compressive strength (approximately 20 to 1) makes design of anchorages for tendons in post-tensioned concrete structures particularly challenging.

Systems currently used for steel reinforcement are unsuitable for use with FRP tendons and could, indeed, lead to failure. As the above brief case histories indicate, there is no current anchorage system that is universally suitable for all types of FRP reinforcement. Systems developed to date are mostly grout and wedge-type anchorages. Grout-type anchorages suffer the disadvantage that the tendon and its anchorage have to be supplied as a preassembled unit. This makes service life replacement difficult but, nevertheless, they have been used extensively for both pre and post-tensioning. Wedge-type anchorages are a more recent development used for carbon and aramid fibre tendons. Proprietary systems of both types are described and reviewed in reference 5.

There is no doubt that further development of anchorage systems for post-tensioned prestressed concrete members must take place if FRP reinforcement is to receive greater use in these types of members.

OTHER APPLICATIONS OF FRP

Marine Structures

FRP reinforcement has excellent corrosion resistance and, for this reason, is beginning to find applications in the marine environment. Possible uses are as reinforcing material for fenders, tensioning material for precast members, lateral tie material for precast members and in off-shore floating breakwaters and other floating structures.

Shotcrete

Glass fibre FRP products are increasingly being used instead of welded wire mesh for shotcrete. FRP has excellent corrosion resistance as well as being lightweight and easy to handle. It can also be moulded to the profile of the shotcrete surface more easily than is possible with wire mesh.

Rock Bolts and Embankments

Although more expensive than steel, GFRP and AFRP rock bolts are beginning to be used in tunnelling works. FRP products are also finding a use as soil reinforcement in embankment construction.

USE OF FRP REINFORCEMENT IN NORTH AMERICA

As might be expected, FRP reinforcement has generated considerable interest in North America. As yet there is no design standard for this reinforcement, but developments are reviewed and co-ordinated by the American Concrete Institute through its Committee 440 "Fiber Reinforced Plastic Reinforcement". This committee was established as recently as 1991 and has a large and active membership drawn from universities and industry, although the UK representation is very small. It has produced a state-of-the-art report for FRP [6].

A number of bridges have been constructed in North America using FRP reinforcement, particularly in Canada where winters can be extremely harsh thus making the corrosion resistance of FRP particularly attractive. One of the first to be equipped with FRP tendons was in Calgary, Alberta, which opened in November 1993 [7]. This bridge used both carbon fibre composite cables and Leadline rods (see Table 1). It is a two-span continuous skew bridge of 22.83 m and 19.23 m spans consisting of 13 prestressed concrete beams in each span. However, what is claimed as the world's longest span bridge to use CFRP as prestressing and shear reinforcement was opened in October 1997 in Headingly, Manitoba, Canada. This bridge is described in detail in reference 8. However, as it could be considered as representing the current state-of-the-art use of FRP, a brief description is given below.

Taylor Bridge, Headingly, Manitoba, Canada

Taylor Bridge has a total length of 165 m divided into five equal spans. Each span consists of eight 1.8 m deep I-shaped prestressed concrete beams. A mixture of FRP and steel reinforcement was used. Two beams were pre-tensioned with CFCC composite cables, 15.2 mm diameter, whilst another two beams were pre-tensioned with indented Leadline bars, 10 mm diameter. Two of these beams used both CFCC and Leadline stirrups for the shear reinforcement whilst the other two used epoxy coated steel bars. The deck slab was reinforced with 10 mm diameter Leadline bars and GFRP reinforcement, 15 mm diameter, was used in part of the barrier wall. Notable features of this bridge are the use of FRP as shear reinforcement and the first use in a field application of draped FRP prestressing tendons.

The absence of design codes led to research being conducted at the University of Manitoba to assist with the design process. A scale model of a beam was tested [9] as was a full-scale model of the deck slab [10]. Straight and draped CFRP reinforcement was tested under axial tension and the performance of CFRP reinforcement as shear reinforcement was also investigated [11]. In addition, transfer and development lengths were evaluated [12] and further work was carried out by the Ministry of Transportation of Ontario [13]. Figure 2 compares the moment-curvature relationships for the beams prestressed with CFRP with that for a beam prestressed with steel wire. Before cracking the relationships are identical, but the ultimate strength of the CFRP beams was 50% higher than that for the steel beams. However, the reduced curvature exhibited by the CFRP beams should be noted.

Thirty two straight and 14 draped cables were used in beams prestressed with CFCC and 38 straight and 18 draped bars in beams prestressed with Leadline. Typically, "steel" beams were prestressed with 26 straight and 14 draped steel strands. The cross diaphragms in the bridge were designed to support the dead load of the bridge should there be a failure of the two CFRP beams. Non-prestressed reinforcement was also provided in these beams to develop catenary action should the prestressing fail.

The strain in the CFRP reinforcement was monitored while the beams were transported to site but was found to be very small. However, monitoring is continuing on a long-term basis. The bridge is instrumented with fibre optic sensors and conventional electric resistance strain gauges embedded in the main beams, deck slab and barrier wall. Data are transmitted through telephone lines allowing continuous monitoring of the effects of traffic loads and extreme environmental conditions. These readings can also be correlated with visual records from a video camera making this an example of the new generation of "smart" structures.

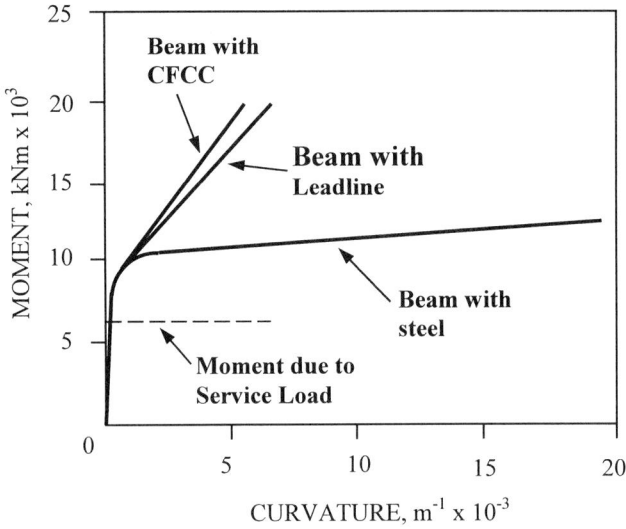

Figure 2 Flexural Capacity of Bridge Beams
(after Rizkalla et al [8])

CONCRETE REPAIR

The use of steel plates for the repair and/or retro-fitting of reinforced concrete structures is well established and has been used very effectively in many countries. The technique consists of bonding steel plates to the surface of a concrete member using epoxy resin. It is a fairly simple technique which can be a very economical way of achieving both increased strength and stiffness. An additional application is to provide confinement to concrete subjected to compression. This can be a very effective means of upgrading structures which have inadequate resistance to seismic effects, and interest in this area has increased dramatically in recent years, particularly in the USA.

A potential problem, however, is the danger of corrosion at the epoxy-steel interface leading to degradation of the bond strength. This can be eliminated by replacing the steel plates with FRP composites. Steel plate bonding has been in used for at least 30 years but the use of FRP composites is much more recent. The largest research and development activity for FRP composites has, to date, been in Japan which has also seen the greatest number of field applications. However, there is growing interest in Europe and the USA (and elsewhere) with ACI Committee 440 again providing a focus for the preparation of design documentation.

The main advantage of using FRP composites in preference to steel is, once again, their resistance to corrosion. Disadvantages are their low moduli of elasticity in tension and low compressive properties.

The composites are anisotropic and have high tensile strength in the direction of the fibres, a feature which can be used to advantage in field applications. Field application of fibre composites to a concrete structure has the following basic steps, as described by Nanni [14]: concrete preparation (cleaning, sealing cracks, rust proofing existing steel reinforcement, smoothing etc.); application of a primer coat; application of the resin undercoat; adhesion of the sheets; application of resin; curing; application of the finish coat.

As stated earlier, Japan has led the field in the development and use of FRP composites and a review of field applications there has been published by Nanni [14]. Nanni has also described work which has been recently completed in Italy [15]. The following examples, to illustrate the field use of FRP composites, have been drawn from this work.

Applications

FRP reinforcement was used in 1994 to strengthen the soffit of a bridge deck at Hiyoshikura, Japan, to increase the load rating of the structure. The bridge had a reinforced concrete deck supported on steel beams. After sealing extensive map cracking in the soffit, two plies of dry carbon fibre sheet were applied, placed parallel and perpendicular to the roadway. A total area of 164 m^2 was treated. Strain gauges installed on some of the existing steel reinforcing bars indicated that tensile strains were reduced by 30-40% as a result of the strengthening work.

CFRP sheets were used in 1996 to repair some columns on a viaduct near Spoleto in Italy. The columns transferred load from the reinforced concrete road deck to a reinforced concrete arch which formed part of the fifty year old viaduct. Corrosion of the steel reinforcement had led to concrete spalling from the face of the columns. Damaged concrete was removed and the original cross-section of the columns reinstated with no-shrinkage mortar. The columns were then wrapped with a single ply of CFRP sheets. Since the column ends were designed to act as hinges, only transverse fibres were used in the wrapping so that confinement was provided to the column without adding moment capacity at the hinge locations.

CONCLUSIONS

The use of FRP reinforcement in reinforced and prestressed concrete structures is a recent development, but it is one which is growing fast. It is an exciting development which is generating increased interest as work moves from the laboratory into field applications. More widespread use continues to be hindered by its high cost and by material properties which can complicate the design process. However, its excellent resistance to corrosion will ensure that research into its use will continue apace and the recent serious problems with epoxy coated steel bars make the use of FRP reinforcement the natural alternative. FRP reinforcement has already been used in a remarkable range of applications. What are now needed are formal design guidelines to assist with its more widespread adoption.

ACKNOWLEDGEMENTS

The assistance of the Fibermesh Division of Synthetic Industries and permission by ACI for the use of their published material are both gratefully acknowledged.

REFERENCES

1. NEVILLE, AM AND BROOKS, JJ. Concrete Technology, Longman, 1990, pp 438.

2. MAIDL, B R, Steel Fibre Reinforced Concrete, Ernst & Sohn, 1995, pp 292.

3. ERKI, M A AND RIZKALLA, S H. FRP Reinforcement for Concrete Structures, Synthetic and Other Non-Metallic Fiber Reinforcement of Concrete, American Concrete Institute, Compilation 28, pp 26-31.

4. MINOSAKU K. Using FRP Materials in Prestressed Concrete Structures, Synthetic and Other Non-Metallic Fiber Reinforcement of Concrete, American Concrete Institute, Compilation 28, pp 17-20.

5. ERKI, M A AND RIZKALLA, S H. Anchorages for FRP Reinforcement, Synthetic and Other Non-Metallic Fiber Reinforcement of Concrete, American Concrete Institute, Compilation 28, pp 33-37.

6. ACI COMMITTEE 440. State-of-the-Art Report on FRP for Concrete Structures, ACI 440R-96.

7. RIZKALLA, S H AND TADROS, G. A Smart Highway Bridge in Canada, Concrete International, Vol. 16, No. 6, June 1994, pp 42-44.

8. RIZKALLA, S H, SHEHATA, E, ABDELRAHMAN, A AND TADROS, G. The New Generation, Concrete International, Vol. 20, No. 6, June 1998, pp 35-38.

9. FAM, A, RIZKALLA, SH AND TADROS, G. Behaviour of CFRP for Prestressing and Shear Reinforcements of Concrete Highway Bridges, ACI Structural Journal, Vol. 94, Jan-Feb 1997, pp 77-86.

10. CHARLESON, K, ABDELRAHMAN, A, RIZKALLA, S H AND SALTZBERG, W. Behavior of a Model Concrete Bridge Reinforced by CFRP, Proc FRPRCS-3, Vol 2, Sapporo, Japan, 1997, pp 575-582.

11. MORPHY, R, SHEHATA, E AND RIZKALLA, S H. Bent Effect on Strength of CFRP Stirrups, Proc FRPCS-3, Vol 2, Sapporo, Japan, 1997, pp 19-26.

12. MAHMOUD, Z AND RIZKALLA, S. Bond of CFRP Prestressing Reinforcement, Proc ACMBS Conference, Montreal, Canada, 1996, pp 877-884.

13. MAHEU, J AND BAKHT, B. New Connection Between Barrier Walls and Bridge Decks, Proc 22nd CSCE Annual Conference, Vol 2, Winnipeg, Manitoba, Canada, 1994, pp 224-229.

13. NANNI, A. Concrete Repair With Externally Bonded FRP Reinforcement, Concrete International, Vol 17, No. 6, June 1995, pp 22-26.

14. NANNI, A. CFRP Strengthening, Concrete International, Vol 19, No. 6, June 1997, pp 19-23.

LOCAL STRENGTH REDUCTION AT BOUNDARIES DUE TO NON-UNIFORMITY OF STEEL FIBRE DISTRIBUTION

P Stroeven

Delft University of Technology

Netherlands

ABSTRACT. The steel fibre reinforcement in concrete generally reveals anisotropy and segregation resulting from the production process. The paper presents a brief survey of the modelling concept for such partially oriented fibre structures. It is demonstrated that this geometrical-statistical approach can be extended to the boundary layer of fibre reinforced concrete elements. Solutions are presented in terms of a product of fibre factor and fibre efficiency. The orthogonal components of fibre efficiency in boundary layers are presented, involving for the `random' portion a decline upon approach of the external surface. The result would be a local reduction in crack resistance, and strength. Approximate expressions are derived

Keywords: Anisotropy, Boundary layer, Concrete, Dispersion, Fibre efficiency, Local strength, Steel fibres, Stereology.

Professor Piet Stroeven is Head of the Group on Materials Engineering of the Division of Mechanics of Materials and Structures, Faculty of Civil Engineering and Geosciences, Delft University of Technology, The Netherlands. He has published widely on subjects in concrete technology relevant for industrialized as well as developing countries, such as micromechanics, mechanisms of damage evolution, bond, fibre reinforcement, cement blending. He serves as a member in RILEM committee TC162-TDF. He is a member of the Advisory Board of two International Journals.

INTRODUCTION

The production process of steel fibre reinforced cementitious materials will in general lead to partial orientation and segregation in the reinforcement structure. This has been demonstrated by quantitative image analysis approaches [1-4]. Anisotropic behaviour in splitting tension and disproportionally improved bending behaviour constitute additional evidences for the occurrence of such phenomena [3]. External surfaces, such as the mold, also influences on the particulate and fibre reinforcement structures in concrete [5,6]. It is found that surface layers are mostly under-reinforced [3]. Such reinforcement characteristics control the local crack driving force, which should be considered in fracture mechanical experiments. It also influences outcomes in studies of damage evolution when based on surface observations.

Models have been developed for the steel fibre reinforcement structure in concrete. A practical solution is provided by taking mixtures of portions of 1-D, 2-D and 3-D fibre systems. Anisotropy is governed by the percentage of 1-D and 2-D fibres. Gradients in the gravity field direction of volume density will reflect segregation [3]. Also the under-reinforcement of boundary layers can be estimated correctly by applying a similar stereological concept [5]. Estimates of global composite mechanical parameters are derived from a law-of-mixtures type of model [7]. The contribution of the steel fibres to these parameters is expressed by the product of 'fibre factor' and 'fibre efficiency'. The fibre factor contains the major fibre features contributing to mechanical behaviour, i.e. volume fraction, aspect ratio and interfacial friction resistance. The appropriate components of the efficiency factor are determined with the help of a first order stereological concept. Next, it is extended to also cover the reinforcement structure near boundaries. An exact solution based on stereological notions has been published earlier [5]. Here the main lines in this approach will be highlighted, whereupon the mechanical implications will be indicated.

FIBRE EFFICIENCY IN BULK

The reinforcement efficiency of small volume fractions of short steel fibres becomes significant only in *cracked* concrete, hence, beyond the so called onset of major cracking at, say, three-quarters of the structural element's ultimate capacity. Mechanical improvements due to the fibrous reinforcement can as a consequence only come from the fibres *intersecting with a crack*, thereby inhibiting it to open up under increasing loadings by gradually slipping out of the matrix and by shearing over the crack edges.

The *active part* of the cracks in concrete subjected to uniaxial tensile stresses is assumed to approx-imately constitute a parallel array. A similar simplifying assumption could govern the case of direct compression. The density of active cracks will be much higher in the second case, however. In the first case, the crack formation will be concentrated for the major part in the *weakest area* of the structural element. This will lead to an additional stochastic component. The crack bridging force in compression is governed by the global value of fibre density, but in the tensile case a 'minimum' value should be considered [9]. This will introduce a size-dependence (among other things).

The distribution characteristics of the *sub-set of fibres intersecting with this crack array* are readily obtained from those in bulk. Stereological estimates for fibre efficiency under these conditions have been based on such notions [7,8]. When a large number of fibres is involved in the process of stress transfer over cracks, the stress component due to the fibres transferred perpendicular to the crack plane will be given by

$$\sigma_f = \frac{\pi}{4} dl\tau_f N_A \overline{(\cos\theta + f\sin\theta)} \qquad (1)$$

in which θ is the angle between the fibre and the norm vector on the crack plane, d and l are the fibre diameter and length, respectively, τ_f is the interfacial friction resistance and N_A is the density of the fibres in the crack plane. It is demonstrated in [8], that for non-plain fibres the plouging action of hooks or anchors at the end of the fibre can be incorporated in τ_f. Actual distribution of fibres can be assumed to approximately form a mixture of 1-D, 2-D, and 3-D portions. Obviously, N_A as well as $\overline{\cos\theta + f\sin\theta}$ depend on composition characteristics, i.e, on the relative contributions from these components. For N_A the following equations are readily obtained [8]

$$N_A(x) = \frac{1}{2} L_{V3} + \frac{2}{\pi} L_{V2} + L_{V1} \qquad (2)$$

$$N_A(y) = \frac{1}{2} L_{V3} + \frac{2}{\pi} L_{V2} \qquad (3)$$

$$N_A(z) = \frac{1}{2} L_{V3} \qquad (4)$$

Here it is arbitrarily assumed that the 1-D component is aligned in the direction of the x-axis (say, the axial direction of a slender structural element), and the $\{x,y\}$ plane is the 2-D orientation plane perpendicular to the gravity field. Further, L_V is the lineal fibre fraction. The trigonometric functions in Equation (1) can be determined for the various composing parts. However, volume averaging should be restricted to the *active* fibres alone. For the 3-D case

$$\overline{\cos\theta} = \overline{\sin\theta} = \frac{\int_0^{\frac{\pi}{2}} \cos^2\theta \sin\theta d\theta}{\int_0^{\frac{\pi}{2}} \cos\theta \sin\theta d\theta} \qquad (5)$$

For the 2-D case

$$\overline{\cos\theta} = \frac{\int_0^{\frac{\pi}{2}} \cos^2\theta d\theta}{\int_0^{\frac{\pi}{2}} \cos\theta d\theta} = \frac{\pi}{4} \quad \text{and} \quad \overline{\sin\theta} = \frac{\int_0^{\frac{\pi}{2}} \cos\theta \sin\theta d\theta}{\int_0^{\frac{\pi}{2}} \cos\theta d\theta} = \frac{1}{2} \qquad (6)$$

Herewith, Equation (1) can be specified for the elementary fibre portions. Further, using $L_V = 4V_f/\pi d^2$ in these expressions, and adding up the respective components yields

$$\sigma_f(x) = \left[\frac{1}{3} V_{f3}(1+f) + \frac{1}{2} V_{f2}\left(1 + \frac{2}{\pi} f\right) + V_{f1}\right] a\tau_f \qquad (7)$$

$$\sigma_f(y) = \left[\frac{1}{3} V_{f3}(1+f) + \frac{1}{2} V_{f2}\left(1 + \frac{2}{\pi} f\right)\right] a\tau_f \qquad (8)$$

$$\sigma_f(z) = \left[\frac{1}{3} V_{f3}(1+f)\right] a\tau_f \qquad (9)$$

in which a and V_f are the fibre's aspect ratio and volume fraction, respectively. The 1-D and 2-D components give the fibre structure its partial orientation. The degree of partially linear orientation is therefore expressed by $\omega_{1,3} = L_{V1}/L_V$ and the degree of partially planar orientation by $\omega_{1,3} = L_{V2}/L_V$. Stress transfer capacity is a product of the fibre factor, $aV_f\tau_f$, and a fibre efficiency, η. The latter components are

$$\eta(x) = \frac{1}{3}(1 - \omega_{1,3} - \omega_{2,3})(1 + f) + \frac{1}{2}\omega_{2,3}\left(1 + \frac{2}{\pi}f\right) + \omega_{1,3} \qquad (10)$$

$$\eta(y) = \frac{1}{3}(1 - \omega_{1,3} - \omega_{2,3})(1 + f) + \frac{1}{2}\omega_{2,3}\left(1 + \frac{2}{\pi}f\right) \qquad (11)$$

$$\eta(z) = \frac{1}{3}(1 - \omega_{1,3} - \omega_{2,3})(1 + f) \qquad (12)$$

The efficiency components according to Equations (10) to (12) explicitly reflect the contributions by shearing over the crack edge (f), and by anisotropy (ω). Particularly for a correct interpretation of materials research it would be advisable to use realistic values for the fibre efficiency, as demonstrated by Table 1.

Table 1 Orthogonal efficiency factors for fibre reinforcement

REINFORCEMENT MECHANISM	$\eta(x)$	$\eta(y)$	$\eta(y)$
a) shearing + pull out + anisotropy: $f = 0.33$; $\omega_{1,3} = 0.10$; $\omega_{2,3} = 0.25$	0.54	0.44	0.29
b) no shearing + pull out + anisotropy: $f = 0.33$; $\omega_{1,3} = 0.10$; $\omega_{2,3} = 0.25$	0.44	0.34	0.22
c) no shearing + isotropy (design): $f = 0$; $\omega_{1,3} = \omega_{2,3} = 0$	0.33	0.33	0.33

It should be noted, that use has been made in Table 1 of an average value of $\omega_{2,3}$. Compositional parameters and compaction procedure influence the degree of orientation. The actual value in materials research can therefore be (considerable) higher [10]

FIBRE DENSITY NEAR BOUNDARIES

Upon approach of an element's surface, more and more fibres of the developed model would start violating the boundary conditions. In a random generation process of coordinates and orientations for fibres, a fibre violating the boundary conditions can be rejected or replaced by one in the same location but with an orientation fulfilling the boundary conditions. Upon rejection, the model would provide a reinforcement structure in the boundary zone of which the fibre density declines in the direction of the specimen's surface. This conforms to experimental data [3]. So, this concept is elaborated, although it can easily be adapted to cover other cases.

The following imaginary experiment is performed (see Figure 1). Fibres with a length l are 'randomly' generated and positioned with one of their ends in the origin of a cartesian coordinate system $\{x, y, z\}$. The other ends of the fibres will cover in a uniformly random way the surface of a sphere with radius l. Thereupon, these fibres are disposed to random positions inside a prismatic element, maintaining their original orientation. The element's surfaces are for the time being assumed to be *permeable*. Figure 1 (b) shows the sphere for $l = 1$, the so called *unit sphere model*. It is used for determination of volume averages of trigonometric functions relevant for the solution of fibre efficiency problems. Next, a crack is considered to run in the boundary zone parallel to the $\{y,z\}$ plane. Fibres will intersect with this crack at a variable distance t from the surface. Fibre and z-axis enclose a sharp angle φ. Next, the boundary zone is subjected to serial sectioning leaving a stack of infinitely thin slices. All fibres intersecting the crack at the same distance t from the external boundary are associated with the local slice and form an *ensemble*. The orientation distribution of these fibres is similar for all ensembles provided they are remote from the external surface (*i.e.* $t \geq l$).

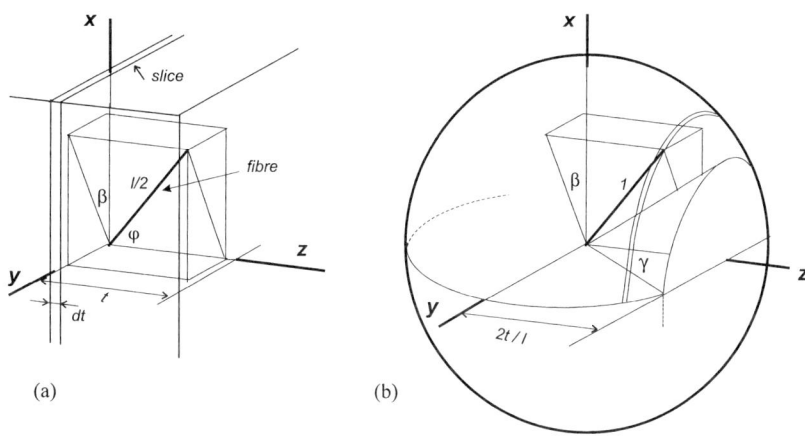

Figure 1 (a) Boundary zone of SFRC element and (b) unit sphere of the (normalized) fibre ensemble associated with the infinitely thin slice shown at the left

At $\varphi \leq \alpha$, with $\cos \alpha = t/l$, a portion of the fibres will penetrate the element's surface for a particular ensemble. This portion is 'removed' from the model. 50 % of the fibres is removed when $\varphi = \gamma$, with $\gamma = 2t/l$. The angles α and γ can be associated with unit sphere sectors, one of which is indicated in Figure 1. Figure 2 presents a $\{x, z\}$-cross section of the boundary zone shown in Figure 1, at the left. The removed portion can be demonstrated proportional to $\cos \alpha / \cos \varphi$. The 2-D fibre component is taken parallel to the $\{x, y\}$-plane. In this case, the external surface will have no influence on the fibres. The fibres will not contribute to stress transfer when the 2-D portion would have been taken parallel to the $\{y, z\}$ plane. For the two-dimensional case, where the orientation plane of the 2-D portion coincides with the $\{x, z\}$-plane, see [13]. Hence, in analogy with Equations (3) to (4)

$$N_A(x) = \eta_3 L_{V3} + \frac{2}{\pi} L_{V2} + L_{V1} \tag{13}$$

$$N_A(y) = \eta_3 L_{V3} + \frac{2}{\pi} L_{V2} \tag{14}$$

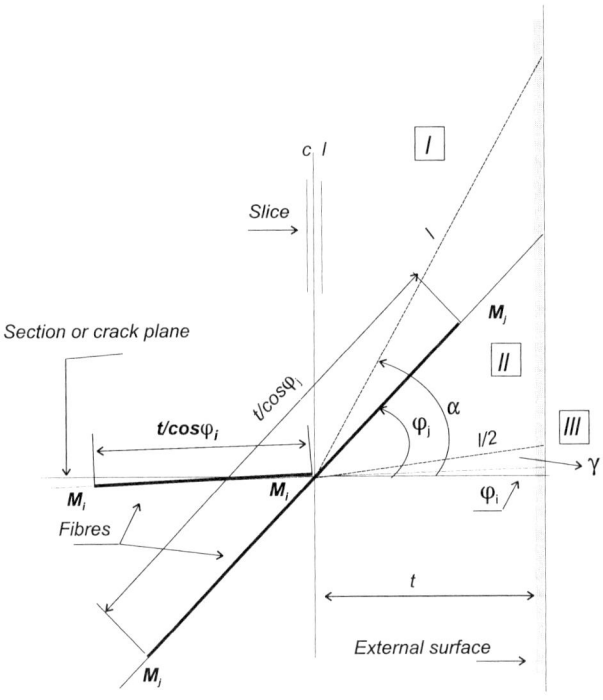

Figure 2 Fibre ensemble in cross section {x, z} perpendicular to external boundary of SFRC element of Figure 1, at the left

$$N_A(z) = \eta_3 L_{V3} \tag{15}$$

in which [5]

$$\eta_3 = \frac{2}{\pi} \frac{\int_0^\alpha \frac{\cos \alpha}{\cos \varphi} \sin^2 \varphi d\varphi + \int_\alpha^{\frac{\pi}{2}} \sin^2 \varphi d\varphi}{\int_0^\alpha \sin \varphi d\varphi}$$

$$= \frac{2}{\pi} \cos \alpha \, \ln \tan\left(\frac{\alpha}{2} + \frac{\pi}{4}\right) - \frac{1}{2\pi} \sin 2\alpha + \frac{1}{2} - \frac{\alpha}{\pi} \tag{16}$$

This yields a value of 0.5 in bulk *(t≥l)* and a to zero declining value in the boundary zone. The partially ordered structure has a finite efficiency factor at the exterior surface according to Equations (13) to (15).

Substitution of the degrees of orientation yield for the orthogonal components of the reinforcement density

$$N_A(x) = \left[\eta_3 + \omega_{2,3}\left(\frac{2}{\pi} - \eta_3\right) + \omega_{1,3}(1 - \eta_3)\right]L_V \quad (17)$$

$$N_A(y) = \left[\eta_3 + \omega_{2,3}\left(\frac{2}{\pi} - \eta_3\right) - \eta_3\omega_1\right]L_V \quad (18)$$

$$N_A(z) = \eta_3(1 - \omega_{2,3} - \omega_{1,3})L_V \quad (19)$$

with $\omega_{2,3} = L_{V2}/L_V$ and $\omega_{1,3} = L_{V1}/L_{V3}$. The average value of η_3 over a boundary layer with a thickness of half the fibre length equals 0.305 [11]. Table 2 presents the average values of the orthogonal components of the fibre reinforcement density in boundary layers with a thickness of half the fibre length.

Table 2 Average orthogonal values of fibre reinforcement density in boundary layers

STRUCTURAL CONDITIONS	$N_A(x)/L_V$	$N_A(y)/L_V$	$N_A(y)/L_V$
Partially linear-planar: $\omega_{1,3}=0.10$ $\omega_{2,3}=0.25$	0.46	0.36	0.20
Partially planar: $\omega_{1,3}=0$ $\omega_{2,3}=0.25$	0.39	0.39	0.23
Isotropic: $\omega_{1,3}=\omega_{2,3}=0$	0.30	0.30	0.
design $(t/l \geq 1)$	0.5	0.5	0.5

The ratio of the two extreme values of the orthogonal components of the reinforcement density defines the degree of *anisometry*, Ω. By good approximation, it is found that

$$\Omega = 1 + 3.2\,\omega_{2,3} + 5\,\omega_{1,3} \quad (20)$$

Using average material research data $\omega_{2,3} = 0.25$ and $\omega_{1,3} = 0.1$, the degree of anisometry is 2.3 (=0.46/0.20). When the linear component is negligible - other conditions being similar – Table 2 yields $\Omega = 1.7$ (=0.39/0.23). The anisometry in reinforcement underlies the mechanical response which is denoted by *anisotropy*. Anisometry in bulk (where $\eta_3=0.5$) will amount only 1.8 for the partially linear-planar, and 1.4 for the partially planar systems (under the same conditions as for the boundary zone). Hence, anisometry is enhanced near the boundaries of structural elements of SFRC. Since the impact of the boundary is on the 3-D fibre component, inevitably, $N_A(z)$ will be the most seriously affected (*i.e. reduced*).

FIBRE EFFICIENCY NEAR BOUNDARIES

The efficiency factors for *stress transfer* can be determined along similar lines. From Figure 2 it can be derived that the embedment length, l_e, is given by either one of the following expressions

$$l_e = \frac{l}{4}; \quad l_e = \frac{l}{4}\left(-1 + 4\frac{\cos\alpha}{\cos\varphi} - 2\frac{\cos^2\alpha}{\cos^2\varphi}\right); \quad l_e = \frac{l}{4}2\frac{\cos^2\alpha}{\cos^2\varphi}\bigg);\quad (21)$$

whereby the first equation holds for $\varphi \geq \alpha$, the second for $\gamma \leq \varphi \leq \alpha$, and the third one for $0 \leq \varphi \leq \gamma$, with $\cos\gamma = 2t/l$. The appropriate projection factor is $\sin^3\varphi\cos^2\beta$, in which β is an angular parameter in the $\{x,y\}$ plane. In the outer half of the boundary zone, fibres can be categorized in three groups, in the inner half of the boundary, domain III in Figure 2 is nonexistent. Therefore, integration of the projection factor over all angles φ should be accomplished separately for these two half zones. In doing so, it is found that the efficiency factor for stress transfer at a crack is given by [5]

$$(t/l \leq 1/2) \quad \eta_{c3} = \cos^3\alpha - 2\cos^2\alpha + 2\ln 2\cos\alpha \quad (22)$$

$$(t/l \geq 1/2) \quad \eta_{c3} = -\frac{1}{3}\cos^3\alpha + 2\cos^2\alpha - \cos\alpha - 2\cos\alpha \ln\cos\alpha - \frac{1}{3} \quad (23)$$

These curves have a common value of 0.95 at $t = 1/2$. Again, the 1-D and 2-D components are not affected, òr they are not contributing to stress transfer.

For reasons of simplicity, a partially-planar system is considered. The contribution of L_{V1} can always be added. It is not affected by nearby boundaries, because the fibres are assumed parallel to the specimen axis. When it is assumed for convenience reasons, that $(1+2f/\pi)/(1+f)$, Equations (7) to (9) can be specified for the boundary zone by

$$\sigma_f(x) = \frac{1}{3}aV_f\tau_f\eta_{c\|} = \frac{1}{3}aV_f\tau_f(1+f)\left(\eta_{c3} + \left[\frac{3}{2} - \eta_{c3}\right]\omega_{2,3}\right) \quad (24)$$

$$\sigma_f(z) = \frac{1}{3}aV_f\tau_f\eta_{c\perp} = \frac{1}{3}aV_f\tau_f(1+f)\eta_{c3}(1-\omega_{2,3}) \quad (25)$$

The orthogonal fibre efficiency components in Equations (24) and (25) are displayed in Figure 3.

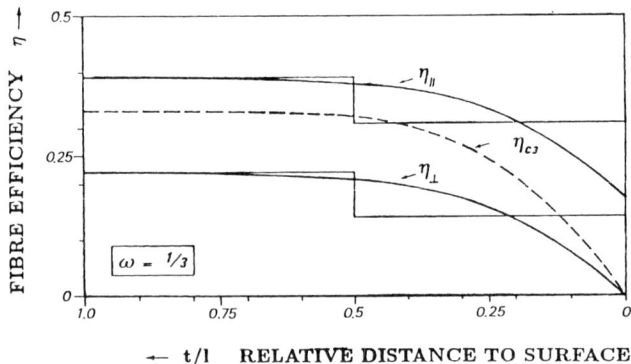

Figure 3 Fibre efficiency factors parallel and perpendicular to external surface

STRENGTH AT BOUNDARY LAYER

The average fibre contributions to the orthogonal *composite strength* components are given by Equations (24) and (25) $\overline{\eta}_{c3} = 0.64$ is substituted [11]. Table 3 presents the resulting average orthogonal values of fibre efficiency for various concepts (anisotropic or isotropic composite; shearing and/or pull out mechanism).

Table 3 Average values of fibre efficiency for stress transfer in orthogonal directions in boundary zone and in bulk

STRUCTURAL CONCEPT	$\eta_{c\|}$	$\eta_{c\perp}$
Partially planar ($\omega_{1,3} = 0$) +		
Shearing and pull out ($f = 0.33$)	0.37	0.21
Anisotropy + pull out ($f = 0$)	0.28	0.16
Isotropy + pull out ($f = 0$)	0.21	0.21
Design: isotropy + pull out	0.33	0.33
Bulk: anisotropy + pull out	0.38	0.25
Bulk: anisotropy +		
Pull out and shearing ($f = 0.33$)	0.50	0.33

Obviously, the fibre contribution to the relevant strength component strongly depends on the structural evaluation concept. The first line of Table 3 is the most realistic one for a boundary zone. Compared with the design concept - which neglects the boundary effect - a dramatic reduction in fibre efficiency can be observed in the direction perpendicular to the 2-D fibre orientation plane (a drop from 0.33 to 0.21). The last line in Table 3 represents a realistic solution for bulk properties. The significant under-estimation by the design concept of reinforcement efficiency in the direction of the 2-D fibre orientation plane can be observed.

For composite strength properties, also the matrix contribution has to be accounted for. The average strength decline in a boundary zone of half the fibre length is for the two orthogonal components of a partially planar system given by

$$R(x) = \frac{\sigma_e(x)}{\sigma_i(x)} = \frac{1 + 0.21A(1 + \frac{4}{3}\omega_{2,3})}{1 + 0.33A(1 + \frac{1}{2}\omega_{2,3})} \tag{26}$$

$$R(z) = \frac{\sigma_e(z)}{\sigma_i(z)} = \frac{1 + 0.21A(1 - \omega_{2,3})}{1 + 0.33A(1 - \omega_{2,3})} \tag{27}$$

with indices e and i denoting the 'exterior' (boundary layer) and 'interior', respectively. $A = a.V_f/(1-V_f).\tau_f^*/\sigma_m$ This factor contains all relevant reinforcement parameters (the fibre factor), and the tensile strength of the cementitious matrix, σ_m. It should be noted further, that $\tau_f^* = (1+f)\tau_f$ provides for the pull out resistance due to interfacial friction resistance (steel shaft and matrix), and to the ploughing mechanism of the hook or anchor at the fibre's end [8]. So, it can be applied for plain and non-plain fibres alike.

For a plain steel fibre type, a value of 6 was experimentally found [8]. In that case, $R(x)=0.82$ and $R(z)=0.78$. Both values of the strength drop will be reduced to 0.76, when the anisometry of the fibre reinforcement is neglected. The strength drop at the boundary will somewhat increase at larger values of A (hooked fibres). The component perpendicular to the external surface will drop in the boundary layer to about three quarters of its bulk value in the range from zero to moderate degrees of fibre orientation. This might be a consideration with respect to spalling off risks. The degree of anisotropy in bulk (defined by the ratio of largest and smallest strength components) is about 1.3 (for $\omega_{2,3}=0.25$). This will rise to about 1.4 for the average orthogonal strength components in the boundary zone.

CONCLUSIONS

The presented methodological stereological framework allows an economic and reliable solution to the assessment of global fibre dispersion characteristics. The most common situation of a partially linear-planar system is described. Boundary effects are successfully analysed along similar lines. Herewith, relevant mechanical properties (particularly anisotropic ultimate strength values) are estimated. The presented concept is (at least) relevant when assessing properties of SFRC in materials research.

REFERENCES

1. STROEVEN, P AND SHAH, S P. Use of radiography-image analysis for steel fibre concrete. Testing and Test Methods of Fibre Cement Composites, Ed. R N Swamy, Lancaster, Construction Press, 1978, pp 345-53.

2. KASPERKIEWICZ, J, MALMBERG, B AND SKARENDAHL, A. Determination of fibre content, distribution and orientation in steel fibre concrete by X-ray technique. Testing and Test Methods of Fibre Cement Composites, Ed. R N Swamy, Lancaster, Construction Press, 1978, pp 298-305.

3. STROEVEN, P AND BABUT, R. Fracture mechanics and structural aspects of concrete, Heron, 1986, 31(2), pp 15-44.

4. GRANJU, J-L AND RINGOT E. Amorphous iron fibre reinforced concretes and mortars, comparison of the fibre arrangement. Acta Stereol., 1989, 8(2), pp 689-694.

5. STROEVEN, P. Effectiveness of steel wire reinforcement in a boundary layer of concrete. Acta Stereol., 1991, 10(1), pp 113-122.

6. STROEVEN, M AND STROEVEN, P. Computer-simulation approach to the ITZ; effect of fine particle additions. Brittle Matrix Composites 5, Eds. A M Brandt, V C Li and I H Marshall, Warsaw, Woodhead publ, 1997, pp 310-319.

7. STROEVEN, P. Structural characterization of steel fibre reinforced concrete. Brittle Matrix Composites 2, Eds. A M Brandt and I H Marshall, Elsevier Appl. Sc., London, 1989, pp 34-43.

8. STROEVEN, P AND DE HAAN, Y M. Structural investigations on steel fibre concrete by stereological methods. High Performance Fibre Reinforced Cement Composites, Eds. H W Reinhardt and A E Naaman, E&FN Spon, 1992, pp 407-418.

9. STROEVEN, P AND GUO, W. Structural modelling and mechanical behaviour of steel fibre reinforced concrete. Recent Developments in Fibre Reinforced Cement and Concretes, Elsevier Appl. Sc., London, 1989, pp 345-354.

10. STROEVEN, P AND BABUT, R. Structural variations in steel fibre reinforced concrete and its implications for materials behaviour. Brittle Matrix Composites 1, Eds. A M Brandt and I H Marshall, London, Elsevier Appl. Sc., 1986, pp 421-434.

11. STROEVEN, P. Boundary effects in fibre reinforced composites. Report 25-89-01, Delft University of Technology, 1989.

12. STROEVEN, P. Steel fibre reinforcement at boundaries in concrete elements. Workshop on High Performance Fiber Reinforced Cement Composites, Mainz, May 16-19, 1999 (to be published by RILEM)

EXPERIMENTAL INVESTIGATION OF POLYPROPYLENE FIBRE REINFORCED CONCRETE

A Al-Robaidi

M R Resheidat

Al-Balqa' Applied University

Jordan

ABSTRACT. This paper presents the results of experimental study to develop concrete composite of selective individual ingredients and their compatible mixtures by using fibrous PP materials aiming at producing lightweight PP concrete parts for a wide range of applications. Selection of Polypropylene fibres was made due to its superior mechanical properties and its lower cost. In a typical process of tailoring the specific combination of properties, physical mixing of PP fibres and reinforcing additives to concrete were developed.

The binary and multi component mixture of PP-fibres and concrete obtained was investigated. The evaluation of PP- fibre concentrations, fibre length/practical size, fibre orientation and its effect on the composite properties was carried out to develop relationship of phase morphology and product performance. The fibres were quickly and easily mixed with the concrete mass creating a very effective multi-directional secondary reinforcement. Conclusions were derived in conjunction with the orientation of fibres in the concrete mix.

Keywords: Concrete, Composite, Reinforcement, Polypropylene, Fibres, Orientation, Tensile strength, Elongation, Morphology.

Dr Amin Al-Robaidi is a an Assistant Professor at the Faculty of Engineering, Al-Balqa' Applied University. He is lecturing Material Science and Polymers. He is specialized in polymeric material and its applications.

Professor Musa R Resheidat is a Professor and Dean of Engineering. His research interests are concerned with the structural design of reinforced concrete structures in general, assessment and development of material in particular. Professor Resheidat has published some forty papers in his field. He is an ASCE fellow, an ACI member, and registered Expert with Jordan Engineers Association.

INTRODUCTION

Polypropylene fibers are being used through the USA and Canada in concrete constructions to control unsightly and troublesome cracking in concrete. Polymer blends are a potentially inexpensive route to the formulation of new products without the need for exhaustive research and development cost[1, 2]. Plastic materials including Polypropylene PP-fibres are now one of the fastest growing materials for different applications worldwide with average growth rates per year of 5% [3, 4]. The increasing concern about environmental pollution highlights the needs for new materials and requires innovative solutions [5, 6]. This research demonstrates the feasibility of reinforcing concrete by using PP-fibre reinforced plastics. These fibres are attractive for this use due to their high tensile strength, light weight, and resistant to corrosion. With the conventional PP-fibres an outstanding mechanical properties of this material represented with its high elastic modulus up to 5100 MPa is realized. At this expense a lightweight system could be developed. The main drawback is the material cost. The primary area of interest was the magnitude of increasing strength and stiffness of the composite provided by the PP-fibres and the effect that the differing strength of elastic modulus contributed to this increase. Accurate predictions are needed to develop suitable procedures for strengthening. Other interest includes the theoretical analysis of prediction of strength and stiffness and investigation of failure modes will be investigated in a later stage.

EXPERIMENTAL DETAILS

Materials

LADENE PP homo-polymer were supplied by SABIC / Saudi Arabia. To produce fibres on a Brabender spinning line. PP continuous fibres were produced with a low denier monofilament and then cut in 50 mm length. The materials properties were tested and are listed in Table 1.

Table 1 Material properties

MELTING TEMPERATURE (°C)	E- MODULUS (MPa)	TENSILE STRENGTH (N/mm²)	ELONGATION AT BREAK (%)
135-140	2400 – 4800	90 - 140	15-40%

Tensile strength and elongation at break were measured along the machine direction at a crosshead speed of 10 mm/min.

Casting of Composite Samples

Eight cylinders and sixteen sample bars were produced for this program with a maximum aggregate size of 5 mm and a minimum compressive strength of 30 N/mm² were specified. In order to fulfill these objectives crushed limestone and basalt were mixed with 6% cement by total weight of the mixture. The samples produced in this work were classified into four main groups according to the PP Fibre concentration, as shown in Table 2. Fibre length was 50 mm and fibre diameter 50 microns.

For compression test, two cylinders of each batch were produced with the dimensions ϕ152.4 x 305 mm. The strength of the cylinder in unconfined compression is shown in Table 2. As it can be seen in Figures 1 and 2, the actual strength some what deteriorates and did not over exceeded the non reinforced cylinders.

Table 2 Concrete compression tests, ϕ152.4mm .x 305mm cylinders

CYLINDER	BATCH	ULTIMATE STRENGTH (N/mm²)	E-MODULUS (N/mm²)	FIBRE CONCENTRATION
A	1	35.88	3700	0% ref.
A	2	34.4	3200	0% ref.
B	1	36.7	3500	2.5%
B	2	33.8	3300	2.5%
C	1	29.9	3450	5%
C	2	37.6	3600	5%
D	1	36.2	3100	10%
D	2	39.3	3400	10%

Samples Used

Four 150mm x 30mm x 10mm thickness concrete samples-bars were cast for each batch. The PP-fibres in Table 3 were randomly mixed in the first batch. In a second batch, the PP-fibres were oriented in axial direction.

Table 3 Concrete tensile samples

SAMPLES	BATCH	ULTIMATE STRENGTH(N/mm²)	FIBRE CONCENTRATION	E-MODULUS (N/mm²)
A		34.88	0%	3700
Er	Random	40.2	2.5%	3800
Eo	Oriented	47.1	2.5%	3845
Dr	Random	43.5	5%	3850
Do	Oriented	52.1	5%	4050
Fr	Random	47.7	10%	3825
Fo	Oriented	54.3	10%	4090
Gr	Random	38.7	25%	3750
Go	Oriented	57.2	25%	4080

RESULTS AND DISCUSSIONS

All the bars Eo (o for oriented), Do, Fo, Go, and Er (r for random), Dr, Fr, and Gr were tested on an Instron Tensile Testing machine. The samples were subjected to longitudinal tensile test (according to ASTM D 3039-76) to determine elastic modulus and tensile strength Figures 3 through 5.

The tensile strength of the miscible blends frequently strength vs. fibre concentration exhibited an increase with increasing PP concentration up to a maxima at 10% in randomly mixed samples. After that, a decrease in tensile values was observed.

This was explained due to phase separation occurred due to partial immiscibility of PP in the concrete matrix. The mechanical properties were adversely influenced by phase separation. Since PP is a polar material, the bond between fibre and matrix was not very strong.

Fibre treatment with acid could increase the bond to the matrix, but on the other hand will badly affect the mechanical properties of the fibres.

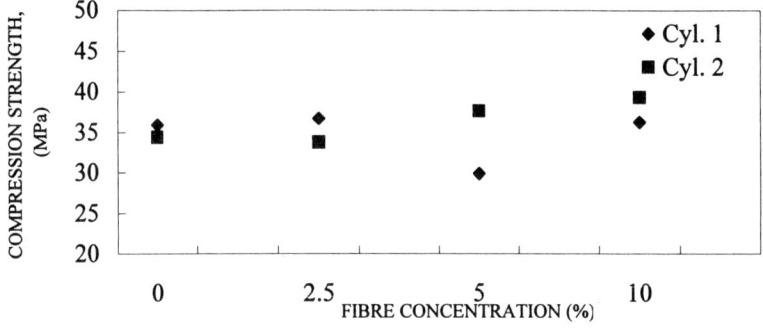

Figure 1 Compression strength vs. fibre concentration

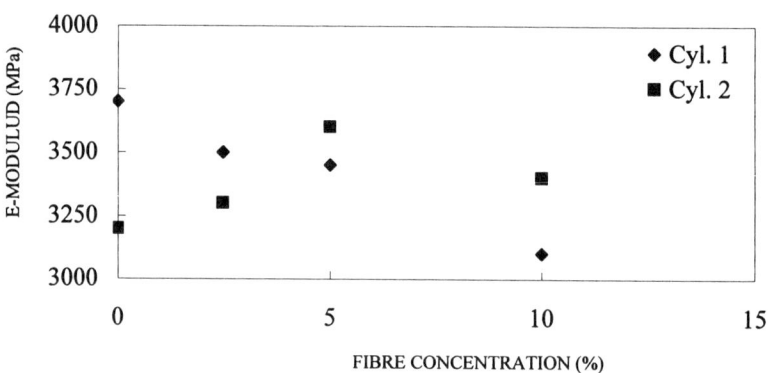

Figure 2 Modulus of elasticity vs. fibre concentration

Polypropylene Fibre Concrete 393

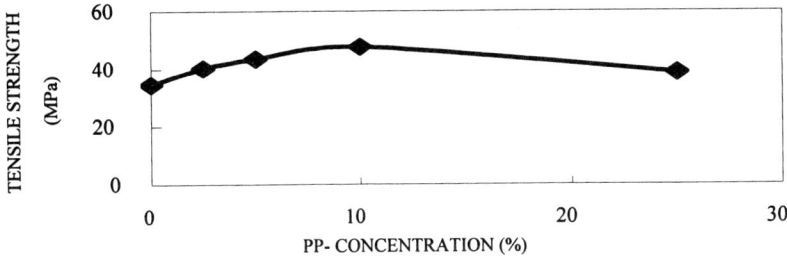

Figure 3 Tensile strength of random PP- fibre reinforced samples. vs. fibre concentration

As mentioned earlier, the presence of fibre in the concrete matrix resulted in an increased tensile strength, Figures 3 and 4.

It is therefore essential to optimize first the PP concentration otherwise an adverse effect would be observed. The available data indicate that 10 % is the best concentration, Figure 3.

Samples with oriented fibres the phase separation was clearly observed. The tensile values were proportional to the fibre concentration, Figure 4. Whereby it is expected that a concentration of more than 25% will have similar effect to the non oriented ones.

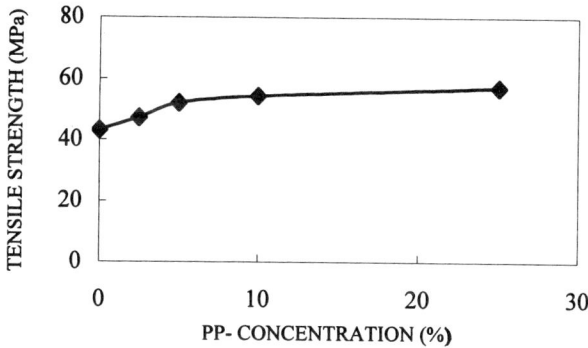

Figure 4 Tensile strength of oriented PP-fibre reinforced concrete samples

Phase separation of oriented fibres was very clear present and observed in three-point binding test. Accumulation of PP-fibres did accelerate phase separation It should be highlighted that fibres did not contribute to improve the extensibility of the bars. Since an accurate elongation measuring device was not available, visual inspection did not indicate any improvement related to elongation.

We want the best of both materials without having along with it the worst of both materials. This is the challenge on optimizing the mixture concentration and to increase the bond between fibre and matrix. The result indicates that it's possible to develop such formulations. More optimization and development work has to be carried out.

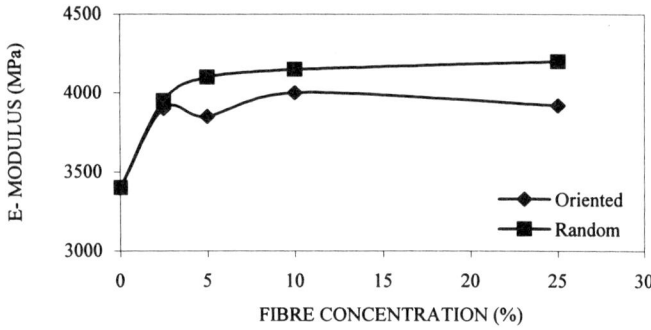

Figure 5 Modulus of elasticity of random and oriented PP- reinforced concrete samples

CONCLUSION

The presence of Polypropylene fibre in a concrete mixture and its contribution to improve the composite mechanical properties has been investigated. The fibre concentration was varied to optimize the composite mechanical properties. Slight increase in tensile strength and modulus of elasticity could be observed. Typical composites are blended to produce lightweight systems with competitive cost benefit relation.

Since Polypropylene material is a polar organic material, delaminating or phase separation was observed, especially in unidirectional fibres. However bond effectiveness between PP-fibres and concrete should be verified to achieve best results. More research and development work have to be conducted to optimize the bond between fibre and matrix.
Randomly mixed PP-fibres assist in relieving the stress that is developed during shrinkage of the concrete mixture. This energy is distributed to the fibres and will result in the reduction of plastic and drying shrinkage cracking. The study indicated that more effort is needed to avoid delaminating and to optimize composition ratios.

REFERENCES

1. KOHUDIC, M A. Advances in polymer blends and alloys technology; Techonomic Publishers, Lancaster, PA., 1988.

2. FYFE, E R, DUANE, J GEE, AND MILLGAN P B. Composite system for seismic applications. Concrete International; June, 1998.

3. AL- ROBAIDI AMIN, M. HAMMDAN. Unpublished work; SABIC R&D Center.

4. RITCHE, P A, AND THOMAS D A, ET. AL. External reinforcement of concrete beams using fibre reinforced plastics. ACI Structural Journal, July –August, 1991, Vol.88, No.4.

5. JENG, Y S AND SHAH, S P. Crack propagation in fibre-reinforced concrete. Journal of Structural Engineering, Vol.112, No.1, January, 1986, pp 19-34.

6. BARR, B AND MOHAMAD NOOR, M R. The toughness index of steel fibre reinforced concrete. ACI Journal, No.5, September-October, 1985, pp 622-629.

MECHANICAL PROPERTIES OF CARBON FIBRE REINFORCED CONCRETE (CFRC)

Y Sato
C Kiyohara
Oita University

H Sakai
M Nakamura
Mitsubishi Chemical Corporation

K Ueda
Sato-Benec Corporation
Japan

ABSTRACT. It is well known that carbon fibre reinforced cement composites (generally, carbon fibre reinforced mortar, CFRM) have their high strength and toughness. Recently, the new type of carbon fibre, like a steel fibre, was developed in order to produce carbon fibre reinforced concrete (CFRC). Since the best mix proportion of CFRC with a volume content of 0.5% has been proposed through many trial mixes, in this research project, an investigation was performed into the mechanical properties of CFRC. It was found that CFRC showed the increase in direct tensile and flexural strength. The occurrence of cracking due to drying shrinkage of CFRC was remarkably delayed compared to the plain concrete, although the drying shrinkage and creep strain were almost same as those of the plain concrete. Drying shrinkage stress generated in CFRC was calculated by use of F.E.M., and the mean tensile stress subject to the restraint concrete was obtained. It was found that the development of tensile stress occurred in CFRC can be roughly predicted.

Keywords: Carbon fibre reinforced concrete (CFRC), Mechanical properties of CFRC, Creep, Drying shrinkage, Shrinkage cracking, Drying shrinkage stress.

Dr Yoshiaki Sato is a Professor in Director of Concrete Technology of the Department of Human Welfare Engineering, Oita University, Japan. His research interests include the creep and drying shrinkage properties of concrete, the repair of concrete, and the effective use of concrete sludge.

Miss Chizuru Kiyohara is a Research Associate of the Department of Architecture Engineering, Oita University, Japan. She received her Master's degree from Kyusyu University, Japan. Her research interests are in the area of high strength and high performance concrete.

Mr Kenji Ueda is working at the Research and Development Centre of Sato-Benec Corporation, Japan. He received his Master's degree from Oita University, Japan, and is presently working at Oita University as a Doctor's Courses Student.

Dr Hiromichi Sakai is a General Manager of Mitsubishi Chemical Corporation, Japan.

Mr Moriyasu Nakamura is a Chief Manager of Mitsubishi Chemical Corporation, Japan.

INTRODUCTION

Since carbon fibres (CF) exhibit an excellent characteristic of strength and stiffness, they can lead to high flexural and tensile strength, and toughness of composite materials. CF may be considered as one of the optimum reinforcement in many structures. In general, CF is applied to cement mortar based materials, and they can be called carbon fibre reinforced mortar (CFRM). A large number of studies on the properties of CFRM have been made [1,2], and the applications of CFRM have been also reported.

Recently, the strand type of CF obtained by bonding CF elements together has been developed as a reinforcement material for concrete, such as a steel fibre, in order to produce carbon fibre reinforced concrete (CFRC). Studies on the properties of CFRC have just started [3-5], and the best mix proportion of CFRC has been proposed through many trial mixes. This research project was designed to collect the fundamental data on the mechanical properties of CFRC, such as strength, creep, and drying shrinkage cracking resistance.

OUTLINE OF EXPERIMENT

Experimental Program

The experiments consist of the following three series; (1) Series I on the various strength tests (compressive, direct tensile and flexural strength), (2) Series II on the compressive creep tests under the basic and total creep, (3) Series III on the drying shrinkage cracking test. In these series, the effects of the age of CFRC and the volume fraction of CF on the mechanical properties of CFRC were investigated. The outline of experiments is shown in Table 1.

Materials and Mix Proportions

Materials used for the production of CFRC were as follows; high-early-strength Portland cement, silica fume, blended sand, crushed stone, air-entrained water reducing agent and carbon fibre. Physical properties of these materials are shown in Table 2.

Table 1 Outline of experiments

	CONTENTS OF EXPERIMENTS	VOLUME FRACTION OF CF	SIZE OF SPECIMEN	AGE AT TESTING
Series I	Compressive strength Direct tensile strength Flexural strength	0, 0.5 & 1.0 %	φ10x20 cm 10x10x60 cm 10x10x40 cm	7, 28, 56 & 240[*1] days
Series II	Creep tests[*2] (basic & total creep)		φ10x20cm	7 & 28 days
Series III	Cracking tests		See Figure 1	7 & 28 days

[*1] flexural strength test only, and deflection curve was measured at the same time.
[*2] concretes containing 0 & 0.5 volume % of CF with the age of 28 days were used.

Table 2 Physical properties of concrete materials used

Cement	High-early- strength Portland cement, Specific gravity : 3.16
Silica fume	Specific gravity : 2.20
Fine aggregate	Blended sand, Maximum size: 2.5mm, Specific gravity:2.45, Absorption: 2.32%
Coarse aggregate	Crushed stone, Maximum size:13mm, Specific gravity: 2.66, Absorption : 0.43%
Chemical admixture	Air-entrained water reducing agent
Carbon fibre	Diameter: 0.8mm, Length: 40mm

The basic concept in the mix proportion of CFRC is that the slump and air content of concrete containing 0.5 volume % of CF is 18cm and 4.5%, respectively. Mix proportions of concrete with CF content of 0 and 1.0% were adjusted based on the basic mix proportion, so that the slump is totally different from the basic one. Mix proportions of CFRC obtained through many trial mixes are shown in Table 3.

Mixing and Curing

A pan-type mixer with a capacity of 100 ℓ was used for the production of CFRC. The mixing procedure was of importance. All the solid materials except CF were fed uniformly and simultaneously into the mixer, and water and admixture were fed after 60 seconds of dry mixing, and mixed for 80 seconds. Carbon fibres were last to be charged into the mixer because of the prevention of damage of CF from mixing. The volume of one batch was 50 ℓ, and several batches in each mix proportion were needed.

Specimens were demolded at the age of 1 day, and cured in the thick plastic bag (sealed curing) until the age at testing.

Procedures

Strength tests were carried out at the age of 7, 28 and 56 days (240 days for only flexural strength test). Compressometer was used for the measurement of stress-strain curve in compression test. In direct tensile strength test, the new test method developed by the authors [6], (a kind of friction grip type), was applied. In flexural strength test, the third-point loading method was applied, and the maximum tensile fibre strain was measured by use of polyester strain gauge. Load deflection curve was also recorded only at the age of 240 days.

Table 3 Mix proportions

MIX	Vf, %	W/C, %	S/a, %	UNIT WEIGHT, kg/m³						SLUMP, cm	AIR, %	UNIT MASS, kg/ℓ
				C	SF	W	CF	S	G	A			
1	0	50	70	400	44	200	0.0	1097	485	6.66	24	6.8	2.16
2	0.5	50	70	400	44	200	5.4	1089	481	6.66	16	4.2	2.20
3	1.0	50	70	400	44	200	9.0	1085	480	6.66	2	4.9	2.18

C: cement, SF: silica fume, W: water, CF: carbon fibre, S: fine aggregate, G: coarse aggregate
A: chemical admixture

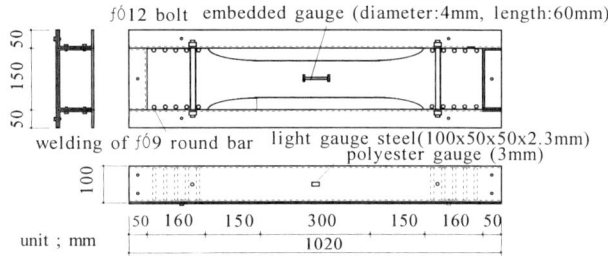

Figure 1 Apparatus for the drying shrinkage cracking test

Creep tests were performed at the age of loading of 7 and 28 days under the basic and total creep condition. Here, basic creep is defined as creep occurring under conditions of no moisture exchange with the ambient; total creep as creep occurring under conditions of drying. Specimens for basic creep were completely sealed with a laminate film, paraffin and moisture prevention tape. The stabilised hydraulic load system was used, and four specimens to be tested simultaneously were held in series in the frame (two specimens for basic creep and two for total creep). The measurement of creep strain was performed by use of the small-embedded gauge developed by the authors [7].

In the drying shrinkage cracking tests, testing apparatus prescribed in Regulation JIS (draft), as shown in Figure 1, was used. Ages at the start of drying were 7 and 28 days. Mean drying shrinkage strain and water loss of companion specimen were measured. Small-embedded gauge was also used to measure the concrete strain. Both the strains of the restraint concrete and steel were measured. Cracking tests were carried out in the room maintained constant temperature and humidity (20±0.5°C, 60±5%R.H.).

RESULTS AND DISCUSSION

Strength Tests (Series I)

Figure 2 shows the development of various strengths with age, and also that of modulus of elasticity under each strength test. It seems that the effects of CF addition on compressive strength (σ_c) are not remarkable, while CF has contribution to the increase in the direct tensile (σ_t) and flexural (σ_f) strength. The direct tensile strength of CFRC showed 1.2 to 1.3 times that of plain concrete, and incorporation of 1.0 volume % of CF increased the flexural strength by approximately 50%, although CFRC with 0.5 volume % of CF showed almost the same strength of plain concrete.

The modulus of elasticity under compressive stress decreases as the CF volume fraction increases. The addition of CF shows a slight increase in the modulus of elasticity under the direct tensile and flexural stress; especially, it seems to show the maximum value at the volume fraction of 0.5%. For the production of concrete specimen with 1.0 volume % of CF, sufficient compaction can not be done because of its low workability, and this is considered to influence the strength and the modulus of elasticity.

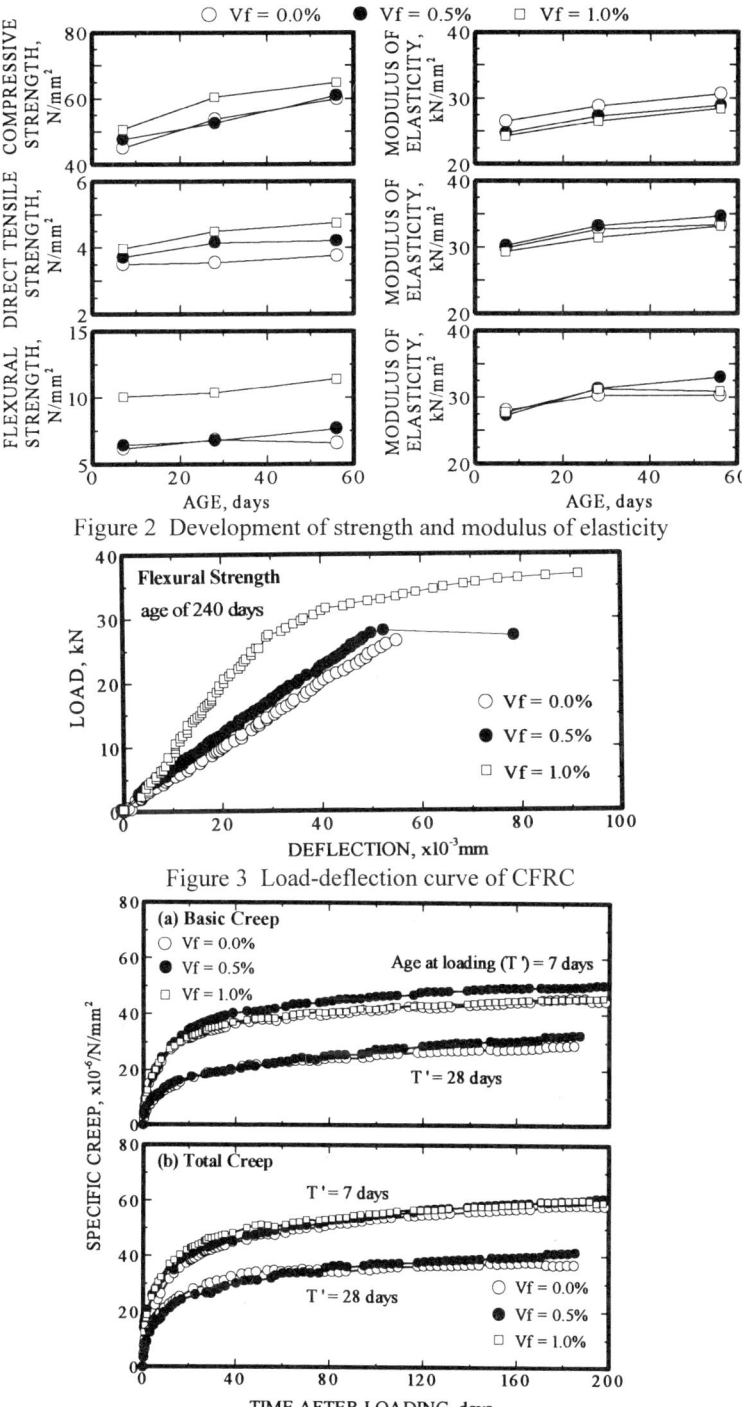

Figure 2 Development of strength and modulus of elasticity

Figure 3 Load-deflection curve of CFRC

Figure 4 Relationship between time after loading and specific creep strain

The greatest advantage in fibre reinforcement of concrete is the improvement in flexural toughness. Figure 3 shows the load-deflection curve of CFRC at the age of 240 days. From this figure, the influence of fibre reinforcement is remarkable; CFRC continues to sustain considerable loads even at deflections considerably in excess of the fracture deflection of plain concrete.

Creep Test (Series II)

Figure 4 shows the relationship between time after loading and specific creep under the basic and total creep condition. Specific creep is defined as creep strain per unit stress, as obtained by the following equation;

$$C_B = (\varepsilon - \varepsilon_e - \varepsilon_a)/\sigma_0, \quad C_T = (\varepsilon - \varepsilon_e - \varepsilon_{sh})/\sigma_0$$

where, C_B = specific basic creep strain, C_T = specific total creep strain, ε = measured strain, ε_e = instantaneous strain at the time at application of load, ε_a = autogeneous shrinkage strain, and ε_{sh} = drying shrinkage strain. ε_a and ε_{sh} with time are shown in Figure 5, and ε_a is started to measure at the age of 7 days. There are no difference due to the addition of CF in both autogeneous and drying shrinkage strain. As can be seen from Figure 4, the inclusion of CF in concrete has little effect on both basic and total creep. For the comparison between basic and total creep, the amount of total creep strain at the age of 28 is about 1.5 times lager days than that of basic creep strain.

Cracking Test (Series III)

Figure 6 shows the results of mean water loss and mean drying shrinkage strain. The effects of CF addition are not recognized in both mean water loss and mean drying shrinkage strain, because these concrete have the same water cement ratio.

The results of cracking test (cracking time and maximum width of cracking) are shown in Table 4. It is obvious that the inclusion of CF has a great contribution to the occurrence of cracking and the width of cracking. In the case of concrete with 1.0 volume % of CF, there were no specimens showing the occurrence of cracking. Regarding the effects of the age at the start of drying, the higher the age at the start of drying the latter the time at cracking.

Figure 7 shows the results of mean drying shrinkage stress subjected to restraint concrete. These tensile stresses were obtained using the measured restraint steel strain. Calculated stresses are also plotted in this figure, and these stresses were obtained by use of F.E.M. program for drying shrinkage stress analysis developed by the authors [8]. It is found that the calculated value agrees with the measured one in the case of the age of 7 days at the start ofdrying, while there is big difference in the case of K=28 days. Although the stress analysis is not always sufficient and further discussion for F.E.M. program should be required, it seems that the tendency of the development of drying shrinkage stress can be approximately predicted.

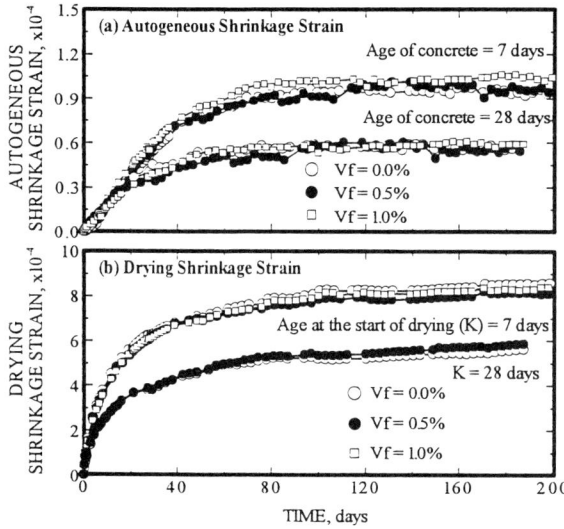

Figure 5 Relationship between time and shrinkage strain

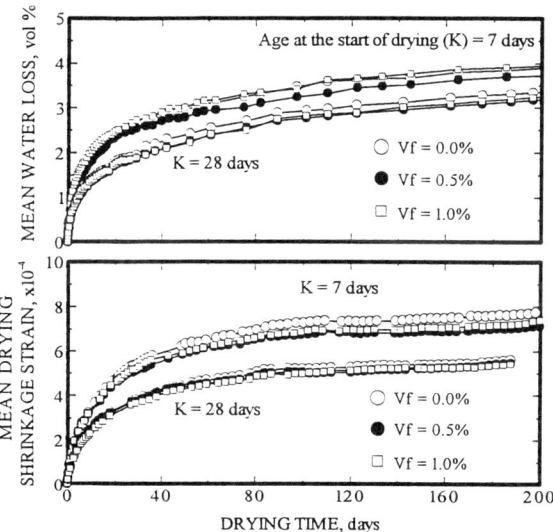

Figure 6 Mean water loss and mean drying shrinkage strain

Table 4 Results of cracking test

AGE AT TESTING	VOLUME % OF CF	DRYING TIME AT CRACKING		MAXIMUM WIDTH OF CRACKING	
7 days	0 %	7 days	8 days	0.35 mm	0.40 mm
	0.5 %	62 days	–	0.06 mm	–
28 days	0 %	42 days	43 days	0.30 mm	0.35 mm
	0.5 %	–	–	–	–

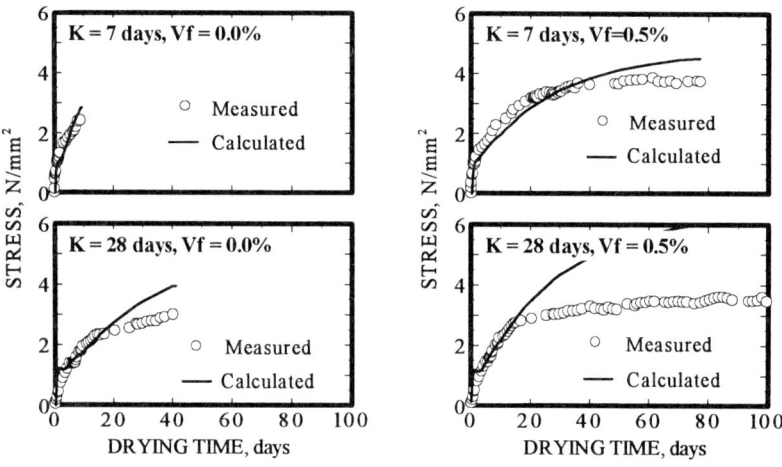

Figure 7 Results of drying shrinkage stress analysis

CONCLUSIONS

Concrete containing the strand type of CF recently developed was produced, and it's mechanical properties such as strength, creep and drying shrinkage cracking were investigated. It was found that the direct tensile and flexural strength increase as the volume fraction of CF increases. The remarkable cracking resistance was also recognized from the comparison of test results of the plain concrete. However, the studies on CFRC have just started. The fibre volume content of CFRC mixed in this research is smaller than the conventional steel fibre reinforced concrete with 1~2 volume % of steel fibre. Further investigation on the mix proportion of CFRC, the size and shape of CF, etc. should be required.

REFERENCES

1. SOROUSHIAN, P, NAGI, M, AND MUSTATA, E. Drying shrinkage characteristics of carbon fibre reinforced cement composite, Creep and Shrinkage of Concrete, ACI SP135, pp 65-76.

2. KUCHARSKA, L, AND BRANDT, A, M. Cracking of carbon fibre reinforced materials, Proc. of Int. Conf. Concrete In The Service Of Mankind, Appropriate Concrete Technology, Edited by R. K. Dhir and M. J. McCarthy, 1996, pp 421-429.

3. HAYASHIDA, J, MITSUI, Y, MURAKAMI, K, SAKAI, H, AND HISABE, N. Experimental investigation on mix proportions and physical properties of carbon fibre reinforced concrete using coarse aggregate (part 1 Mix design of CFRC), AIJ Kyusyu Chapter Architectural Research Meeting (Japanese), 1996, pp 133-136.

4. IWATA, T, SATO, Y, KIYOHARA, C, SAKAI, H, NAKAMURA, M, AOYAGI, H, AND TEZUKA, M. Mechanical properties of carbon fibre reinforced concrete (part 1), AIJ Kyusyu Chapter Architectural Research Meeting (Japanese), 1997, pp 45-48.

5. TAIRA, S, SATO, Y, KIYOHARA, C, SAKAI, H, NAKAMURA, M, AOYAGI, H, AND TEZUKA, M. Mechanical properties of carbon fibre reinforced concrete (part 2), AIJ Kyusyu Chapter Architectural Research Meeting (Japanese), 1998, pp 97-100.

6. UEDA, K, SATO, Y, KIYOHARA, C AND, NAGAMATSU, S. Effect of drying on direct tensile strength of concrete, JCA Proceedings of Cement & Concrete, Japan Cement Association (Japanese), 1997, pp 828-833.

7. KIYOHARA, C, SATO, Y, UEDA, K AND, NAGAMATSU, S. Creep properties of cement paste under compressive load, JCA Proceedings of Cement & Concrete, Japan Cement Association (Japanese), 1997, pp 812-817.

8. UEDA, K, SATO, Y, AND, KIYOHARA, C. Analysis of Drying Shrinkage Stress Occurring in Concrete Specimen, Proceedings of Japan Concrete Institute (Japanese), 1998, Vol. 20, No. 2, pp 637-642.

STRUCTURE AND MECHANICAL BEHAVIOUR OF CONCRETE REINFORCED WITH HYBRID STEEL-CARBON FIBRES

Z H Shui
Wahun University of Technology
China

Y R Cheng
Northern Jiaotong University
China

P Stroeven
D H Dalhuisen
Delft University of Technology
The Netherlands

ABSTRACT. Concrete composites were designed by incorporating various combinations of steel and carbon fibres into plain concrete. Only low-volume fraction of such hybrid fibre systems were considered in this study, because of possible application for railway sleepers where high electrical conductivity should be prevented.

The effects of such hybrid fibre reinforcement systems on structural and mechanical properties of the concrete composites were investigated. In comparison to the plain concrete, compressive strength and flexural strength of the hybrid fibres reinforced concrete composites increased by 5~15%, and 45~80%, respectively. The fracture energy increased 10~20 times. SEM analysis showed that the difference of surface status between PAN and Pitch carbon fibres is one of the reasons that leads to different strengthening effects. Porosity measurements shown that the increased porosity in the composites caused by the addition of carbon fibres can be controlled with a special mixing procedure.

Keywords: Composite, Steel fibre, Carbon fibre, Structure, Mechanical properties.

Associate Professor Z H Shui is head of Building Materials & Products Group, Wuhan University of Technology, Wuhan, China. His main research interests include the properties and uses of binding materials, fibre-reinforced cementitious composites.

Dr P Stroeven is head of Group Materials Engineer, Faculty of Civil Engineering, Delft University of Technology, Delft, The Netherlands. He specialises in micro-mechanics, material structure and properties, computer simulation of material behaviour.

Professor Y R Cheng is Professor of Northern Jiaotong University, Beijing, China. He specialises in fracture mechanics and in the use of cementitious composite in railway engineering.

D H Dalhuisen is staff of Faculty of Civil Engineering, Delft University of Technology, Delft, The Netherlands.

INTRODUCTION

It is well known that fibres can improve the behaviour of cement based composites. Many attempts have been made to develop new and novel FRC materials. With the studying progress, the hybrid fibre reinforcement systems were gradually considered as an effective way to improve the mechanical properties of the FRC composites [1], [2].

In the hybrid fibre system consisting of one macro fibre and one micro fibre, the larger and smaller fibres can control macro-cracks and micro-cracks of matrix, respectively. When the hybrid fibre system is a combination containing a stiffer fibre and a more ductile fibre, the system will simultaneously improve the first crack strength, ultimate strength and strain capacity of the composite. In general, the microstructure around fibres is quite different for different kind of fibres. There is interfacial transition zone (ITZ) between macro-fibre and matrix, whereas ITZ is largely eliminated in micro-fibre cement composite [3]. In some cases, the hybrid fibre systems are more useful than the mono-fibre systems for strengthening and toughening of FRC.

However, the strengthening and toughening effects of the hybrid fibre system depend on fibre combinations, fibre contents, particularly the dispersion degree of the fibres, as well as *w/c*. For example, in the carbon-steel fibre reinforced concrete, negative effects of the hybrid fibre systems on composite mechanical properties were observed due to the improper technological procedure during specimen preparation [4].

The importance of proper fibre dispersion was gradually recognized. New methods were built to disperse fibres properly in matrix [5], [6]. In this study, the relationship between dispersion degree of carbon fibre and composite density was taken accounted, and the porosity of the composites was determined and calculated for various mixes. The relationship was found useful for analysing the structures and properties of the composites.

EXPERIMENT DETAILS

Raw Materials and Mix Proportions

The used cement was produced by ENCI, The Netherlands, with a characteristic strength 52.5 MPa. The properties of short carbon fibres are described in Table 1.

Table 1 Properties of PAN and Pitch carbon fibres

TYPE	NOMINAL LENGTH (mm)	FILAMENT DIAMETER (μm)	TENSILE STRENGTH (MPa)	ELASTICITY MODULUS (GPa)
PAN-based	5	7	2500	250
Pitch-based	12	14	485	45

The length and diameter of hooked steel fibres are 30mm and 0.5mm, respectively. For plain steel fibres, 25mm and 0.4mm, respectively. The river sands had a fineness modulus of 3.23. The coarse aggregate had the maximum size of 16mm. By the way, the superplasticizer, Tillman OFT 4 was used. The mix proportion of the concrete matrix is shown in Table 2.

Table 2 Mix proportion of the concrete matrix

CEMENT	SAND	GRAVEL	SUPERPLASTICIZER	W/C
1.0	1.92	2.35	0.015	0.40

The fibre contents in volume fraction are presented in Table 3.

Table 3 Fibre volume content in composites (%)

MIX CODE	PAN FIBRES	PITCH FIBRES	HOOKED FIBRES	PLAIN FIBRES
M01	0	0	0	0
M02	0	0	0	0.5
M03	0	0	0.5	0
M04	0.25	0	0	0
M05	0	0.3	0	0
M06	0.25	0	0.5	0
M07	0.25	0	0	0.5
M08	0	0.3	0.5	0
M09	0	0.3	0	0.5
M10	0	0	0	0.8

Preparation of the Specimens

To properly disperse the hybrid fibres into the concrete mixture, a "Two-stage" mixing procedure was executed.

In the first stage, carbon fibres were dispersed in cement paste, and then, in the second stage, the fibre-cement paste was mixed with the aggregates and other components. The fibre concrete mixtures were shaped into 500×400×100mm slabs and 100×100×100mm cubes. Concrete beams for bending testing were obtained by sawing the slabs after hydrating one week. The size of concrete beam was 500×100×100mm.

The concrete beams were notched at the middle of the bottom surface. The depth of notch was about one third of the beam.

Physical and Mechanical Properties Testing

Density determination was performed for the cubic and beam specimens. Compressive strength of the fibre composites was determined with the cubic specimens. Flexural strength and fracture energy of the composites were determined and calculated based on four-point bending tests. The surface status of the two kinds of carbon fibres was observed with SEM

RESULTS AND DISCUSSIONS

Density and Compressive Strength

The density and compressive strength of the composites are presented in Figure1.

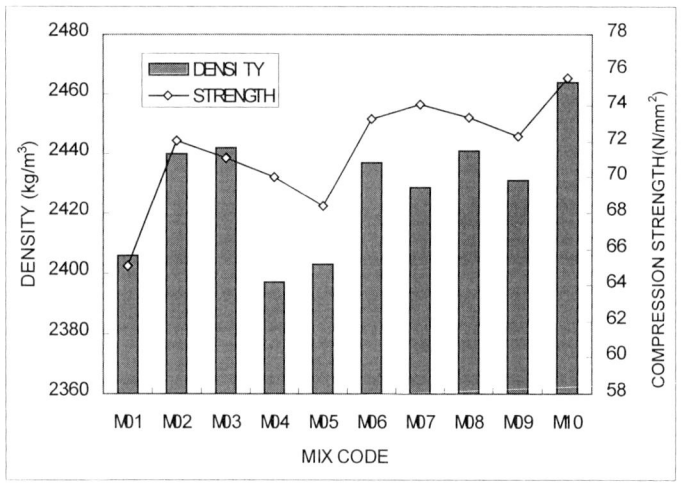

Figure 1 Density and compression strength of the composites

The results show that the density depends on fibre types and amount. Steel fibres lead to increased material density while carbon fibres lead to decreased material density. Steel fibre plays stronger strengthening effect than carbon fibre when fibres were added in single form.

Based on the four-point bending tests, load-displacement curves were plotted and fracture energy was calculated. The results are shown in Figure 2. The flexural strength of the composites was determined based on the assumption that there is no notch sensitivity [7].

In comparison to the plain concrete, flexural strength of the hybrid fibres reinforced concrete composites increased by 45~80%. The addition of carbon and steel fibres dramatically increases fracture energy of the composite. Further increased fracture energy, up to 20 times of plain concrete, was achieved as the composites reinforced with hybrid fibre systems.

Flexural Strength and Fracture Energy

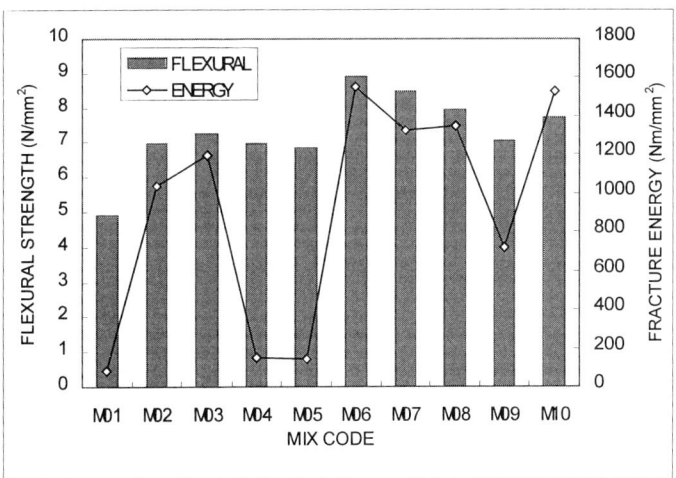

Figure 2 Flexural strength and fracture energy of the composites

The Effects of Porosity

The properties of fibre composite not only depend on the compositions, but also depend on the interactions between fibre and matrix. In some cases, the interactions may lead to negative effects on mechanical properties of the composites. A case of negative effects caused by carbon fibres is shown in Figure3 [4].

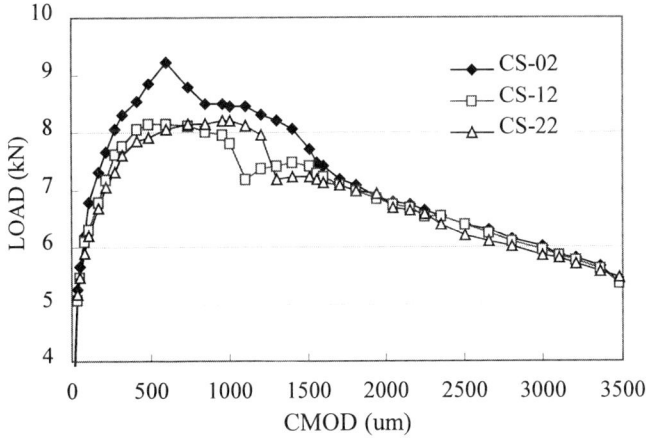

Figure 3 A case of fibres leading to negative effects

The possible reason is the improper dispersion of carbon fibres leading to multifilament structure in the composite [8].

For a cementitious composite system with water saturated, the bulk density of the composite can be approximately expressed as:

$$d_c = \frac{\sum W_i + W_w'}{\sum V_i + V_w'} \qquad (1)$$

where, W_i is the weight of a component and V_i is relevant volume of the component. W_w' and V_w' represent the weight and volume of the filling water, which fills the pores caused by air during mixing.

W_w' should be obtained from (1):

$$W_w' = \frac{\sum W_i - d_c \cdot \sum V_i}{(d_c / \gamma_w) - 1} \qquad (2)$$

Accordingly, additional porosity is defined as:

$$P' = \frac{V_w'}{\sum V_i + V_w'} \% = \frac{(W_w' / \gamma_w)}{\sum V_i + (W_w' / \gamma_w)} \% \qquad (3)$$

Additional porosity of hardened composite reflects the interactions between fibres and matrix. For the case shown in Figure3, the additional porosity, P' was presented as Table 4.

Table 4 The additional porosity of hybrid fibre concrete based on the previous study

MIX CODE	CARBON FIBRE (v.%)	STEEL FIBRE (v.%)	BULK DENSITY (kg/M³)	ADDITIONAL POROSITY (%)
CS-02	0.0	1.0	2419	5.80
CS-12	0.1	1.0	2381	8.27
CS-22	0.2	1.0	2377	8.48

The results showed that the additional porosity increased due to the improper dispersion of carbon fibres. It was verified by SEM image analysis

However, when the carbon fibres were properly dispersed in the matrix by "Two-stage" mixing procedure, the additional porosity of hybrid fibre composites dramatically decreased. This shows in Table 5.

Table 5 Additional porosity (P) for hybrid steel-carbon fibre composites

CODE	M01	M02	M03	M04	M05	M06	M07	M08	M09	M10
(%)	2.82	2.31	2.12	3.28	2.85	2.36	2.88	2.02	2.68	1.73

The Effects of Surface Status of Carbon Fibres

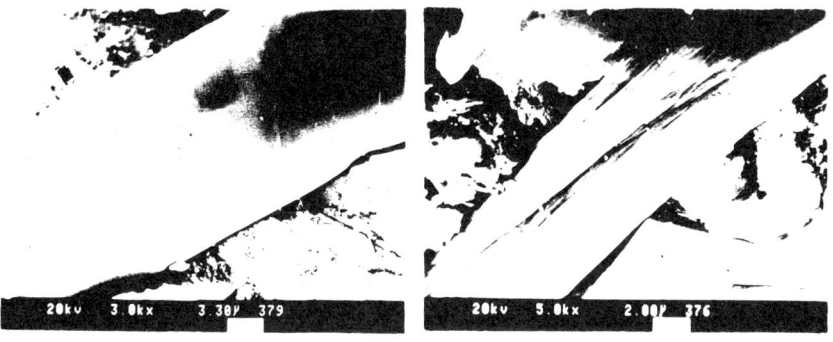

Figure 4 Surface of Pitch fibre   Figure 5 Surface of PAN fibre

SEM analysis was executed for the hardened fibre composites. Figure4 and Figure5 show that the pitch carbon fibre is of very smooth surface while PAN carbon fibre is of coarse surface which contains a lot of twisty veins. SEM images showed that the pitch carbon fibres were easily removed from matrix implying poor bond between fibre-matrix interfaces. This is one of the reasons why the PAN carbon fibres are more effective than pitch carbon fibres in strengthening and toughening.

Relatively, the surfaces of steel fibres are coarse. Hence the higher friction stresses occur between the fibres and matrix. The larger friction stress conduces to improvement of toughening effect on composites.

CONCLUSIONS

1. Low-volume content carbon and steel fibres can enhance flexural strength and fracture energy of concrete composites in mono and hybrid styles. In particular, the composites reinforced with the hybrid fibre systems lead to double strengthening and toughening effects.

2. Improper carbon fibre dispersion will lead to high porosity in composites. Hence, the dispersion degree of carbon fibres strongly affects the strengthening and toughening results of reinforced systems on composites. Uniform dispersion of carbon fibres can be approached by a specific procedure.

3. SEM analysis revealed that the surface of the pitch carbon fibres is very smooth, resulting in poor interfacial bonds. The surface of PAN fibre is coarse, hence, it is more effective in strengthening and toughening.

REFERENCES

1. ARNON, B, AND SIDNEY, M. Fibre Reinforced Cementitious Composites, Elsevier Science Publishers Ltd., Essex, England, 1990, pp 418-420.

2. HUA, Y, ZHANG, S, AND JIANG, Y. Experimental Study on the Flexural and Fatigue Behaviour of Hybrid Fibre Concrete, China Concrete and Cement Products, 1997, No.4, pp 40-43.

3. NAAMAN, A, E, AND, REINHARDT, H,W. (Edited), High Performance Fibre Reinforced Cement Composites, Volume 2: HPFRCC-95, Michigan, June 11-14,1995, pp 140-146.

4. SHUI, Z, STROEVEN, P, AND, CHENG, Y. Study on the Technology and Properties of Cementitious Composites Reinforced with Hybrid Steel-Carbon Fibre Systems, Report 03.21.1.32.11, Delft University of Technology, July, 1997.

5. STROEVEN, P. On Simulation, Image Analysis and Structural Modeling of Steel Fibre Concrete, Advanced Technologies, Elsevier Science Publishers, 1993, pp 399-404.

6. SHUI, Z, STROEVEN, P, GAO, Q, AND, CHENG, Y. Method for Evaluation the Uniformity of Carbon Fibres Dispersion in Cement Paste, ICFRC, Guangzhou, Nov. 1997, pp 89-93.

7. ZHOU, F, P, BARR, B, I, G., AND LYDON, F, D. Fracture Properties of High Strength Concrete with Varying Silica Fume Content and Aggregates, Cement and Concrete Research, Vol. 25, No. 3, pp 547.

8. ARNON, B. Microstructure, Interfacial Effects, and Micromechanics of Cementitious Composites, Conference on Advances in Cementitious Materials, Gaithersburg, 1990, Vol. 16, pp 523-547.

BEHAVIOUR OF CARBON FIBRE SHEET AS FLEXURAL REINFORCEMENT IN RC BEAMS WITH ARAMID FRP RODS

Y Takahashi	**C Hata**
Hokkai-Gakuen University	Hokkaido Development Bureau
Y Sato	**T Maeda**
Hokkaido University	Shimizu Corporation
Japan	Japan

ABSTRACT. The simple-supported concrete beams reinforced with aramid FRP (AFRP) rods and carbon FRP (CFRP) sheets were tested to failure using a symmetric two-point concentrated static loading system. AFRP rods were used instead of steel reinforcing bars, and CFRP sheets were epoxy bonded to the tension face of the concrete beams to enhance their flexural strength. Moreover, 5-cm wide strips of CFRP sheet in some places were wrapped around the web (hereafter, called "U-jacket") after the CFRP sheets were bonded. The strain distributions on AFRP rods and CFRP sheets, and flexural behavior of the beams with AFRP rods and CFRP sheets were examined experimentally. The results showed that; 1) peeling of CFRP sheets occurred near the maximum flexural moment region, 2) ultimate load and deflection of the beam with U-jackets were higher, and 3) the U-jacket was a significant factor affecting the ductile behavior of beams with CFRP sheets.

Keywords: Carbon FRP sheet, Aramid FRP rod, Flexural reinforcement, U-jacket of CFRP sheet, Deflection ability

Yoshihiro Takahashi is a professor in the Department of Civil Engineering, Hokkai-Gakuen University, Sapporo, Japan. He received his D.Eng. degree from Hokkaido University, Sapporo, Japan. His research interests include the use of FRP in reinforced concrete structures.

Chihiro Hata is a civil engineer at Hokkaido Development Bureau, Japan. He is a graduate of Hokkai-Gakuen University and obtained his B.Eng. degree.

Yasuhiko Sato is a research associate in the Department of Structural and Geotechnical Engineering Graduate School of Engineering at Hokkaido University, Japan. He obtained his D.Eng.degree from Hokkaido University. His research interests include the application of FRP reinforcements in concrete structures.

Toshiya Maeda is an engineer of Civil Engineering Division at Shimizu corporation. He received D.Eng. degree from Hokkaido University, Japan. His research interests include the rehabilitation and strengthening of concrete structures.

INTRODUCTION

High expectation has been placed on continuous fiber reinforced sheets as a strengthening and rehabilitation material for existing reinforced concrete structures, and the number of existing concrete structures using these reinforcement sheets is increasing. The purpose of reinforcing with continuous fiber sheets is chiefly to increase flexural and shear strength. A large deflection ability is also required in the case of strengthening of a bridge pier. The CFRP sheet is excellent in terms of high tensile strength, lightness, and resistance to corrosion and chemical attack. In addition, the CFRP sheet can be handled easily because of its flexibility.

In our previous study(1997), we performed flexural tests on reinforced concrete beams in which CFRP sheets had been bonded to the bottom surface and a deformed steel bar used as a main tension reinforcement. The strain behavior of the CFRP sheet used in that study was very complex because the deformed steel bar had a yielding phenomenon[1].

In the present study, in order to obtain the basic information needed to establish a design method, we experimentally examined the flexural behavior of a concrete beam reinforced with CFRP sheets on the bottom surface and the strain behaviors of CFRP sheets and AFRP rods. The AFRP rods were used as main tension bars. Since the AFRP rod has no yielding phenomenon, there is no effect of yielding on strain behavior of the CFRP sheet.

OUTLINE OF EXPERIMENT

The experimental study consisted of casting six concrete beam specimens. Each specimen had a 150mm\200mm cross section. All specimens were reinforced for shear by closed stirrups with D10(deformed steel bar with a nominal diameter of 9.53mm) equally spaced at 50mm. The beams were reinforced in this manner to prevent shear failure and to isolate the flexural behavior from shear behavior. The flexural reinforcement ratio(AFRP rod ratio) p=0.008588 used for the specimens. All beams were tested as simply supported beams under two symmetrical load with a total span of 1200mm and shear span of 500mm. The beams were loaded by monotonic static loading to failure at increments of 5kN. The flexural reinforcement(AFRP rods) in these beams consisted of three K96s AFRP rods and two D6 steel bars(deformed steel bar with a nominal diameter of 5.635mm) in the top. The shape, size, reinforcement details and loading of flexural specimens are shown in Figure 1.

Reinforcement method of CFRP sheets was used as a variable in these tests. A sketch of these test specimens is shown in Figure 2. Except for specimen F0, a CFRP sheet was wrapped around the web along the entire left span of all specimens, and the right span was made the examination span. Test specimens consisted of: specimen F0 was not reinforced with CFRP sheets in the beam soffit; F1 was reinforced with only one CFRP sheet attached to the beam soffit in the longitudinal direction; F2 was reinforced with three CFRP sheets; F3 was reinforced with three CFRP sheets attached to the beam soffit, and additionally wrapped up to 10cm in height by U-jackets in three places at 35cm, 45cm, and 55cm from the center line in examination span; F4 which was the same as F3 except that U-jackets were applied in six places at 5, 15, 25, 35, 45 and 55cm from the center line in examination span; and F5 which was the same as F4 except that the wrapped up to 20cm in height (full height of the web) by U-jackets.

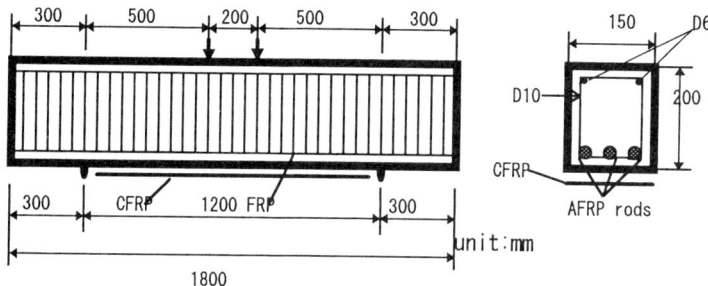

AFRP : Aramid FRP rod , CFRP: Carbon FRP sheet

Figure 1 Beam dimensions, internal reinforcing and load configuration

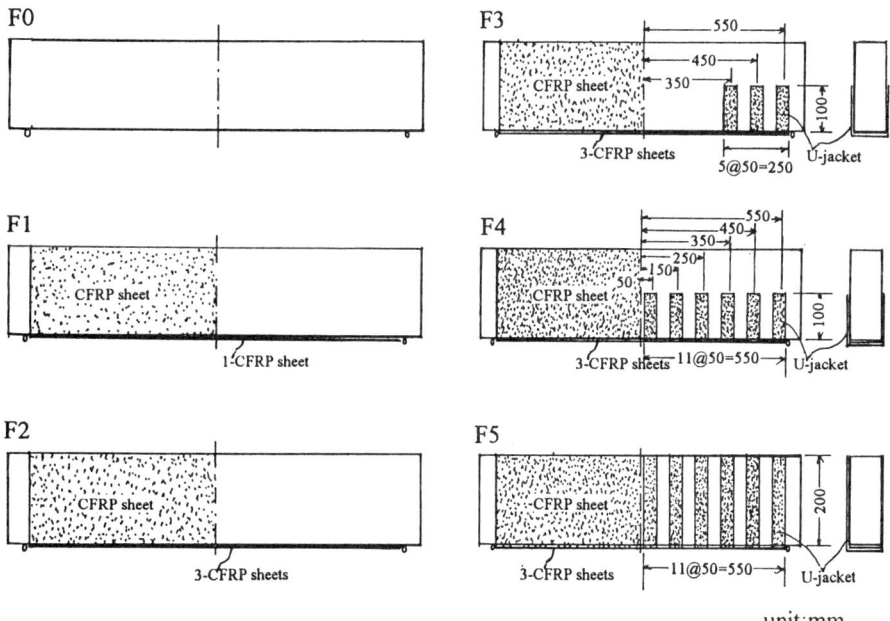

Figure 2 Configuration of test specimens

The mechanical properties of the material used for the test specimens are shown in Table 1. Strain gauges were attached to the CFRP sheet and AFRP rod. In specimens F3, F4, and F5, some strain gauges were attached to the U-jacket of CFRP sheet. These gauges were attached at a height of 3.0cm from the bottom surface.

Table 1 AFRP rod, steel bar and CFRP sheet properties

TYPE	CROSS SECTION (mm²)	YOUNGS MODULUS (Gpa)	YIELD STRENGTH (Mpa)	TENSILE STRENGTH (Mpa)
AFRP K96S	73.0	66	--	1280
Steel D10	71.3	200	377	548
CFRP	1.65	230	--	3480

The concrete was composed of high early-strength Portland cement, sea sand and river gravel. The water-cement ratios and fine-coarse aggregate ratios were kept at 45% and 38%, respectively. The target compressive strength of concrete was 30Mpa.

In this study, the failure modes of test specimens were examined, and the strains of AFRP rod and CFRP sheet and the deflection of midspan of specimen and that just under the loading point were recorded at each load increment. Deflection and strength of the test specimens were also evaluated using the layer-by-layer procedure.

EXPERIMENTAL RESULTS

Ultimate Strength and Failure Modes

The experimental results are shown in Table 2. In general, when the concrete beam reinforced with a CFRP sheet attached to the beam soffit is subjected to both flexural moment and shear force, shear stress is generated at the concrete-adhesive interface (adhesive stress) due to the movement in the interface between the CFRP sheet and concrete in the direction of the member axis.

Table 2 Experimental results

SPECIMEN	COMPRESSIVE STRENGTH (MPa)	TENSION BAR	CFRP A_{CFS} (mm²)	ULTIMATE LOAD (kN)
F0	39.15	K96S x 3	0	141.3
F1	38.70	K96S x 3	198	142.3
F2	38.33	K96S x 3	594	138.5
F3	37.90	K96S x 3	594	156.8
F4	44.20	K96S x 3	594	160.0
F5	44.20	K96S x 3	594	178.0

It seems that if this adhesive stress will reach a certain limit, peeling of the CFRP sheet occurs first from the region of the limit stress value and then propagates over the whole beam until failure. An increase in ultimate strength with an increase in the number of layers of

CFRP sheet can be expected; however, differences in ultimate strength due to a difference in the number of layers was hardly seen in this experiment. More detailed study is need to explain this result. On the other hand, in the case of specimens F3, F4 and F5, which had three CFRP sheets and U-jackets, peeling of the CFRP sheets was restrained by the U-jackets, and the ultimate strength was found to be higher than that of the specimen without U-jackets.

The failure modes were flexural compression failure in specimen F0 and breakage of the CFRP sheets just under the loading point after partial peeling of the CFRP sheets in specimen F1. The failure mode in specimen F2 was peeling of the CFRP sheets. Specimen F3, with three CFRP sheets and U-jackets, failed due to breakage of the U-jacket at 55cm after peeling of the CFRP sheets and U-jackets at 35cm and 45cm. In specimen F4, which was the same as F3 except that there are U-jackets in six places, peeling occurred first in the CFRP sheets in the central region (in the constant moment region), and the U-jacket suppressed the peeling up to a certain load. However when the U-jacket could no longer suppress the peeling, the beam reached ultimate failure due to debonding in the anchorage of the U-jackets. In specimen F5, which was the same as F4 except for the wrapped height of the U-jacket, peeling occurred first in the constant moment region as it did in specimen F4. Then debonding of the U-jackets occurred toward support from the center span of the beam, and finally the U-jacket at 55cm was torn.

Deflection Behavior and Stress Distribution of AFRP rods

Load-deflection relations of all specimens just under the loading point are shown in Figures 3(a) and (b). The values calculated by the layer-by-layer procedure for the beam without CFRP sheets, the beam with one CFRP sheet, and the beam with three CFRP sheets are also shown in these Figures [2],[3].

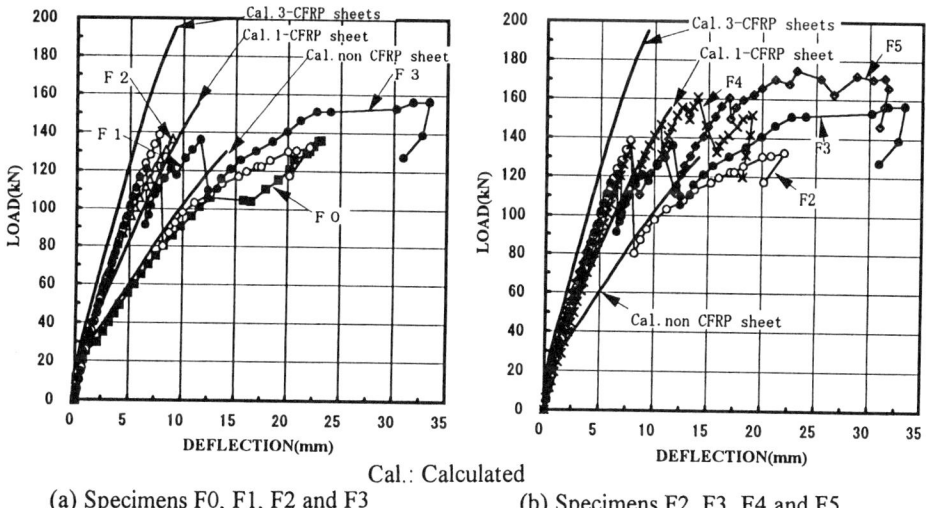

(a) Specimens F0, F1, F2 and F3   (b) Specimens F2, F3, F4 and F5

Cal.: Calculated

Figure 3 Load vs deflection under load point deflections

These figures show that the inclinations of deflection curves in the specimens with CFRP sheets are steeper than those of the specimen without CFRP sheets in the beam soffit. This indicates that the flexural rigidity of specimens with CFRP sheets is higher. However, there was little difference in flexural rigidity between specimen F1 with one CFRP sheet and F2 with three CFRP sheets. After the peeling of CFRP sheets, the deflection behavior of the specimen with CFRP sheets became similar to that of specimen F0 that has no CFRP sheets. This indicates that after the peeling of CFRP sheets, the AFRP rods take all of the load. Moreover, an increase in the ultimate displacement can be confirmed by the use of U-jackets, and the ultimate displacement is higher than that of the specimen without CFRP sheets.

Load-strain relations of the AFRP rod in all specimens just under the loading point are shown in Figures 4(a) and (b). As shown in Figure 3, the calculated strain value for the specimen without CFRP sheets, with one CFRP sheet, and with three CFRP sheets are also shown in these figures. These figures show that the strain of the AFRP rod in the specimen with CFRP sheets is smaller than that in the specimen without CFRP sheets before the peeling of CFRP sheets, but the strain approaches that of the specimen without CFRP sheets after peeling. This result indicates clearly that the CFRP sheets take the load until they peel.

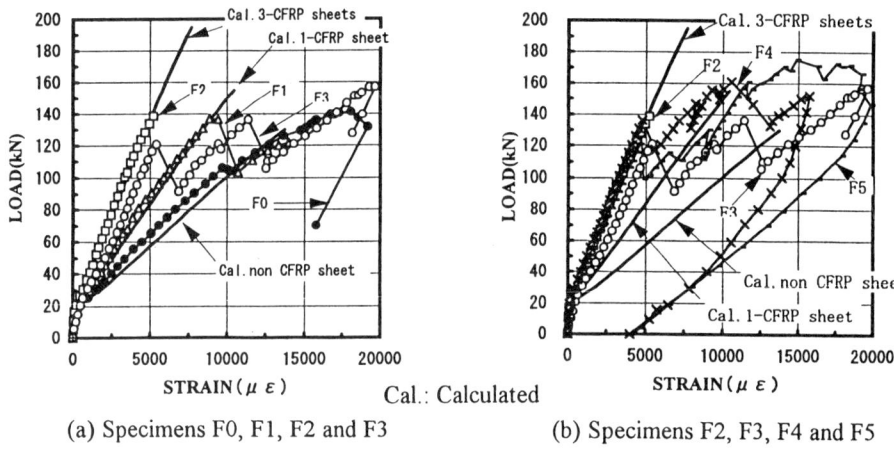

(a) Specimens F0, F1, F2 and F3   (b) Specimens F2, F3, F4 and F5

Cal.: Calculated

Figure 4 Load vs strain in AFRP rod (under loading point)

Strain Distribution of CFRP sheets

Load-strain relations of the CFRP sheet on the bottom surface of specimen F1 at 0cm, 24cm and 44cm in the right span from the center line are shown in Figures 5(a), (b) and (c), respectively. The values calculated by the layer-by-layer procedure for the beam with one CFRP sheet are also shown in these figures. In Figure 5(a), the strain at 0cm, which is the constant moment region, remains constant at about 4000 beyond 60kN. This result shows the occurrence of partial peeling of the CFRP sheet in this region. On the other hand, the flexural and shear region (at 24cm and 44cm), there is no peeling and strain increases with increases in load. When the load increased to at about 140kN, complete peeling occurred in the constant moment region and the strain in this region suddenly decreased to almost zero. The strain at 24cm decreases to 1500 from 8000 with increases in load. According to behavior of strain at 44cm, it is seen that adhesiveness in the CFRP sheet near this region remains until the load reaches the ultimate load.

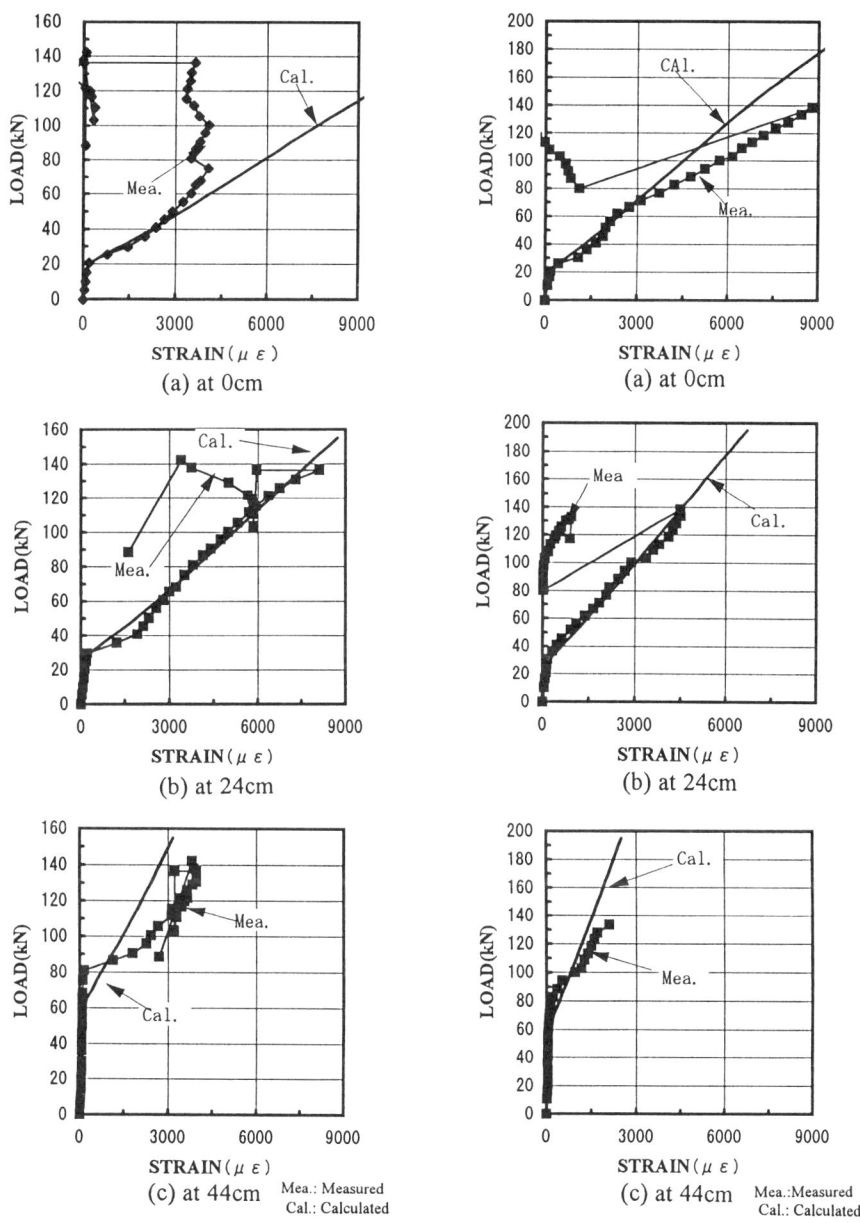

Figure 5 Load vs strain in CFRP sheet (F1)   Figure 6 Load vs strain in CFRP sheet (F2)

Figures 6(a), (b) and (c) show strain distributions in specimen F2 at 0cm, 24cm and 44cm, respectively. The calculated values for the beam with three CFRP sheets are also shown in these figures. Peeling of the CFRP sheet in the constant moment region occurred suddenly at a load of about 140kN, as it did in specimen F1. On the other hand in the flexural and shear region (in Figure 6(b)), though the strain decreased suddenly at a load of 140kN, deflection progressed and the strain increased with increases in load. This result indicates incomplete peeling. As shown in Figure 6(c), at 44cm, there was no peeling until the beam failed at the ultimate loading.

Figures 7(a), (b) and (c) show the strain in specimen F3 at 0cm, 24cm and 44cm on the CFRP sheet, respectively. The calculated values for the beam with three CFRP sheets are also shown these figures. Specimen F3 has three CFRP sheets and U-jackets in three places. In this specimen, the effect of reinforcement with the U-jackets is verified by an increase in the ultimate load. As shown in Figure 7(a), when the load decreased to 90kN from 120kN, the strain decreased to 4000 from 6000 . However, when the load was later increased to 140kN from 90kN, the strain remained at almost 4000 . This result indicates that peeling of the CFRP sheet in the constant moment region occurred at this time. Though the strains in the flexural and shear regions (in Figures 7(b) and (c)) decreased momentary, the strains later increased with increases in load. This result indicates that the bonding remains. The U-jacket suppresses the peeling of CFRP sheets on the bottom surface. The load increased to at about 140kN, then decreased to at about 100kN, and later increased to at about 160kN. This result indicates that the tension force on AFRP rods shifts to CFRP sheets by debonding of the U-jacket. This result was also verified from the strain distribution in Figure 4.

Figures 8(a), (b) and (c) show the strain in specimen F4 at 4cm, 24cm and 44cm on CFRP sheets, respectively. As in Figure 7, the calculated values are also shown in these figures. This specimen is the same as F3 except that there are U-jackets in six places. Since the strain of CFRP sheets remains constant at about 4000 in the constant moment region (at 4cm) and remains constant at about 2000 in the bending and shear region, the peeling of sheets occurred at about 140kN. However, the load increased to at about 140kN due to the effect of reinforcement with U-jackets.

Figures 9(a), (b) and (c) show the strain in specimen F5 at 0cm, 28cm and 44cm on CFRP sheets, respectively. As in Figure 8, the calculated values are also shown in these figures. This specimen is the same as F4 except for being wrapped up to 20cm in height (full height of the web) by the U-jacket. The peeling of CFRP sheets on the beam soffit occurred at about 120kN in the constant moment region due to the strain of CFRP sheets reaching about 6000 . Although both the strain in the constant moment region and the load value are almost the same as those in specimen F4 when the first peeling occurred, the strain of the CFRP sheets after peeling occurred remained constant at about 3000 4000 due to the reinforcement by U-jackets. The strain at 28cm on CFRP sheets remained constant at about 1400 , and the load increased to about 180kN. The increment in load was 20kN higher than that in specimen F4. This increment in load was due to the effect of the wrapped height of the U-jacket. The CFRP sheets at 44cm did not peel completely at about 120kN. This is understood from an increase in strain with an increase in load. However, the peeling progressed considerably at loads beyond 160kN, the strain decreased with an increase in load, and the load reached ultimate load and the strain decreased to zero.

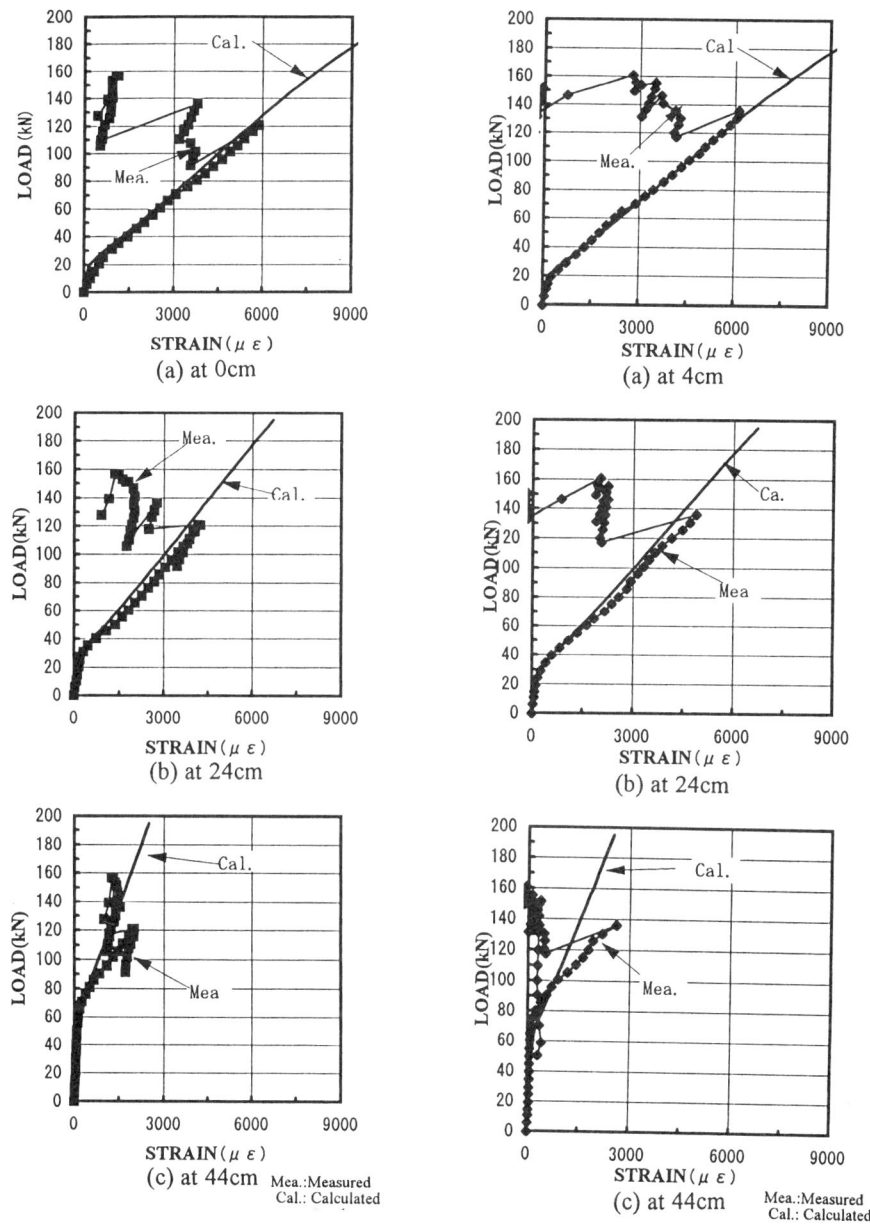

Figure 7 Load vs strain in CFRP sheet (F3)   Figure 8 Load vs strain in CFRP sheet (F4)

From Figures 7, 8 and 9, it is seen that the resistance strength to peeling failure can be improved by making the increment of the place of the U-jacket and the development length of the U-jacket sufficiently long.

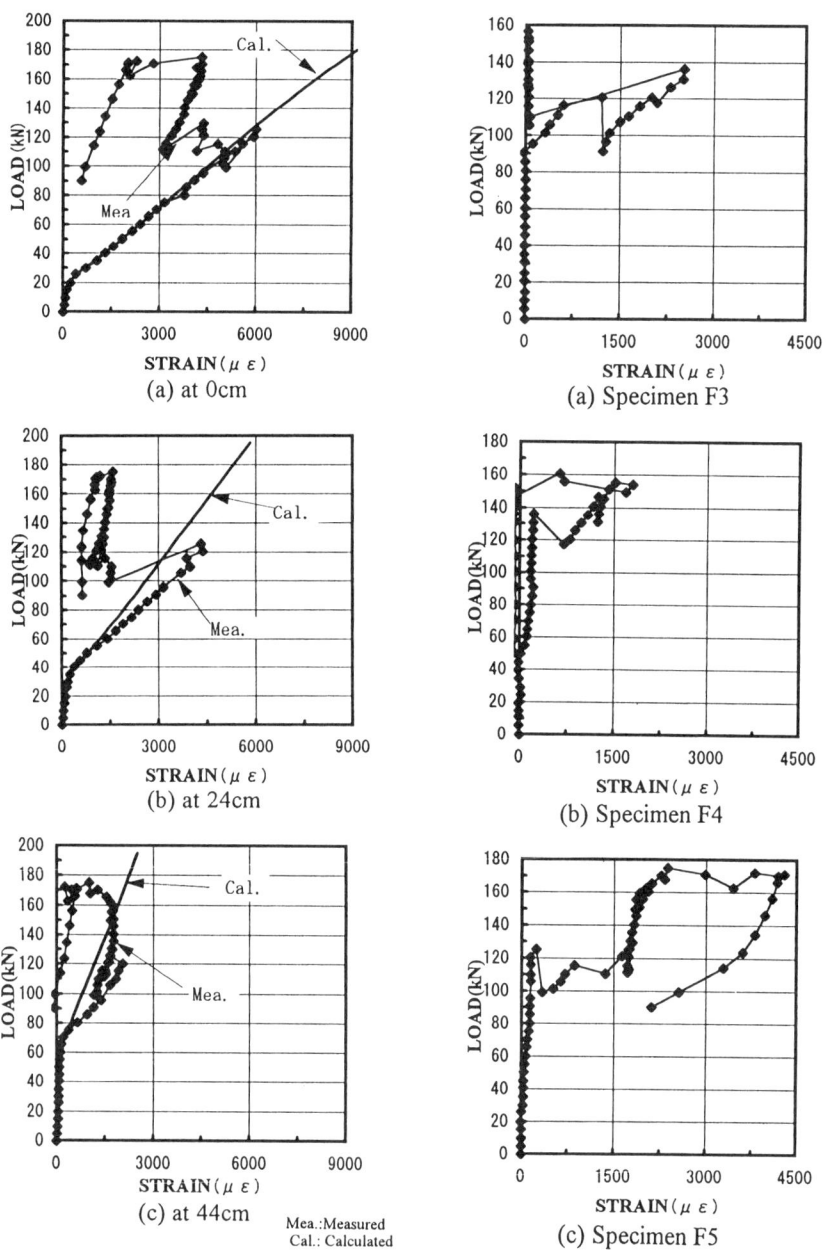

Figure 9 Load vs strain in CFRP sheet (F5)   Figure 10 Load vs strain in CFRP sheet (F6)

Figures 10(a), (b) and (c) show the relationships between load and strain on the U-jacket at 45cm in specimens F3, F4 and F5, respectively. These strains were measured by a strain gauge attached at 3cm from the bottom surface. Figure 10(a) shows that the strain in specimen F3 increased gradually beyond a load of about 90kN and the load decreased suddenly to 90kN from 120kN, while the strain remained constant at about 1200 and later increased. The strain became almost zero at a load of about 140kN, at which peeling of the U-jacket occurred. Figure 10(b) shows that strain increment in specimen F4 was hardly seen until the load reached about 140kN. After that, the load decreased to about 120kN, but the strain increased . In that loading stage, peeling of CFRP sheets on the bottom surface occurred. Since the strain on the U-jacket increased, the effect of reinforcement by the U-jacket was verified. Figure 10(c) shows that the load in specimen F5 decreased to about 100kN from 130kN and that both the strain and load later increased. This phenomenon indicates that the first peeling of CFRP sheets on the bottom surface occurred at this load stage. In the ultimate condition, the strain on the U-jacket at 45cm remained at about 2000 . This result shows that this U-jacket was not debonded completely and that the load was increased by this U-jacket.

SUMMARY

In this study, using concrete beams reinforced with CFRP sheets on the bottom surface and AFRP rods as the main tension bars, we experimentally investigated the flexural behavior of the concrete beams and the strain behavior of the CFRP sheets. The conclusions were:

1. Differences in ultimate strength due to a difference in the number of layers of CFRP sheet was hardly seen in this experiment.

2. Deflection near maximum load is larger in specimens with U-jackets than in those with no U-jackets. Therefore, there was a considerable increase in ultimate deflection. Ultimate strength increased and the effect of reinforcement by U-jackets was confirmed.

3. The process leading to failure in specimens with U-jackets is as follows. First, peeling occurs in the CFRP sheets. The U-jacket suppresses this peeling up to a certain load, but when the U-jacket can no longer suppress the peeling, the beam reaches ultimate failure due to debonding in the anchorage of the U-jackets. However, it seems that the resistance strength to peeling failure can be improved by making the development length of the U-jacket sufficiently long.

REFERENCES

1. TAKAHASHI,Y, SATO,Y, UEDA,T, MAEDA,T and KOBAYASHI,A. Flexural behavior of RC beams with externally bonded carbon fiber sheet. Non-Metallic(FRP) Reinforcement for Concrete Structures. Proceedings of the Third International Symposium, Vol.1,Oct,1997,pp 327-334

2. JAPAN SOCIETY OF CIVIL ENGINEERING (JSCE). Standard specification for design and construction of concrete structures (Design), (in Japanese),1996,pp 24

3. OKAMURA,H and MAEKAWA,K. Nonlinear analysis and constitutive models of reinforced concrete. Giho-do Inc.,1991, pp.36-38

STUDY ON MECHANICAL BEHAVIOUR OF MORTAR REINFORCED WITH DISCONTINUOUS CARBON FIBRES

N Koshiishi

Waseda University

Japan

ABSTRACT. One of the methods to improve the brittleness of cementitious materials is to reinforce this material with discontinuous discrete fibres. The short fibre reinforced cement composites (FRCs) are superior in resistance to cracking, tensile strength and ductility. Ductility would be expected to increase by inducing the multiple cracks. However, it is difficult to measure accurately the mechanical behavior of FRCs by experiments, because the strain gauge attached on the surface of specimen restrains the development of cracks, and also cracks resulting disturb the operation of the strain gauge. Furthermore, it is not clear how to develop the multiple cracks into the composites, which are loaded in flexure. In this study, the experiments were carried out to investigate the relationship between the stress and deformation on FRCs containing carbon fibres, when these FRCs were individually loaded in tension, compression and flexure. Next, the relationships between the flexural stress and deflection were calculated by using the results of tension and compression tests. As a result of calculations, the deflection at the maximum stress point of FRCs, which were loaded in flexure, could be predicted by using the ratio of the tensile strength to the first crack strength, and the average width of the cracks, both obtained by tension test, along with other relationships experimentally obtained.

Keywords: Fibre reinforced cementitious composite, Silica fume, Carbon fibre, Tension test, Flexural strength, Ductility, Crack density, Crack opening mouth.

Dr Naoyuki Koshiishi is an Associate Professor in the Department of Architecture, at Waseda University, Tokyo, Japan. His main research interests include the mechanical properties of short fibre reinforced cementitious composites and the rheological properties of cement paste using mineral and chemical admixtures.

INTRODUCTION

For randomly distributed short fibre reinforced cement composites (FRCs), the pseudo strain-hardening behavior has been modeled and comparisons have been made between model predictions and experimental measurements, especially for the effect of strain-hardening on the flexural/tensile-strength ratio [1]. The present paper examines the characteristics influencing the deflection of FRCs which are loaded in flexure, by analytical study based on the experimental data.

EXPERIMENTAL DETAILS

Materials

Blended cement in which ordinary Portland cement (C) is partially replaced with silica fume (Sf) was used as the cementitious material. Siliceous sand (S) (particle size : under 300μm) was used as fine aggregate. Pitch-based carbon fibre (Cf) (diameter : 18μm, length : under 6mm, tensile strength : 590MPa, modulus of elasticity : 30GPa) was incorporated into the mortar. Naphthalene sulfonate based superplasticizer (Sp) was also used.

Mix Proportions

The mixes prepared in this study are shown in Table 1. The sand to cementitious materials ratio (S/(C+Sf)) and fibre volume content (Vf) are respectively constant among all mixes (0.5 by volume and 2%). The amount of superplasticizer in each mortar matrix, which was produced from different combinations of the water to cementitious materials ratio (W/(C+Sf)) and Sf/(C+Sf), was decided according to prior mixing tests, so as to achieve uniform fibre dispersion and ensure the operations of placing and compacting.

Specimen Preparation

Mixing was carried out by using an ordinary mortar mixer. The actual mixing time was 7 minutes in total. Freshly mixed FRCs were cast into metal molds. Specimens were stored in a humid room after casting, and were demolded 26 hours after mixing. Following demolding, they were immediately cured in water at a temperature of 20±3° C for 4 weeks before testing.

RESULTS OF THE EXPERIMENT

Tensile Stress (σ_t) - Elongation (δ_t) Curve

All of the experimentally obtained $\sigma_t - \delta_t$ curves were characterized by the three points A, B and C in Figure 1. Point C was defined for reasons of expediency, as the point at which the stress was reduced to 90% of the maximum stress. The characteristics of some FRCs given by each point are shown in Table2.

Table 1 Mix proportions

MARK	WATER TO CEMENTITIOUS MATERIALS RATIO W/(C+Sf)		REPLACEMENT RATIO OF SILICA FUME Sf/(C+Sf)		SAND TO CEMENTITIOUS MATERIALS RATIO S/(C+Sf)	
	by Weight	by Volume	by Weight	by Volume	by Weight	by Volume
A2	0.32	1.0	0.00	0.00	0.41	0.50
B2	0.33	1.0	0.07	0.10	0.42	0.50
C2	0.34	1.0	0.15	0.20	0.43	0.50
D2	0.35	1.0	0.23	0.30	0.45	0.50
E2	0.36	1.0	0.32	0.40	0.46	0.50
F2	0.38	1.2	0.00	0.00	0.41	0.50
G2	0.39	1.2	0.07	0.10	0.42	0.50
H2	0.40	1.2	0.15	0.20	0.43	0.50
I2	0.42	1.2	0.23	0.30	0.45	0.50
J2	0.43	1.2	0.32	0.40	0.46	0.50
K2	0.46	1.4	0.07	0.10	0.42	0.50
L2	0.47	1.4	0.15	0.20	0.43	0.50
M2	0.49	1.4	0.23	0.30	0.45	0.50
N2	0.50	1.4	0.32	0.40	0.46	0.50
O2	0.52	1.6	0.07	0.10	0.42	0.50
P2	0.54	1.6	0.15	0.20	0.43	0.50
Q2	0.56	1.6	0.23	0.30	0.45	0.50
R2	0.58	1.6	0.32	0.40	0.46	0.50
S2	0.59	1.8	0.07	0.10	0.42	0.50
T2	0.61	1.8	0.15	0.20	0.43	0.50
U2	0.63	1.8	0.23	0.30	0.45	0.50
V2	0.65	1.8	0.32	0.40	0.46	0.50

The total length of all cracks which developed within the gauge length (L_g) was measured after the tension test, and the number of cracks (N_{cr}) was calculated by dividing the total length by the width of the specimen. The average width of the cracks (δ_{tp}) at the maximum stress point was calculated under two assumptions. One was that the total elongation could be divided into the elastic strain and the total width of all cracks. The other was that all of the widths at the maximum stress point were almost the same.

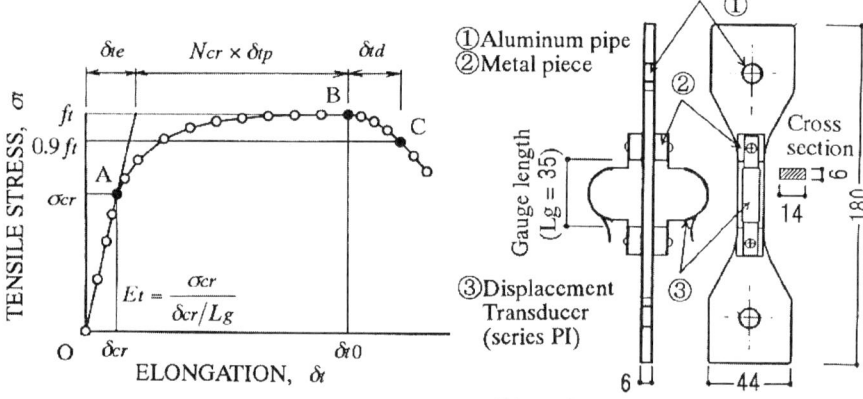

Figure 1 Characteristics of obtained tensile stress - elongation relation

[Note] Shape and dimension of tension test specimen (Unit : mm)

Compressive Stress - Strain Curve

The compression test was conducted with the prism-shaped specimens ($20 \times 20 \times 40^H$ mm), and the strains in the direction of compressive load were measured by using a pair of strain gauges that were 20mm long. Regardless of mix proportions, all the curves were realized by three characteristics. Some of the results are shown in Table 2.

Table 2 Characteristics of some FRC's obtained from tension and compression tests

	TENSION TEST					COMPRESSION TEST		
	E_t	σ_{cr}	f_t	δ_{tp}	δ_{td}	E_c	f_c	ε_{c0}
MARK	(N/mm²)	(N/mm²)	(N/mm²)	(mm)	(mm)	(N/mm²)	(N/mm²)	(x10⁻⁶)
B2	19000	3.57	5.68	0.026	0.024	26700	101	4560
H2	13700	3.13	5.34	0.031	0.024	22300	87	4740
M2	11700	2.44	4.36	0.05	0.028	18700	73	5100
R2	12100	2.09	3.67	0.035	0.045	1500	57	5050

E_t = modulus of elasticity, σ_{cr} = first crack strength, f_t = tensile strength, δ_{tp} = average crack width at maximum stress, δ_{td} = increment of elongation in stress decending portion, E_c = modulus of elasticity, f_c = compressive strength, ε_{c0} = compressive strain at maximum stress

Flexural Stress - Deflection Curve

The third-point loading flexure test (support distance : 120mm) was conducted with plate-shaped specimen ($12^D \times 40^W \times 160^L$ mm). The deflections at the loading points were measured with Displacement Transducers. Some of the results of the flexural strength (fb) and the deflection at the point of maximum stress ($\delta b0$) are shown in Table 3.

As shown in the note of this table, the portion of the specimen within the supporting points was divided into 12 blocks which were the same length of 1cm. The total length of all cracks in each block was measured and the crack density (Dcr) was calculated by dividing the total length by the width of the specimen. The crack density of each block was proportional to the maximum stress caused at the center of the block. Therefore, the crack density of a specimen can be shown by the single number (Dcr^*) which is the average crack density on the areas of B2, B1, B1' and B2'.

Table 3 Results of flexural tests

MARK	f_b (N/mm²)	δ_{b0} (mm)	D_{cr} (1/cm)
B2	14.7	0.53	1.5
H2	13.5	0.76	1.9
M2	12.6	1.12	2.4
R2	11.5	1.55	3.0

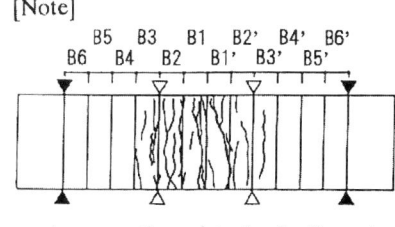

[Note]

▲ : supporting point, △ : loading point

ANALYSIS METHOD

The flexural stress (σ_b) - deflection (δ_b) curves were calculated by using the experimentally obtained $\sigma_t - \delta_t$ and $\sigma_c - \varepsilon_c$ curves, and Dcr^* of the flexure test specimen reserved for analysis. The flexural rigidity of the FRC beam was gradually reduced according to the increase of the external force. Therefore, the inside of the supporting points was divided into 12 blocks of the same length, and the $\sigma_b - \delta_b$ curve was calculated by using the apparent flexural rigidity which corresponded to the resulting moment, by the external force, in each block.

Stress - Strain Relations in Cross Section of Flexural Test Specimen

The tensile stress (σ_t)-strain (ε_t) relation in the small portion of the beam was obtained by replacing the Ncr shown in Figure 1 with the Dcr^* shown in Table 3. And the compressive stress-strain relation was indicated by the experimentally obtained $\sigma_c - \varepsilon_c$ relations without any treatment.

Moment - Curvature Relation and Flexural Rigidity

The moment (M) - curvature (κ) relation was computed by using the $\sigma_t - \varepsilon_t$ and $\sigma_c - \varepsilon_c$ relations, increasing the strain on the extreme tension fibre at an interval of 50 micro-strains, under the assumption that the strain distribution maintained linearity in the cross section. The apparent flexural rigidity (EI) of the small block at a certain bending moment was indicated by the inclination of the straight line which passed through each pair of adjacent points in the obtained $M - \kappa$ relation.

Calculation for Flexural Stress – Deflection Curve

The load limits, at where EI of each blocks was changed, were calculated. And all the sets of values of the EI, which corresponded to each interval between the two adjacent load limits, were listed. The increment of deflection ($\Delta \delta b$) corresponding to the load increment (ΔPb) can be computed by putting the relation between the ΔPb and the EI into the differential equation for the relationship between the stress and deflection of the elastic beam, and by fitting the boundary condition between all of the pairs of adjacent blocks. Finally, the $\sigma b - \delta b$ relation was obtained by accumulating the ΔPb and $\Delta \delta b$.

RESULTS AND DISCUSSION

Influencing Factor Upon Deflection

The analytically predicted $\sigma b - \delta b$ curves corresponded exactly with the experimentally obtained curves. Thus, the data calculated can be used for the purpose of investigating the characteristics, which can hardly be obtained by experiment.

The $\delta b 0$ can be obtained by also calculating the bending moment of the imaginary beam loaded in the curvature of the actual beam as the distributed load [2]. The curvature of point C in Figure 2 is given in equation (1). Considering $\varepsilon_c(\sigma b \cdot max)$ is proportional to $\varepsilon_t(\sigma b \cdot max)$, κmax is given in equation (2). Assuming that the curvature is indicated in the function of the moment as shown in equation (3), the area of imaginary load (S) and the distance from a supporting point to the center of gravity ($x0$) are given in equation (4) and (5) respectively. Finally, $\delta b 0$ is given in equation (6), and it was found that $\delta b 0$ is proportional to $\varepsilon_t(\sigma b \cdot max)$. Actually, $\delta b 0$ was closely proportional to $\varepsilon_t(\sigma b \cdot max)$ in this study, with a correlation coefficient of 0.878.

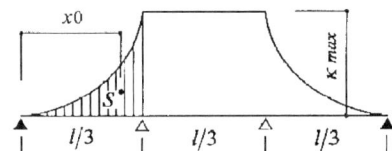

Figure 2 Imaginary beam loaded in the curvature of the actual beam as the distributed load

$$\kappa_{max} = \frac{\varepsilon_t(\sigma b \cdot max) + \varepsilon_c(\sigma b \cdot max)}{h} \quad (1)$$

$$\kappa_{max} = \alpha \cdot \varepsilon_t(\sigma b \cdot max) \quad (2)$$

$$\kappa / \kappa_{max} = f(M/M_{max}) \quad (3)$$

$$S = \kappa_{max} \cdot \frac{l}{3} \int_0^1 f(x) dx = \beta \cdot \kappa_{max} \cdot \frac{l}{3} \quad (4)$$

$$x_0 = \frac{l}{3} \cdot \left\{ \int_0^1 f(x) \cdot x dx \Big/ \int_0^1 f(x) dx \right\} = \beta' \cdot \frac{l}{3} \quad (5)$$

$$\delta_{b0} = \frac{l}{3} \cdot \left(\frac{l^2}{18} + \beta\beta' \cdot \frac{l^2}{9} \right) = \gamma \cdot \varepsilon(\sigma_{b \cdot max}) \quad (6)$$

where, h is the thickness of specimen, α, β, β' and γ are constant.

However, $\varepsilon(\sigma_{b \cdot max})$ has no dependence on the values (i.e. E_t, σ_{cr}, f_t or δ_{tp}). Then, investigating the ratio of $\sigma_t(\sigma_{b \cdot max})$ to f_t, it was found that, regardless the value of $\varepsilon(\sigma_{b \cdot max})$, the ratio $\sigma_t(\sigma_{b \cdot max})/f_t$ is almost constant (=0.947). Where, the coefficient of variation is 1.0%. Therefore, the strain at which the ratio $\sigma_t(\sigma_b)/f_t$ decreased to that value (0.9) should be defined the influencing characteristic upon the deflection at the maximum stress.

Characteristics Influencing Flexural Strength and Deflection of FRCs

The crack density (D_{cr}*), which was obtained from the flexure test, was used as the input data for calculation in order to enhance the accuracy in prediction. However, the deflection can be approximately predicted by also using the data obtained from tension and compression tests.

Figure 3 shows the relation experimentally obtained. It was found that the D_{cr} of each block was in proportion to the ratio of the maximum flexural stress (σ_b) caused at the block to the average σ_{cr} of the tension test specimens produced from the same mix.

The ratios of f_b calculated to f_t used as input data were almost constant among all specimens. Therefore, from the facts that the most part of the $\varepsilon(\sigma_{b \cdot max})$ was occupied with the total width of cracks, the deflection at the maximum stress point can be approximately predicted by using the σ_{cr}, f_t and σ_{tp} measured by tension, as shown in Figure 4.

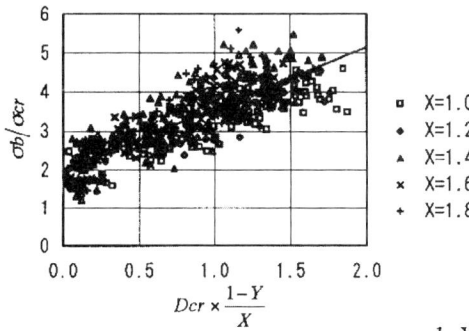

Figure 3 Relation between σ_b/σ_{cr} and $D_{cr} \times \frac{1-Y}{X}$

$X = W/(C + S_f)$, $Y = S_f/(C + S_f)$

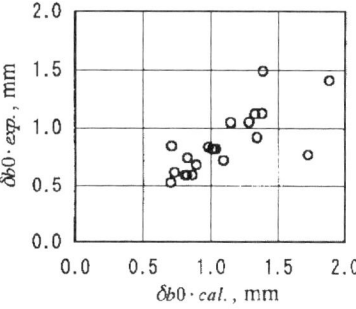

Figure 4 Relation between $\delta_{b0 \cdot exp.}$ and $\delta_{b0 \cdot cal.}$

CONCLUSIONS

It was found that the first crack strength, tensile strength and average width of cracks obtained by tension test were important factor influencing on the deflection of FRCs loaded in flexure.

REFERENCES

1. MAALEJ, M, AND, LI, V, C. Flexural/Tensile-Strength Ratio in Engineered Cementitious Composites, Journal of Materials in Civil Engineering, Vol. 6, No. 4, 1994, pp 513-527.

2. TIMOSHENKO, S. Strength of Materials, Part 1, 1955, pp 154-162.

IMPROVEMENT OF BENDING LOAD-BEARING CAPACITY BY EXTERNALLY BONDED PLATES

R Žarnić

S Gostič

V Bosiljkov

V B Bosiljkov

University of Ljubljana

Slovenia

ABSTRACT. The flexural behaviour of reinforced beams strengthened by externally bonded steel and carbon fibre reinforced plastic (CFRP) is compared using the experimentally obtained data. Two types of strengthened beams of different span to depth aspect ratio have been tested. The different failure mechanisms have been observed in cases of strengthening by steel or by CFRP plates. Two computational methods have been used in the analysis of failure mechanism and failure limit state: FEM and analytical. The FEM model gives promising results, but further development is needed before its introduction in design practice.

Keywords: Reinforced concrete structure, Bending, Steel plate, Carbon fibre reinforced plastic (CFRP) plate, External bonding, Flexural strength, Stiffness, Failure mechanism.

Professor Roko Žarnić holds the Chair for Research in Materials and Structures, Faculty of Civil and Geodetic Engineering, University of Ljubljana, Slovenia. Lately, his main research has been oriented to earthquake resistant masonry structures and strengthening of reinforced concrete structures. Professor Žarnić is also a member of Management Committee of COST C5 Action: Urban Heritage-Building Maintenance.

Samo Gostič MSc is Ph.D. student at the Faculty of Civil and Geodetic Engineering, University of Ljubljana, Slovenia. His main research interest is in the development of experimentally supported computational models of structures.

Vlatko Bosiljkov MSc is Ph.D. student at the Faculty of Civil and Geodetic Engineering, University of Ljubljana, Slovenia. His main interest is experimental and computational research of earthquake resistant masonry and reinforced concrete structures.

Violeta Bokan Bosiljkov PhD is a Teaching Assistant at the Faculty of Civil and Geodetic Engineering, University of Ljubljana, Slovenia. Her main research interest is in testing of polymer modified concrete and experimental research of strengthened structural elements.

INTRODUCTION

The efficient upgrade of flexural strength of reinforced concrete structures by externally bonded steel plates has been introduced to Slovenian practice in 1979 when an insufficiently reinforced newly built reinforced concrete bridge deck was strengthened [1] and [2]. After that more than 100 reinforced structures have been strengthened by steel plate bonding in Slovenia without any deficiency observed by later inspection. Recently, the new generation of plates made of carbon fibre reinforced plastic (CFRP) has been introduced in practice. Although it is well known that carbon fibre reinforced plastic plates can replace those made of steel there are some differences that should be taken into account in strengthening design. For this reason the research project on comparison of steel and CFRP plate bonding has been started. There are many papers reporting the experiences on flexural behaviour obtained by experimental and analytical research [3], [4], [5], [6], [7], [8], [9] that give a good insight in the flexural behaviour of plate bonded structural elements. The intention of the herein-presented research was to verify the ability of Slovenian contractors to introduce the CFRP plates widely in practice. Therefore the familiar strengthening with steel plates is compared with more recently introduced strengthening with CFRP plates.

DESCRIPTION OF SPECIMENS AND TESTS

Two different flexural elements were designed to represent a beam and a strip of the slab. The latter is considered as a flat beam. The dimensions of the beam were 200 mm wide by 300 mm deep by 3250 mm long (Figure 1). The flat beams were of the same length as the beams. Their cross-sectional dimensions were 800 mm wide and 120 mm deep (Figure 1). The design of specimens was carried out according to ENV 1992 (Eurocode 2). The steel reinforcement ratio A_s/bd was chosen in the amount which allows the external reinforcement to be added without over-reinforcing both types of specimens. It avoided premature brittle failure of the concrete in compression. The steel reinforcement ratio of beams was 0.56% and of flat beams 0.40%. The internal flexural reinforcement consisted of ribbed steel bars with 400 MPa minimum yield stress. Shear reinforcement of beams consisted of ribbed steel bar 6 mm in diameter, placed at 100 mm distance in the 1/3 of length at both sides and with 150mm distance in the 1/3 of length in the middle of the beam.

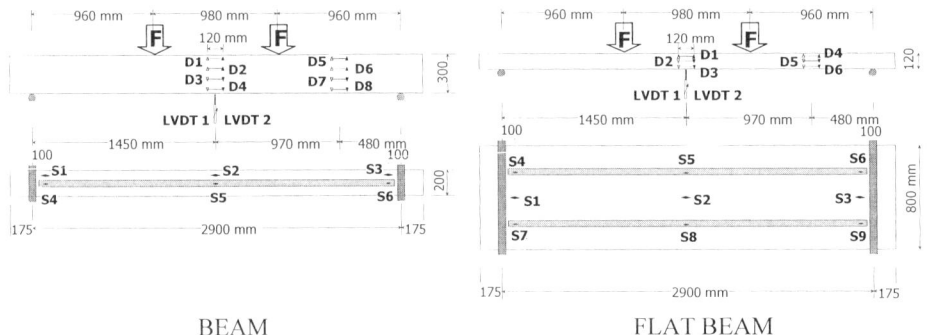

Figure 1 Geometry and instrumentation of test specimens

Altogether fourteen specimens were constructed, seven of each type. The concrete used for the construction of specimens had compressive strength of 25 MPa. Three of each type were strengthened by steel plates and three of them were strengthened by CFRP plates. The mild steel plates (f_y/f_u = 240/360 MPa, E=210 GPa) were 4 mm thick and 50 mm wide. The CFRP plates (f_u = 2400 MPa, E=150 GPa) were 1.2 mm thick and 50 mm wide. The epoxy glue was 2 mm thick in all cases. Each beam was strengthened by one plate (Figure 1). Two plates (Figure 1) strengthened each flat beam. One beam and one flat beam remained unstrengthened and served as reference specimens. Plates were attached on specimens by epoxy resin in the same way as it is case in practice. Since the steel plates are usually nailed every 500 mm on building site to enable fast and efficient pressure on glued surface, the same was done in the case of steel plated specimens. Specimens were tested by displacement controlled hydraulic actuator INSTRON 250 kN. The external load acted on 1/3 of the span as it is shown in Figure 1. Displacements were measured by LVDT-s at midspan of specimens. The distribution of strains over the depth of beams was measured in midspan and at 1/6 of the span by electrical dilatometers (D1 through D8). Strains on lower concrete surface and plate surfaces were measured at midspan and at the ends of plates by strain gauges (S1 through S9). The layout of the measuring points on beams and flat beams are shown in Figure 1. The legends in diagrams refer to measuring points presented in Figure 1.

TEST RESULTS AND DISSCUSSION

Flexural Behaviour of Specimens

The comparison of the deflection curves of strengthened and unstrengthened specimens presented in Figure 2 gives an insight in the level of flexural strength upgrade achieved by plates. Several differences between the effect of steel and CFRP plating can be observed from the diagrams. Using the steel plates, higher increase of stiffness and ductility can be achieved than in case of the CFRP plating. It was also observed that the test results of beams are less scattered, which can be the influence of greater robustness of beam cross-section. The higher ductility of steel plated specimens can be partly influenced by dowel effect of steel nails used for fastening of steel plates during their fitting on the beam surfaces. The diagrams clearly show the occurrence of first cracks and therefore changing of flexural stiffness.

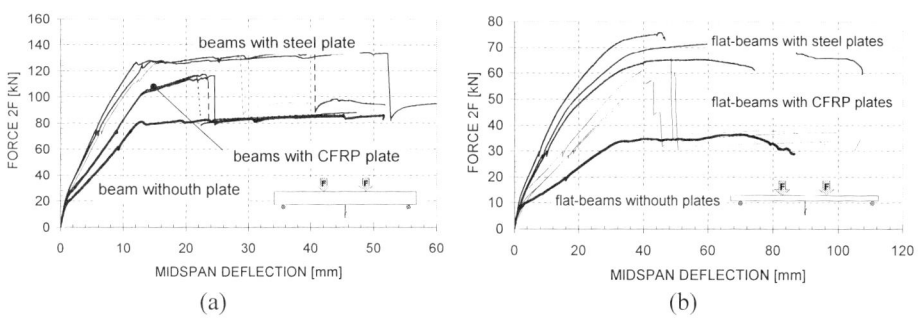

Figure 2 Comparison of deflection curves of beams (a) and flat beams (b)

Flexural strength of beams increased on average by 53% because of the added steel plate, while CFRP plates caused an average increase of strength by 35 %. Unstrengthened beam failed due to action of external force of 86.5 kN. In the case of flat beams steel plates increased their strength on average by 97.6%. CFRP plates increased their strength by 72.5%. Unstrengthened flat beam failed at an external force of 36.5 kN. Reduction of flexural stiffness due to development of flexural cracks occurred at approximately the same external loading in the case of unreinforced and plated specimens. Plates delayed the development of cracks, which resulted in different post-cracking stiffness. Inner reinforcement started to yield at the same magnitude of deflections of all beams. Steel plates were more effective then CFRP ones, which resulted in higher stiffness of steel plated specimens.

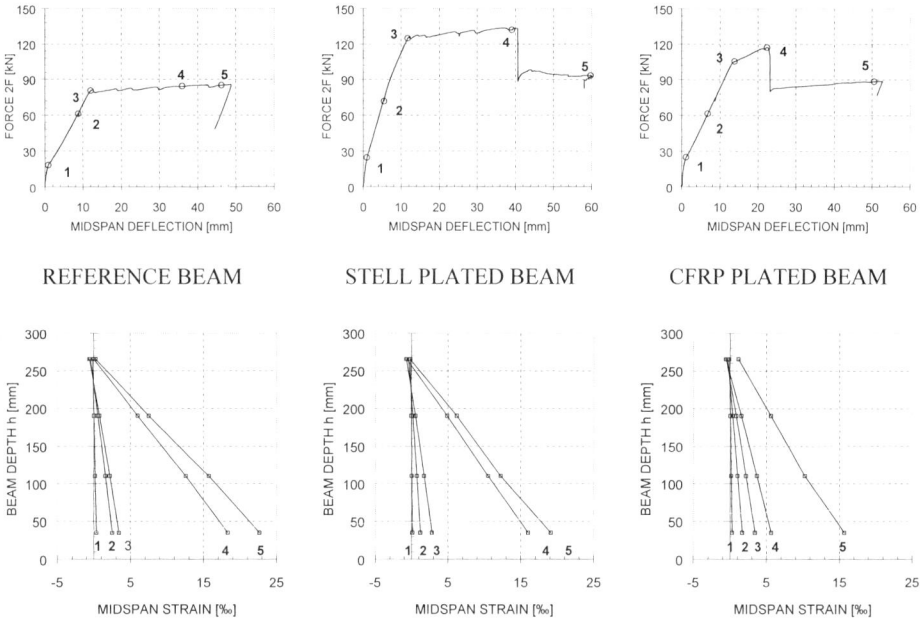

Figure 3 Comparison of strain distribution over the depth of reference beam, steel plated beam and CFRP plated beam

The distribution of strains over the depth of beams was measured at midspan and at 1/6 of span of specimens to obtain the information on validity of Bernoulli's hypothesis. It was confirmed using the set of results presented in Figure 3 both for specimens strengthened by steel and by CFRP plates. The analysis of strain development in plates and on concrete surface helps in understanding different failure modes that developed in the tested specimens. Strains developed on the surfaces of plates and on the lower concrete surface are presented in Figure 4 (a) and (b). It clearly shows the function of plates in delay of crack development. In Figure 4 (c) it schematically shows that the yield limit was achieved in steel plates while CFRP plates remained in elastic range. While for steel plates the stiffness had fallen to almost zero level after reaching the yield point of steel and thus peeling stresses within the concrete were not further increased. The stiffness of the CFRP plates remained unchanged and thus provoked sudden delamination at the critical load.

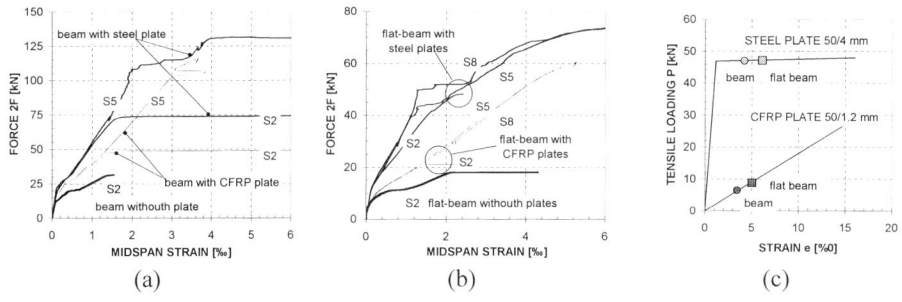

Figure 4 Comparison of strain curves of beam (a) and flat beam (b).
Strains were measured on plates and on concrete on lower surface of beam.

Failure Modes

Three basically different failure modes were observed. All of them are known from experiences of other researcher [5] and [9]. In the case of all three steel plated beams the failure was caused by end-of-plate shear peeling (Figure 5 (a)). It developed due to exceeding strength of the concrete below the plate. The nails that were used for plate bonding limited the area of peeling development, but did not affect the magnitude of failure forces. The failure of CFRP plated beams developed due to debonding or delamination of plates below the concentrated force acting on the upper beam surface (Figure 5 (b)).

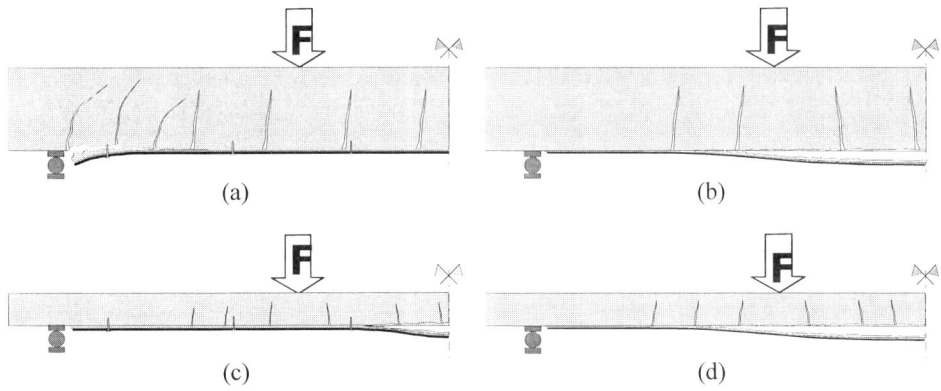

Figure 5 The failure mechanisms that developed in the cases of: beams strengthened by steel (a) and CFRP (b) plates and flat beams strengthened by steel (c) and CFRP (d) plates.

Steel plated flat beams failed due to debonding or due to delamination of plates in relatively limited midspan area. The development of failure (Figure 5 (c)) was influenced by yielding of steel plates (Figure 4 (b) and (c)). In the case of CFRP the failure was caused by delamination or/and debonding of plates (Figure 5 (d)).

The development of failure in the cases of CFRP plated flat beams and flat beams was sudden in contrary to more ductile development of failure in cases of steel plated flat beams.

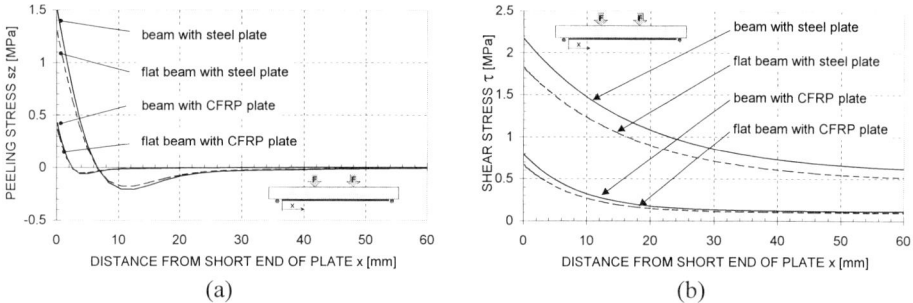

Figure 6 Comparison of the analytically obtained peeling (a) and shear (b) stresses at the end of steel and CFRP plates bonded to beams and flat beams.

The development of strains at the end-of-plate region can be predicted analytically using the expressions that was proposed by Täljsten [7]. The calculated distribution of peeling tensile stresses and shear stresses at the end-of-plates are presented in Figure 6. Stresses were calculated taking into account the geometry of specimens and plates and mechanical properties of materials. The acting force that was taken into account is in every observed case equal to failure force. It can be clearly seen that the combination of peeling and shear stresses gives the highest value in the case of steel plated beams where failure actually occurred due to end-of-plate shear peeling. It is interesting that steel plated flat beams would probably failed in the same way if the failure did not develop by debonding due to midspan yielding of plates. From Figure 6 it can also be concluded that CFRP plated beams and flat beams are not sensitive to end-of-plate failure. In these cases the development of failure due to debonding or delamination in the region of extensive flexural deformations of structure is more critical.

COMPUTATIONAL APPROACH

There are several well-documented and verified analytical approaches to the prediction of global or local flexural behaviour of plated reinforced concrete structures [7], [8], [9] and [10]. Further on in this chapter an alternative approach based on FEM analysis is presented in its early stage of its development. Observing the comparison of the predicted and experimentally obtained deflection curves presented in Figure 8, it is clear that computational results are not yet satisfactory. However, further development of the presented approach is worthwhile. FEM analysis also enables an insight into the development of stresses, strains and cracks as it is shown in Figure 9. Visual presentation of these parameters helps the designer to optimise the layout and the quantity of plates in the cases of real structures, which are to be strengthened by plates.

Computational approach is based on FEM method. The numerical model is based on the homogenisation of cracked material [12]. Hence the tensile cracking is the only non-linearity considered in this model. This allows us to model strengthened materials that exhibit strain-softening behaviour with strong localisation of the strain within their interfaces.

Cracking is judged on the basis of stresses and strengths of each material. After the crack has occurred, the artificial material infills the cracked zone and its further behaviour is governed by its artificial stiffness K_n and K_t. The constitutive model is incorporated in a three-dimensional finite element code. All the models were made from 20-noded solid elements.

Table 1 Material properties for numerical modelling

	E [GPa]	ν	f_c [MPa]	f_t [MPa]
Concrete	32	0.20	25	2.5
Epoxy	12.8	0.35	100	4
Steel bars	210	0.30	400	400
Steel plates	210	0.30	360	360
CFRP plates	150	0.10*	2400	100*

*Assumed values

For the simulation of behaviour of tested specimens two different types of numerical models were made. In the first model (homogenised) the reinforced concrete beam was modelled as a plain material. In the second model (detailed) the concrete and the reinforcement within the beam were modelled separately and each rebar is represented as a set of elements with material properties of steel perfectly bonded to surrounded concrete. Layers of the epoxy and plates were modelled separately with different material properties. In the terms of CPU time, the second model was much more demanding. In the case of beam it numbered 882 elements and 4520 nodes. Material properties for all models are given in Table 1 where f_c and f_t are compression and tension strength, respectively. All materials except for CFRP were modelled with strain softening. For the last one the ideally elastic behaviour is assumed.

Load disposition and boundary conditions for the detailed numerical model are shown in Figure 7 (a). The deflection curves of reference beam that were calculated using both models are presented in Figure 7 (b). It can clearly be seen that better prediction can be achieved using the detailed model. The main weakness of this computational approach is the assumed elastic behaviour of materials instead of more realistic elasto-plastic behaviour. Hence the homogenised model predicts softer behaviour than the detailed model. However, herein presented results of computational modelling should be considered as preliminary ones. Further development is to be carried out especially for those models where extensive cracking has occurred in the single step of loading. In further modelling of strengthened beam and flat beam only detailed models will be considered. The comparison of calculated and experimentally obtained deflection curves of beams (Figure 8 (a)) shows fairly good accordance of experimentally obtained and predicted curves. Significant differences between the calculated and experimentally obtained curves (Figure 8 (b)) are seen in the case of flat beams.

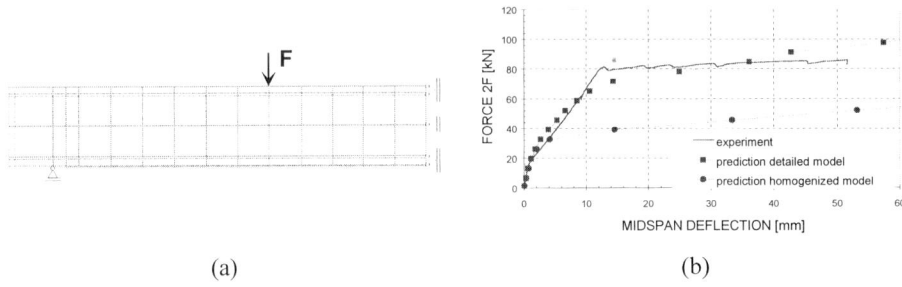

(a)   (b)

Figure 7 Load distribution and boundary conditions of detailed numerical model (a) and comparison of deflection curves obtained by two different FEM models (b).

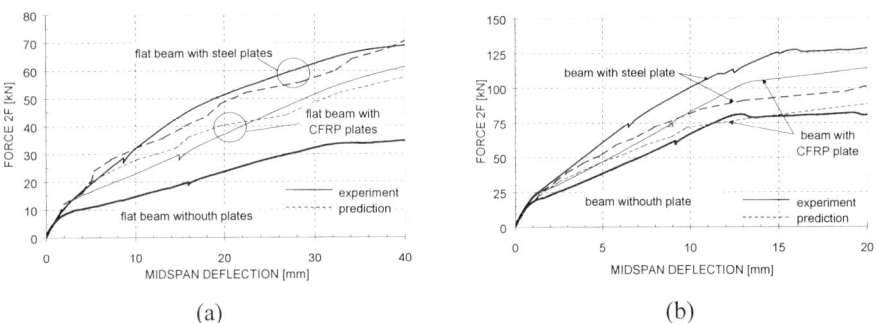

(a)   (b)

Figure 8 Comparison of experimentally obtained and predicted deflection diagrams of beams (a) and flat beams (b)

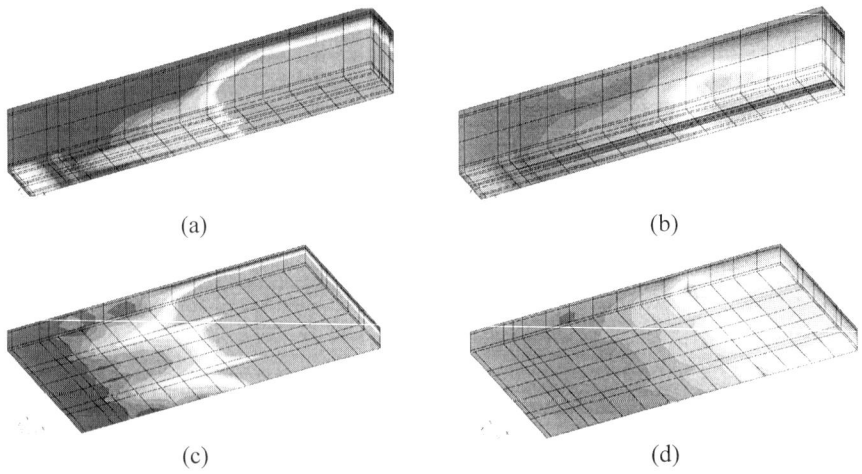

Figure 9 Distribution of cracks in steel plated beam (a), CFRP plated beam (b), steel plated flat beam (c) and CFRP plated flat beam (d)

The distribution of the crack pattern for the beam and the flat beam is presented in Figure 9. Figure 9 (a) and Figure 9 (b) show that strengthening with steel plates induces higher peeling stresses in the concrete within the zone of glued plates. Cracks, which are initiated by those stresses, are more concentrated towards the end of the glued plates. The distribution of cracks was different in the case of CFRP plated beams (Figure 9 (b) and (d)). Their crack pattern was more uniform without any further concentration both in the lower and higher state of stresses. Although the 3D technique enables fairly good prediction of beam flexural behaviour, some further improvements are needed. With the introducition of elasto-plastic material properties in the model, better prediction of elasto-plastic behaviour of strengthened structures is expected. By introducing of shear band [11] it should be possible to model satisfactory the short span strengthened beam. The benefits that are expected from 3D analysis of the strengthened elements are further study of the glued plates, which are closely spaced. This approach can enable successful modelling of mutual influence of closely located plates on the behaviour of strengthened elements.

CONCLUSIONS

The purpose of the described investigation in flexural behaviour of plated beams was to prepare guidelines for practice in Slovenia. Steel plating of reinforced concrete structures is well developed in comparison to recently introduced CFRP plating. The experimental results help in clarifying the differences between both examined plating methods through the comparison of flexural behaviour of tested beams. It is well known that CFRP plates are less sensitive to corrosion and that their application to concrete surface is easier. Our tests showed that steel plated structures are stiffer and more ductile in comparison to CFRP plated ones. It is to be mentioned that our specimens were made in accordance to typical strengthening pattern used in practice. Important information obtained from the tests is the fact that Bernoulli's hypothesis is a correct assumption used in the design procedure of plated structures.

The failure modes are relatively well known due to a number of experimental results available from other researchers. However trying to model behaviour of plated structures up to failure demands the use of non-linear computational models that take into consideration elasto-plastic characteristics of materials. This was demonstrated by the use of known elastic analytical approaches. The behaviour of a structure within elastic range can be successfully predicted by the majority of available computational and analytical models. The computational model that is presented in this paper can be appropriate foundation for the development of a model for the prediction of non-linear behaviour of plated structures. The next step of its development is its upgrading which should enable the use of elasto-plastic material characteristics in the computational process.

ACKNOWLEDGEMENTS

The authors would like to express their appreciation for the research grants made by the Ministry for Science and Technology of Republic of Slovenia and Ministry for Traffic and Communications of Republic of Slovenia to carry out the reported work, which forms part of the research programme on strengthening of reinforced concrete structures.

REFERENCES

1. BOŠTJANČIČ, J, HOČEVAR, B and CAFNIK, F. Design and realisation of strengthening of reinforced concrete bridge deck by externally bonded steel plates, Informacije ZRMK 239, Gradbeni vestnik 7/8, 1982, Ljubljana, Slovenia (in Slovenian) pp 1-4.

2. ŽARNIĆ, R and TERČELJ, S. Testing of reinforced concrete bridge deck after strengthening by externally bonded steel plates, Informacije ZRMK 238, Gradbeni vestnik 6, 1982, Ljubljana, Slovenia (in Slovenian) pp 1-4.

3. SWAMY, R N, JONES, R and BLOXHAM, J W. Structural behaviour of reinforced concrete beams strengthened by epoxy-bonded steel plates. The Structural Engineer, Vol. 65A, Feb. 1987, pp59-68.

4. OEHLERS, D J and MORAN, J P. Premature failure of externally plated reinforced concrete beams. Journal of Structural Engineering, Vol. 116, No.4, April 1990, pp. 978-995.

5. RITCHIE, P A, THOMAS, D A, LU, L W and CONNELLY G M. External reinforcement of concrete beams using fibre reinforced plastics. ACI Structural Journal, V. 88, No 4, July-August 1991, pp 490-499.

6. TRIANTAFILLOU, T C and PLEVRIS, N. Strengthening of RC beams with epoxy-bonded fibre-composite materials. Materials and Structures, RILEM, No. 25, 1992, pp. 201-211.

7. ZHANG, S, RAOOF, M and WOOD, L A. Prediction of peeling failure of reinforced concrete beams with externally bonded steel plates. Proceedings of Instn Civ. Engrs Structs & Bldgs, No. 110. Aug. 1995, pp. 257-268.

8. TÅLJSTEN, B. Strengthening of beams by plate bonding. Journal of Materials in Civil Engineering, Vol. 9, No. 4, Nov. 1997, pp. 206-211.

9. LA TEGOLA, A, MANNI, O and NOVIELLO, G. Flexural reinforcement of concrete beams using FRP plates. Saving Buildings in Central and Eastern Europe, IABSE Colloquium, Berlin, June 1998.

10. KAMISKA, M and KOTYNIA, R. Tests of reinforced concrete strengthened with CFRP plates. Saving Buildings in Central and Eastern Europe, IABSE Colloquium, Berlin, June 1998.

11. BOKAN-BOSILJKOV, V, BOSILJKOV, V, GOSTIČ, S and ŽARNIĆ, R. Critical local failure of strengthened bending concrete elements. Proceedings of International Congress on Creating with Concrete, Dundee, September 1999.

12. PANDE, G N, MIDDLETON, J, LEE, J S & KRALJ, B. Numerical simulation of cracking and collapse of masonry panels subject to lateral loading. Proc. 10th International Brick and Block Masonry Conference, Calgary, Canada, 1994

EFFECTS OF CARBON FIBRE SHEETS ON SHEAR STRENGTH OF REINFORCED CONCRETE COLUMNS ADJOINING WALLS OR SASHES

O Joh
Y Goto
University of Hokkaido
Japan

ABSTRACT. The paper describes an experimental study on the use of carbon fiber sheets for seismic retrofitting of non-ductile reinforced concrete columns with or without adjoining walls or sashes. Eight column specimens designed originally to fail in shear were tested. In the case of independent columns, the column cross-section was modified from a square to a circle, and in the case of columns adjacent to window sashes, various methods of wrapping carbon fiber sheets using lapped joints, separated slits, or a combination of these two at the sheet ends were employed. The test results show that carbon fiber sheets bound to column faces are effective for improving the strength and ductility of columns. A formula for shear strength evaluation, using an effective coefficient for sheet reinforcement, is proposed.

Keywords: Reinforced concrete column, Seismic rehabilitation, Strengthening, Carbon fiber sheet, Shear strength, Ductility, Adjoining sash, Closed wrapping, Separated wrapping, Circular cross-section

Osamu Joh is a Professor of Architectural Engineering, Graduate School of Engineering, Hokkaido University, Sapporo, Japan. He received B.E., M.E. and D. E. degrees from Hokkaido University. His current interests are seismic design of reinforced concrete structures and evaluation of snow loads, particularly shear transfer mechanism in beam-column joints of RC and SRC structures, anchorage of hooked bars and rehabilitation of non-ductile RC members. He is a member of the AIJ Committees on Management of Science, RC Structures and another technical researches.

Yasuaki Goto is an Associate Professor of Architectural Engineering, Graduate School of Engineering, Hokkaido University, Sapporo, Japan. He received B.E., M.E. and D. E. degrees from Hokkaido University. He has teaching and research interests in seismic design of reinforced concrete structures, especially shear transfer mechanism in RC interior beam-column joints.

INTRODUCTION

Carbon fiber sheets have been used in various structures for such purpose as increasing flexural strength of beams, stiffening slabs with bending cracks, and repairing columns that have failed in shear, recently. There have been many studies on the effects of these sheets on flexural performance of beams and slabs, and some studies on shear strength of columns. Studies on shear strength have shown that shear strength of a short column is increased by wrapping carbon fiber (CF) sheets continuously around the column [1-3], including the authors' previous paper [4]. However, it is difficult to wrap CF sheets completely around most columns in buildings, because they are joined to structural/non-structural walls or window/door sashes. Wrapping a CF sheet with vertical slits along the sashes around a column may not be so effective for enhancing shear strength. From the viewpoint of mechanics, the CF sheet must be continuously wrapped around a column to increase the shear strength of the column, and this is usually done by making gaps between non-structural walls and a column and temporarily removing sashes [4]. From a practical viewpoint, however, in order to carry out effective rehabilitation work while the inhabitants remain in the building and to avoid a leak of rain at the gaps after the work, new wrapping techniques that do not require the use of gaps or the necessity for removal of sashes must be established. We therefore conducted an experimental study to determine the effectiveness of new techniques for wrapping a CF sheet around a column that do not require making gaps in adjoining walls or removing sashes in enhancing the shear performance of normal steel-reinforced concrete columns. The experiments focused on obtaining relationships between the shear strength of such columns and the practicability of various new strengthening techniques.

EXPERIMENTS

Configurations of Specimens

The eight column specimens used in this study had a square cross-section of 300 mm x 300 mm on a scale of about half size. Two loading stabs at the top and bottom of the column, column height of 600 mm (shear span ratio of 1.12), axial bars of 12-D16 (p_g =2.65%, SD685), lateral reinforcement of hoop-6 (p_w =0.124%, SR295) and design concrete compressive strength of 20 MPa, as shown in Figure 1(a). The specimens were designed originally to fail in shear before and also after repair. They had the following variables:

1. Shape of column cross-section, which ranged from a square to a circle [-AC] to a rectangle [-RT];

2. Wrapping technique, which included continuous entire wrapping [-A-], continuous partial wrapping [-SA-], separate entire wrapping [-U-] and mixed wrapping [-SU-]

3. Treatment of sheet ends by steel flat bars [-FB] (see Table 1).

The wrapping process of the CF sheet included:

1. Grinding all column surfaces,

2. Making column corners into an arc shape with a radius of 20 mm

3. Undercoating the surface with epoxy adhesive,

4. Cutting out CF bandages of 30-mm-wide from a large CF sheet, and

5. Binding up the column faces with the bandages in every 50mm pitches three times normally. This 20-mm-wide spacing between CF bandages enabled easy observation of the development of cracking on the column surfaces.

The entire surfaces of actual columns were then wrapped without spacing. However, Specimen BS-6SA was wrapped separately with three bandages, each of six layers. Each bandage had a lap joint of about 100 mm in length at the ends for continuous wrapping and was glued to the concrete with epoxy adhesive. The properties of the CF sheet were thickness of 0.111 mm, tensile strength of 3.48 GPa, Young's modulus of 231 GPa, and elongation of about 1.5 %. The concrete and reinforcement properties are shown in Table 2

Table 1 Specifications of specimens and wrapping techniques used

SPECIMEN	VARIABLES			STRENGTH MPa
	Cross Section (mm)	Wrapping	Remarks	
BS	300x300	None	Original column hoop-6 dia @ 150 pw = 0.12%	18.9
BS-3A	300x300	Closed	Standard specimen with normal CF sheet reinforcement	18.2
BS-31-AC	Circular dia = 424	Closed	Changed to circular section by post-cast mortar	21.1
BS-3U-RT	500x300	Separated	Changed to wide column by post-cast concrete	20.8
BS-3U	300x300	Separated	Vertical slits along imaged sashes on transverse faces	18.2
BS-3SU	300x300	Mixed	Arrayed alternately two closed and two separated bandages	24.4
BS-3SU-FB	300x300	Mixed	Jointed all bandage ends with steel flat bars vertically	20.0
BS-6SA	300x300	Partial	Concentrated on 3 bandages of 60mm wide, 3 layers	19.4

The circular column specimen (BS-3A-AC) was made by post-casting mortar in the gaps between the square column and a circular form circumscribed to the column. The rectangular column specimen (BS-3U-RT) was formed by post-casting concrete of 100 mm in thickness on both sides of the column in order to obtain a longer anchorage length of the CF sheet, as shown in Figure 1(d). The post-casting parts were not reinforced. Specimen BS-3A was reinforced by '*closed entire wrapping*', i.e., all column faces were bound continuously with three layers of a CF sheet (CF reinforcement ratio, pf, was 0.13%) and this specimen was used as a standard specimen for comparison with other specimens. The specimens reinforced by '*separated entire wrapping*' had imaged walls or virtual sashes of 60 mm in thickness on

Table 2 Measured properties of concrete, mortar and reinforcement

CEMENT GROUP	σ_B (MPa)	ε_{max} (%)	$E_{1/3}$ (GPa)	$E_{2/3}$ (GPa)
Column (BS)	18.9	0.26	19.3	15.4
Post-cast concrete	22.0	-	-	-
Post-cast Mortar	35.4	-	-	-
STEEL BAR	σ_Y (MPa)	ε_Y (%)	σ_{MAX} (MPa)	Elongation (%)
Column Axial	767	0.376	929	11.7
Shear Rebar	367	0.171	523	20.3
CARBON-FIBRE SHEET	Design tensile strength: Design Young's modulus: Strain at ultimate strength:		3480 MPa 231 GPa 1.5%	

both transverse surfaces of the column, and the CF sheets were pasted with double U-shapes separated at the walls/ virtual sashes as shown in Figure 1(d). Anchorage lengths of the CF sheets on the transverse surfaces were 100 mm for BS-3U and 200 mm for BS-3U-RT. The specimens reinforced with '*mixed wrapping*' and '*closed partial wrapping*' used CF plates as bridges over each vertical slit between both U-shaped sheets separated by imaged sashes. That is, CF plates, which were solidified by epoxy adhesive beforehand and had a total length of sash width plus both lap joint lengths, were inserted into each gap between the transverse column surface and imaged sash and joined both separated CF sheets continuously. For Specimens BS-3SU and BS-3U-FB with mixed wrapping type, half of the CF bandages (sheet) were joined by three layers of CF plates as shown in Figure 1(c), and the other bandages were not joined. In the latter specimen, steel band plates, which were set vertically, joined the ends of all CF sheets in order to transmit stress of all bandages to the CF plates. CF sheets of Specimen BS-6SA with closed partial wrapping were concentrated at three bandages (the intermediate bandage was a horizontal slit to sidestep the central measuring points), and all of the bandages were joined by three CF plates for the purpose of easier repairing work.

Loading and Measuring Methods

The specimens were subjected to no axial load and to static lateral load reversals in drift angles of 1/500, 1/200, 1/100x2, 1/50x2, 1/33x2, 1/25x2, 1/20x2, and finally 1/17 radians. Here, x2 means two time load reversals with the same drift angle, and the loading was supplied in one direction as shown by the arrows in Figure 1. The moment diagram of the columns had point symmetry with respect to the center of column height by N-type loading. This loading supplied large shear at the middle span and an inflection point at the center of the middle span.

The drift angle was obtained as the average of four relative displacements measured between the top and bottom stubs at both column sides. Column shear force, bending and shear deformations on a column face, and strains of axial bars, shear reinforcement and CF sheet were measured.

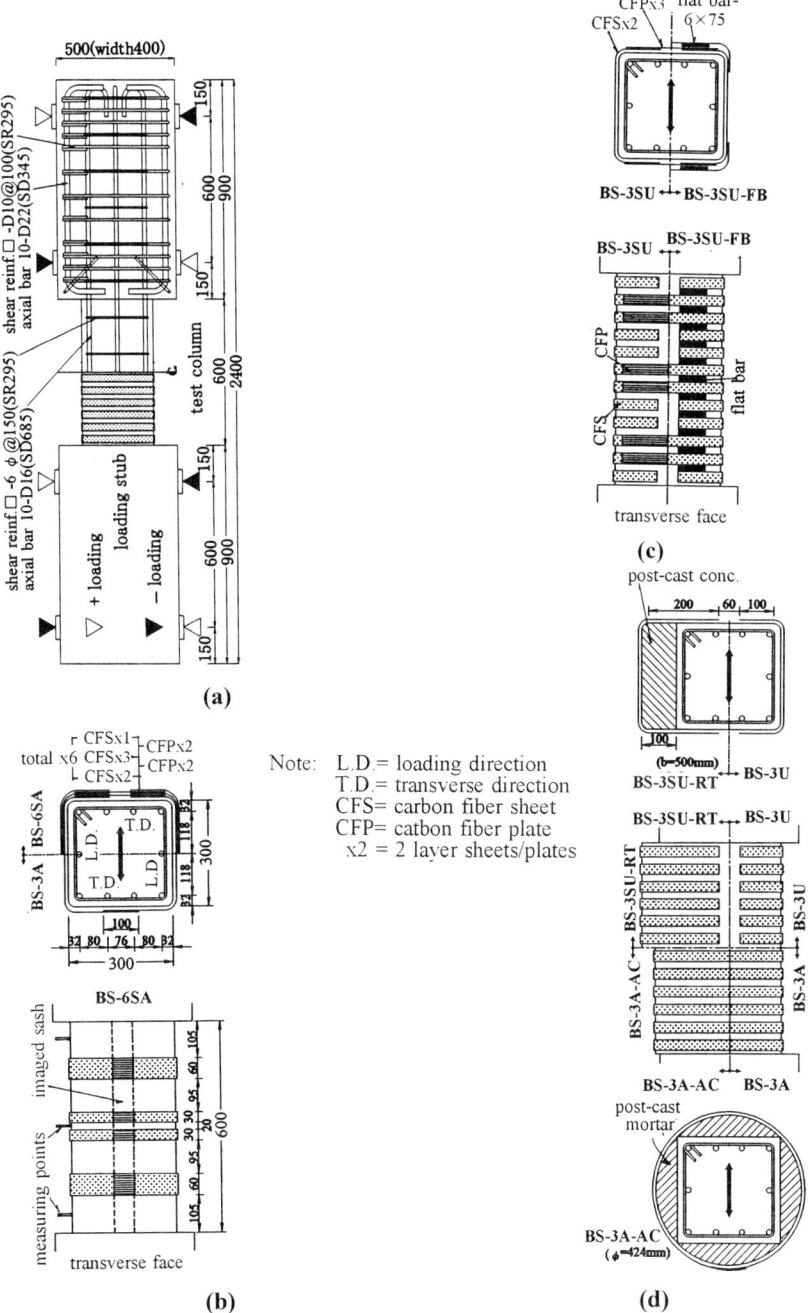

Figure 1 Specimens setup (a) Reinforcement loading (b) Closed partial wrapping (c) Mixed wrapping (d) Closed/separated entire wrap

TEST RESULTS AND DISCUSSION

Circumstances of Failure

The circumstances of some column failures after testing are shown in Figure 2. All of the specimens failed in shear. Initial shear cracks appeared in all specimens except Specimen BS-3U-RT at the drift angle R =1/500 radians. Unrepaired specimen BS reached the maximum load at R =1/75 following yield of shear reinforcement and peeling-off of cover concrete.

Closed Entire Wrapping Type

Specimen BS-3A maintained its original cross-sectional shape: Following the appearance of many shear cracks of 45-degrees, shear reinforcement yielded and shear resistance reached a maximum at R=1/50 (hereafter abbreviated as 1/50). Due to additional drift, the concrete failed in diagonal compression, but the CF sheet did not break. Specimen BS-3A-AC modified to circular cross-sectional shape: Following the appearance of many shear cracks with about 60-degrees parallel to a diagonal line of the column at 1/100, shear reinforcement yielded and shear resistance was increasing still. Axial bars yielded at +1/50 and -1/33, and the resistance was maintained until about 1/20. Finally, extension of the shear crack width and peeling of post-cast mortar became remarkable, but the CF sheet did not break.

Figure 2 Examples of column failure after testing

Separated Entire Wrapping Type

Specimen BS-3U maintained its original cross-sectional shape: On the loading faces of the column (see Figure 1(d)), shear cracks appeared at the top and bottom of column, and shear reinforcement started to yield at 1/200. On the transverse faces, vertical cracks that appeared along each sheet end at 1/300 to 1/200 extended markedly to the whole height of the column at 1/100, and shear resistance decreased rapidly at 1/50 due to anchorage failure, which resulted in the cover concrete under the CF sheets in the transverse direction opening like hinged double doors. Specimen BS-3U-RT modified to rectangular cross-sectional shape:

On the loading faces of the mortar, there was little cracking, and shear reinforcement yielded beyond 1/100. On the transverse faces, although vertical cracks appeared at 1/100, the column showed the same anchorage failure as that of BS-3U at 1/50. Separation of the cover mortar from the original column concrete was checked after the test.

Mixed Wrapping and Closed Partial Wrapping Types

Specimen BS-3SU without steel flat bars: Since the maximum shear resistance resulted from separation of a CF plate from a CF bandage at the middle of column height in approaching 1/50, loading was temporarily stopped. In a reloading test that was conducted after reinforcement by pasting on one additional layer of CF sheet to all joints, the column failed in anchorage at the ends of separated bandages before recovering the maximum shear resistance mentioned above, but the failure was not remarkable in comparison with that of BS-3U. Finally, joints between the CF plates and CF bandages were separated at -1/25. Specimen BS-3SU-FB with steel flat bar anchorage: Maximum shear resistance was reached at 1/50, as in the case of specimen BS-3SU. However, its value was larger and load degradation was clearly less than those of Specimen BS-3SU because stress of separated bandages could be transmitted to CF plates through the steel plates. Finally, joints between the CF plates and the steel flat bars were separated at -1/25, and then shear resistance decreased rapidly. Specimen BS-6SA reinforced by closed partial wrapping: After reaching maximum shear resistance at 1/33, the shear resistance decreased gradually with increase in drift due to crack width expansion and concrete crash at portions without CF bandages. The CF sheet did not break until the final stage.

Load vs Displacement Relationship

Column shear force 'Q' vs column drift angle 'R' relationships for Specimens BS and BS-3A are shown in Figure 3 as examples. In the case of Specimen BS, stiffness on the relation curve degraded at initial shear cracking, and shear resistance almost reached the upper limit at about 1/200 and then decreased drastically after the hoops yielded. In the case of Specimen BS-3A, stiffness degradation appeared at hoop yielding but not initial shear cracking, and the decreasing rate of shear resistance was a little beyond reaching the maximum load. Skeleton curves of the relationships are shown in Figure 4 for all specimens. The vertical axis of these skeleton curves were normalized by the square root off_B, which is compressive concrete strength, in order to reduce the effect of differences in the concrete strengths of specimens.

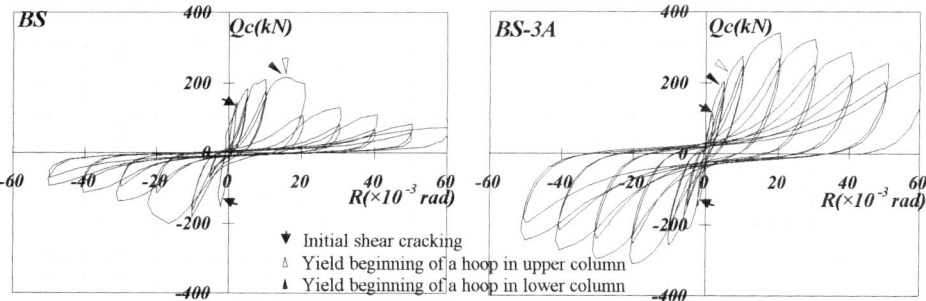

Figure 3   Examples of column shear vs drift-angle relations

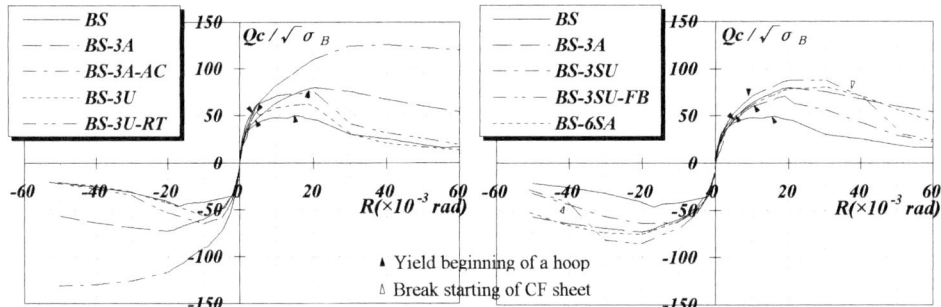

Figure 4 Skeleton curves of column shear vs drift-angle relations

Closed Entire Wrapping Type

Specimen BS-3A: Since the CF sheet shared much of the shear stress due to shear crack extension, the column shear strength increased by 1.7-times more than that of Specimen BS and the degrading rate of shear resistance after reaching maximum shear strength was very small. This indicates that the technique of closed entire wrapping of a CF sheet is effective for enhancing strength and ductility in shear-failing RC columns. Specimen BS-3A-AC: Initial stiffness became larger than that of BS due to enlargement of the cross-sectional area. The shear strength, which was not clear due to premature axial bar yielding, increased more than 2.2 times that of BS and ductility also became greater. The increasing rate of strength was greater than the enlargement rate of the concrete cross-section plus the strengthening effect of the CF sheet; therefore, strength of a column can also be enhanced by modifying the rectangular cross-sectional shape of the column to a circular cross-sectional shape.

Separated Entire Wrapping Type

Specimen BS-3U: Although the skeleton curve until about 1/100 was similar to that of BS-3A, shear resistance reached a maximum at about 1/50, just after peeling of the cover concrete on the transverse faces (mentioned in the last section), and shear resistance decreased quite rapidly. The enhancing effects on shear strength and ductility were less than those in BS-3A. Specimen BS-3U-RT: The skeleton curve within a range of 1/200 to 1/50 could be simulated by addition of the strength increment by post-cast concrete to the skeleton curve of BS-3U, and the observed skeleton curve was similar to that of BS-3U in the range beyond 1/50. The shear strength was enhanced to the same degree as that of BS-3A, but shear resistance decreased after the CF sheet anchorage failed.

Mixed Wrapping and Closed Partial Wrapping Types

Specimen BS-3SU: The skeleton curve before reaching 1/50 at maximum shear resistance was similar to that of BS-3A, but the degradation of shear resistance after 1/50 was slightly greater than that of BS-3A due to separation of the CF sheet from the CF plates at their joints and due to damage to the concrete in unclosed wrapping portions. Specimen BS-3SU-FB: The ascent toward maximum shear strength became steeper in the range of 1/100 to 1/50 than that of BS-3U due to steel plate anchorage; the shear strength at 1/33 was the greatest among all specimens except the circular cross-sectional column. However, separation of the CF sheet from the steel plates resulted in a large decline in shear resistance after 1/33, and the resistance became less than that of BS-3A. Specimen BS-6SA: The relation curve was very similar to that of BS-3A. This indicates that the technique of closed wrapping concentrated at a few parts can improve shear performance of columns to the same degree as that by closed entire wrapping when the same amount of CF sheet is used in both cases.

Strain of CF Sheet

Strain distributions of the CF bandages measured at maximum shear resistance are shown in Figure 5. Except for the top and bottom measuring points on each column face, the average strains of Specimen BS-3A with closed entire wrapping were about 0.7% in the loading direction (hereafter abbreviated as LD) and about 0.4% in the transverse direction (TD). Average of hoop strains, which are not shown in the figure, were 1.3% and 0.1% in LD and TD, respectively. Average strains of Specimen BS-3A-AC in LD and TD were 1.0% and 0.7%, respectively, about 1.5-times higher than those of BS-3A. These values were higher because the drift angle at maximum resistance of the latter was larger than that of the former.

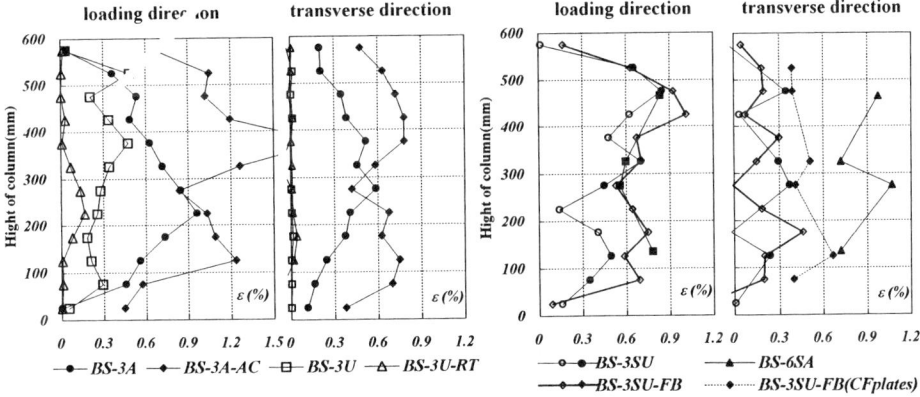

Figure 5 Distributions of CF sheet strains at the ultimate stage

The separated entire wrapping specimens (BS-3U and BS-3U-RT) had lower average values of LD (0.3% and about 0%, respectively). Therefore, it is clear that most of the shear strengthening resulted from the additional concrete cross-section. For the mixed wrapping specimen BS-3SU, the strain of separated CF sheet in LD was smaller than that of BS-3A but larger than that of BS-3U. Consequently, the average strains in the case of BS-3SU were 0.5% and 0.2% in LD and TD, respectively.

In Specimen BS-3SU-FB, the difference in strain between closed and separated CF sheets was so small that the average strains in LD and TD became 0.7% and 0.5%, respectively, which were equal or larger than those of BS-3A due to the effect of the steel plate on the CF sheet anchorage. The average strains of Specimen BS-6SA in LD and TD were 0.7% and 0.9%, respectively; only in BS-6SA the average strain in TD was larger than that in LD. Considering that cross-sectional area of the CF sheet in TD of BS-6SA was two thirds of that in LD, the equivalent strain in TD was estimated to be 0.6%, but this value was still larger than that of BS-3A.

Shear Resistance of RC Columns Wrapped with CF sheet

Experimental column shear forces (expQu) and drift angles (Ru) of all specimens at maximum resistance are shown in Table 3.

Table 3 observed and calculated values at the ultimate stage: Units Q(kN), R ($\times 10^{-3}$ rad)

SPECIMEN	EXP RESULT		CALCULATED SHEAR RESISTANCE						FAIL MODE
	eQu	Ru	CQsu1	e/su 1	CQsu2	e/su2	CQsu3	e/su3	
BS	216	15.7	196	1.10	196	1.10	-	-	S
BS-3A	340	20.3	265	1.28	287	1.18	-	-	S
BS-3I-AC	577	40.3	370	1.56	398	1.45	428	1.35	B->S
BS-3U-RT	343	21.2	394	0.87	421	0.81	310	1.11	A->S
BS-3U	265	18.5	265	1.00	287	0.92	-	-	A->S
BS-3SU	350	21.3	292	1.20	314	1.11	-	-	S
BS-3SU-FB	393	20.2	273	1.44	295	1.33	325	1.21	S
BS-6SA	353	30.7	280	1.26	305	1.16	-	-	S

1) Failure modes: S = Shear failure, A = Anchorage failure, B = Axial bar failure
2) Modified Ohno-Arakawa Equation

$$calQsu = \frac{(0.115 k_p \cdot k_u (180 + \sigma_B))}{((M/Qd) + 0.12)} + 2.7\sqrt{\rho_w \cdot \sigma_{wy} + \alpha \cdot \rho_f \sigma_f + 0.1 \sigma_o} \cdot b \cdot j$$

Where k_u : coefficient for size effect = 1.0
k_p : coefficient for axial bar ratio = 0.80
ρ_w : steel shear reinforcement ratio = 0.124
σ_w : yield stress of steel reinforcement = 367 Mpa
j: moment arm member = 234mm
b: width of member = 300mm
σ_o: axial stress of member = 0
M/Qd: shear span ratio = 1:12

Calculated shear forces (calQsu) by using the Modified Ohno-Arakawa Equation [6], which estimates shear strength of RC columns, are also shown in the table. The first and second terms in this equation are interpreted as the arch action of concrete and the truss action of shear reinforcement, respectively. In order to apply this equation to evaluation of the shear strength of RC columns wrapped with CF sheets, fiber reinforcement ratio (pf) multiplied by its tensile strength (f) and a reduction factor (α) was added to the second term of the steel reinforcement effect in the equation. Here, subscripts '-su1' and '-su2' were used in case of =2/3 and =1, respectively. The reason why an effective coefficient of 2/3 was used in previous papers [4, 5] was because it showed close agreement between experimental and calculated values. However, the experimental shear force of BS-3A in this study was 1.28-times higher than the calculated shear force (used =2/3), although this value was an overestimation because the average strain of the CF sheet in the loading direction (0.6%) was less than the tensile limit strain (1.5%) of the sheet. The ratio of the experimental value to the calculated value with =1 (hereafter abbreviated as 'exp/su2') was 1.18, and became close to 1.0. In the case of BS-3A-AC, the calculated value was obtained after substituting the circular cross-sectional area with an equivalent square one. The ratios exp/su1 and exp/su2 were 1.56 and 1.45, respectively. When the shear force (calQsu3) was calculated by using 1.0 and the equivalent concrete strength which was modified with mortar strength of 35.4 MPa; it was larger than concrete strength of 21.1 MPa, the ratio exp/su3 was 1.35 and was still large. This is thought to be because the effect of circular wrapping on shear strengthening was larger than that of square wrapping.

In Specimen BS-3U, exp/su1 was 0.99, which was the best agreement, but exp/su2 was 0.92, which was an overestimation. The ratio exp/su1 of BS-3U-RT was only 0.87. Considering that the average strain of the CF sheet was nearly zero, as mentioned above, the ratio of experimental to calculated shear force (calQsu3) with=0 was 1.11, and both shear forces became close. The amount of CF sheet reinforcements of BS-3SU and BS-6SA in the loading direction were exactly the same as that of BS-3A, and the only difference among these three specimens was in concrete strength. Therefore, the ratios exp/su1 and exp/su2 were similar among the three specimens because the difference in concrete strength was small. However, the ratio exp/su2 of BS-3SU-FB was 1.33-times higher than those of the above three specimens, despite the fact that its amount of CF sheet reinforcement was the same as those of the above three specimens. The reason for this was thought to be that the steel plates for anchorage had a dowel action against shear distortion like column axial bars. The shear resistance (calQsu3) was calculated by considering the effect of the cross-sectional area of steel plates on the axial bar effective coefficient (k_P) in the equation. Consequently, the ratio exp/su3 became 1.21, which was similar to that of BS-3A.

Judging from the results, as all experimental values except BS-3U were still larger than the calculated values of calQsu2 or calQsu3, it should be considered that the effect of CF sheet wrapping on shear strength of RC columns is estimated with not only the truss action as an increase in lateral reinforcement but also the arch action as an increase in nominal concrete strength due to confinement effect.

CONCLUSIONS

The following conclusions were drawn from the results of experiments conducted on eight different column specimens subjected to lateral load reversals in order to determine effective strengthening methods using carbon fiber sheets for non-ductile R/C columns that have a square cross-section with or without adjoining imaged sashes.

1. A combination of closed entire wrapping of CF sheets with modification of the shape of column cross-section from a square to circle by post-cast concrete is effective for improving shear strength and ductility.

2. CF sheet wrapping with entire separation along sashes on transverse faces of a column is somewhat effective for improving shear strength, but there is no improvement in ductility.

3. The CF sheet wrapping method that consisted of mixing closed and separated bandages or of closed partial bandages can be expected to have the same enhancing effect on shear strength and ductility as that of closed entire wrapping.

4. The useful application of a previously proposed equation for evaluating shear strength of normal RC columns to CF sheet-wrapped columns was derived.

5. The reinforcement work of separated entire wrapping by CF sheets itself can be applied to columns adjoining walls; however the shear resistance of such columns may be different from that of columns that have adjacent sashes due to the confinement effect from the walls. The authors are currently conducting an experimental study on such columns and hope to present the results in a future report.

ACKNOWLEDGEMENTS

The authors would like to express their appreciation for the financial support of Tonen Corporation and for the works of Mr A. Kitano who is an Instructor of Hokkaido University, and K. Yogo and M. Kondo who were graduate students of the university.

REFERENCES

1. XIAO, Y. Seismic retrofit of concrete columns using advanced composite materials. Proc. of International Conference on Composite Construction, Innsbruck, 1997. pp928-929.

2. ONO, K. Strengthening of reinforced concrete bridge piers by carbon fiber sheet. Proc. of International Conference on Composite Construction, Innsbruck, 1997. pp929-930.

3. KATSUMATA, H, and KOBATAKE, Y. Retrofit of existing reinforced concrete columns using carbon fibers. Proceedings of the Third International Symposium on Non-Metric Reinforcement for Concrete Structures. Vol.1, Sapporo, 1997. pp555-562.

4. JOH, O, and KITANO, A. Shear strengthening of RC columns by carbon fiber sheet. Proc. of International Conference on Composite Construction, Innsbruck, 1997. pp926-927.

5. ISO, M, MATSUZAKI, Y. et al. Experimental study on seismic behavior on RC columns with wing walls retrofitted by carbon fiber sheets. Proc. of the 3rd Inter. Symposium on Non-Metric Reinforcement for Concrete Structures. Vol.1, Sapporo, 1997. pp579-586.

6. ARAKAWA, T. Shear resistance of reinforced concrete beams. Transaction of Architectural Institute of Japan. Vol.66, 1960. pp24-33 (in Japanese)

SEISMIC PERFORMANCE OF RC COLUMNS LATERALLY CONFINED BY CARBON FIBRE REINFORCING PLASTIC TUBE

T Yamakawa
P Zhong
University of the Ryukyus
Japan

ABSTRACT. Widely carbon and aramid fibre sheets have been applied to seismically retrofit existing buildings and bridges because of their unique characteristics related to mechanical response and environmental durability. This paper summarizes the recent research developments of a new structural concept for the design of the hybrid RC columns with premanufactured Continuous Fibre Reinforced Plastic (CFRP) tube, where the CFRP tubes have the dual function of stay-in-place formwork and transverse reinforcement for the structural elements. The paper will discuss the elastoplastic behaviour of RC columns with CFRP tube subjected to lateral cyclic load and constant vertical load simultaneously through experimental and analytical research results. The hybrid column specimens illustrated excellent seismic performance in shear force versus lateral displacement hysteresis loops in spite of very slight bond slip in comparison to that of steel plate retrofitting one. And it is significant to develop the technology and application on circular and square tubes employing carbon or aramid fibre sheets.

Keywords: CFRP tube, RC column, Hybrid, Stay-in-place formwork, Transverse reinforcement, Confinement, Elastoplastic behaviour.

Professor Tetsuo Yamakawa is a professor in the Department of Civil Engineering and Architecture, University of the Ryukyus, Okinawa, Japan. He received his Dr Eng. in 1989 from Kyusyu University. His research activities are in the areas of structural mechanics, steel-concrete composite structures, development of new seismic retrofit. He is a member of AIJ, JCI and NZSEE.

Peng Zhong is a graduate student in University of the Ryukyus, Okinawa, Japan. His main research interests include the earthquake-resistance design and the technology with emphasis on seismic retrofit to structures. He is a member of AIJ and JCI.

INTRODUCTION

Widely expended for aerospace and leisure application, carbon and aramid fibre have been manufactured as sheet, wire and bar to seismically retrofit existing buildings and bridges because of their advanced mechanical characteristics and environmental durability. Furthermore, Continuous Fibre Reinforced Plastic (CFRP) has been introduced as reinforcement, instead of steel, to inhibit or protect against corrosion for concrete structure subjected to corrosive conditions or where electrical insulation is required.

In this paper, a new concept is proposed that CFRP tube perform the dual function of stay-in-place formwork and transverse reinforcement, while longitudinal bars can be steel or CFRP bars. And if necessary, the transverse steel, namely, hoops may be arranged further. In this concept, the concrete sustained compressive load while CFRP tube functions as formwork, confinement material to core concrete and shear reinforcement in a high level. Pilot lateral cyclic loading tests on concrete filled square CFRP tube columns simultaneously subjected to the constant axial load are carried out to verify the enhancement in bending and shear strength and ductility of this new hybrid columns.

Column Specimen

Four columns were provided with 250 mm square and 1000 mm height and shear span to depth ratio $M/(VD)$ of 2.0. Details of specimens are listed in Table 1. Gaps of about 10 mm were provided between end of tube/plate and stub so that axial compressive load might not be transmitted through the tube or plate, which is used as retrofitting jacket in this case. The specimens CR97A-S0 and CR97A-DS are reference ones to the CFRP tube specimens CC97-SD and CC97-SS.

All columns have the same vertical bars and transverse hoops, but hoops are not arranged in the CFRP tube specimen CC97-SS only. A concrete cover of 15 mm was provided, which is measured from the outside of the perimeter hoops and enclose the area of concrete core. All the material properties are shown in Table 2.

Table 1 Column specimens

SPECIMEN	REFERENCE	STEEL PLATE	CFRP TUBE	
	CR97A-S0	CR97A-DS	CC97-SD	CC97-SS
ELEVATION (unit in mm)				
SECTION				
CONCRETE STRENGTH f_c', MPa	31.4	31.4	33.3	33.3
STEEL HOOP	D6-@60 ($p_w = 0.43\%$)	D6-@60 ($p_w = 0.43\%$)	D6-@60 ($p_w = 0.43\%$)	-
REBAR	12-D13 ($p_g = 2.44\%$)	12-D13 ($p_g = 2.44\%$)	12-D13 ($p_g = 2.44\%$)	12-D13 ($p_g = 2.44\%$)

Table 2 Mechanical properties of reinforcing material

REINFORCING MATERIAL	THICKNESS OR SECTION AREA	f_y, MPa	ε_y, %	E, GPa
CARBON FIBRE SHEET (1 LAYER)	0.167 mm	3481.4	1.50	230.4
STEEL PLATE	3.2 mm	265.4	0.20	222.7
STEEL BAR (D6)	32 mm²	388.2	0.22	184.1
STEEL BAR (D13)	127 mm²	360.6	0.21	202.7

Table 3 Transverse swell of CFRP wall plate

MEASURE POSITION		MEASURE mm	CALCULATION mm	MEAS./CALC,
CC97-SD	Flange Midheight	4.90	3.01	1.63
	Web Lower end	-	4.51	-
	Flange Midheight	4.70	2.19	2.15
	Web Lower end	6.60	3.29	1.89
CC97-SS	Flange Midheight	1.80	3.94	0.46
	Web Lower end	4.70	6.06	0.78
	Flange Midheight	2.40	4.03	0.60
	Web Lower end	2.85	6.01	0.47

Carbon fibre sheet with fibre content of 300 g/m² and its design thickness of 0.167 mm was used in manufacturing square CFRP tube. Five layers of carbon sheet with lap of 200 mm were impregnated by epoxy resin. As a result, the thickness of square CFRP tube was 2.7-2.9 mm. The CFRP material exhibited a linear elastic behaviour until fracture according to stress-strain curves obtained from tensile test.

Deflection In CFRP Tube Wall When Casting Concrete

Bending stiffness of wall plate in CFRP tube is weaker than that of steel plate, so it is worried that deflection may occur out-plane when subjected to lateral pressure due to fresh concrete. When fresh concrete is being poured, lateral pressure of fresh concrete against the wall of CFRP tube can not be ignored. Here as a trial, both fixed ends beam subjected to equivalent distributed uniform load instead of lateral pressure could be modeled. The maximum lateral pressure of fresh concrete shall be 22.6 kPa/m. The elasticity modulus is 67.7 GPa for specimen CC97-SD, and 72.6 GPa for CC97-SS. The beam span is 250 mm, depth becomes 2.9 mm for CC97-SD, and 2.7 mm for CC97-SS. Twice deflection calculated from the model and measured when casting are listed in Table 3. Measured values to calculating ones ratio varied from 0.46 to 2.15. This is a reason why manufactured tubes had not a flat surface in its inside and outside, so different thickness exists in different positions.

Measurements And Loading Arrangement

Both horizontal and vertical deformations were recorded using the transducers located between stubs. Strains of corner longitudinal bars, hoops and steel tubes were typically

458 Yamakawa, Zhong

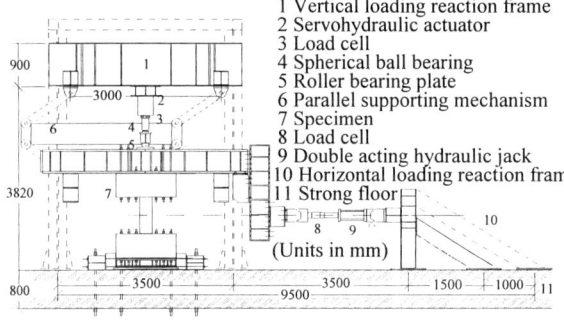

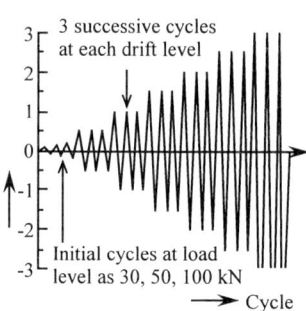

Figure 1 Schematic test setup

Figure 2 Loading programme

monitored using the electric resistance strain gage with length of 5 mm. Relatively high axial load $0.35f_c'A_g$ was applied in this program, where f_c' is concrete cylinder strength and A_g is gross area of section of the columns. The axial load ratio is desirable to be lower than 1/3 in Standards for Structural Calculation of Reinforced Concrete Structures (1991) by Architectural Institute of Japan (AIJ) in order to keep ductility of RC columns.

The loading apparatus with the parallel supporting mechanism developed by the Building Research Institute in Japan, which was called as Ken-ken type, was used in the experiment as illustrated in Figure 1. At first several small cyclic lateral forces were applied to column test specimens simultaneously subjected to constant axial load in order to obtain the initial lateral stiffness curve. Continuously, cyclic lateral forces were referred in the range of drift angle R = ±0.5, ±1.0, ±1.5, ±2.0, ±2.5, ±3.0% at three cycles respectively as depicted in Figure 2. The $R = \delta/h$ is a story drift angle of the column where δ is a story drift and h is the height of the column.

EXPERIMENTAL RESULTS

Shear force V versus story drift angle R, and mean axial strain ε_v versus story drift angle R responses obtained from test are plotted in Figures 3 and 4. The mean axial strain ε_v is given by dividing relative vertical displacement between top and bottom stubs by column height h. Top stub is always forced to drift on a parallel with bottom stub by the parallel supporting mechanism of the test setup.

Reference column specimen CR97A-S0 whose vertical bars yielded at $R = 0.6\%$ reached peak shear force at $R = 1.0\%$. Then numerous bond cracks appeared while lateral force decreased significantly. At $R = 2.0\%$, vertical bars in upper end of column buckled and brittle shear failure occurred. In hybrid RC column specimens confined by CFRP tube, deterioration almost did not occurred in V-R hysteresis loops while a little bond slip cracks appeared along longitudinal bars. When retrofitted by steel plate, just only a little flexural crack was observed in the upper and lower end of column, that is meaning that flexural failure occurred and this is verified by V-R relationship curves.

As seen in ε_v -R hysteresis loops, reference specimen CR97A-S0 illustrates an unsteady axial strain behaviour, and this agrees to that column performed a shear failure after reaching its flexural strength. Specimen confined by CFRP tube or steel plate strained gradually and

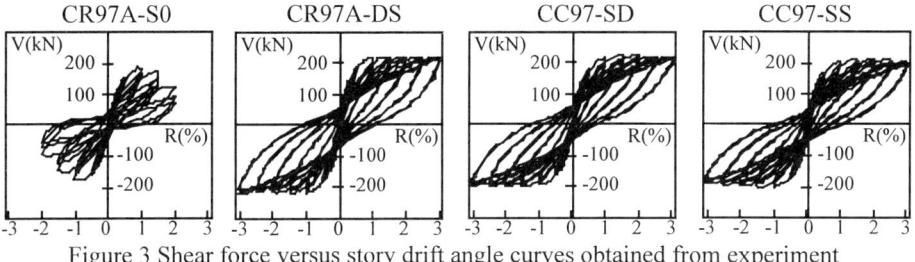

Figure 3 Shear force versus story drift angle curves obtained from experiment

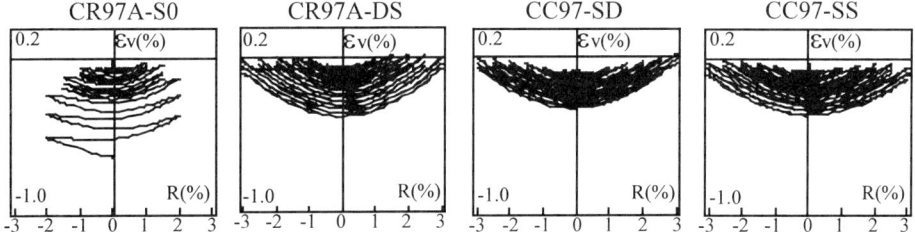

Figure 4 Mean axial strain versus story drift angle curves obtained from experiment

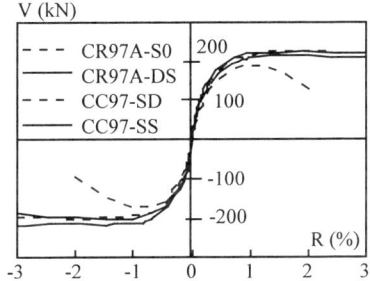

Figure 5 Skeleton curves of V-R responses for columns

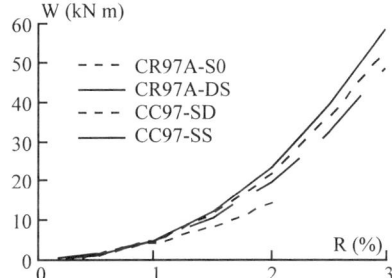

Figure 6 Accumulated energy absorption capacity of columns

stably, which agrees to the experimental results showing excellent ductility and strength. And among those three hybrid columns including the retrofitted RC column with steel plate jacket, little distinction is observed in axial strain, and this could be considered that axial strain of columns also be well suppressed comparing with reference RC column.

The V-R skeleton curves and accumulated energy absorption capacity curves are compared in Figures 5 and 6 respectively. Preventing columns from brittle failure, CFRP tube could well functioned the confining role as same as steel plate jacket.

Theoretical Investigation

The confinement effects of transverse reinforcing material to core concrete were taken into account when evaluating flexural and shear behaviour of column specimens.

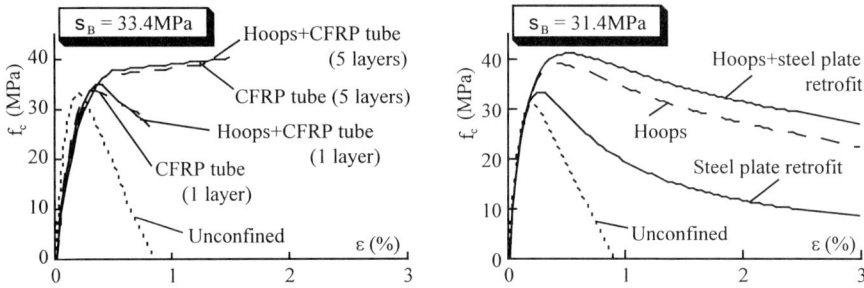

Figure 7 Stress-strain relationship of concrete confined by reinforcing material

For confined concrete by steel hoops, Mander's stress-strain model was employed [1]. By steel tube, compressive strength was suggested by Matsumura. By steel hoops and retrofitting steel plate, the superposed strength was adopted together with Mander's equation, because this constitutive law has been adopted in the hybrid RC columns consisting of concrete and steel [2]. By carbon fibre sheet, Kawashima's law is used [3]. When applying it to CFRP tube, the effect of epoxy may be ignored and just 5 layers of carbon sheet behave confining role. Confinement effect for each case is shown as Figure 7. For carbon sheet, fracture of carbon fibre sheet also be considered in Kawashima's law, and the tangent modulus of elasticity of confined concrete is lower than that in Mander's. It is reasoned that the mechanical properties are different between the two confinement materials, namely, steel and carbon fibre, and this should be investigated deeply in future.

Above-mentioned confinement effect is taken into fibre model to calculate the flexural capacity of column specimens. And the ductility enhancement could also be expected due to the confining reinforcement. On the other hand, modified Arakawa's equation for shear strength is an empirical one, and AIJ method is design law by Architectural Institute of Japan on basis of plastic theory. In AIJ method, shear strength in plastic hinge region decreases when rotation angle becomes large [4].

Shear strength enhancements from retrofit of steel plate and carbon fibre reinforcing sheet could be considered as the truss effect in shear resistance mechanism. And in the arch effect, confined concrete strength is adopted because shear strain is possible to arrive at a greater level when concrete is well confined. This opinion also could be testified by experimental results to a certain degree, where shear failure has not been observed.

Theoretical flexural capacity and shear strength are compared with experimental V-R skeleton curves as depicted by Figures 8 and 9, in which corrugated branch in theoretic curves is caused by the occurrence of carbon fibre tension fracture at the extreme compressive zone according to Kawashima's approach. For reference, extreme fibre compressive strains at cross section of the end of columns versus the drift angle curves are illustrated as Figure 10. But the fracture was not observed during loading test, and the mean compressive strain of specimens CC97-SS and CC97-SD did not exceed 0.4% (see Figure 4) through displacement transducers while in Kawashima's law the ultimate compressive strain is about 1.5% (see Figure 7). Thus, taking stress-strain model of confined concrete under uniaxial compressive loading into flexural capacity analysis is rather conservative.

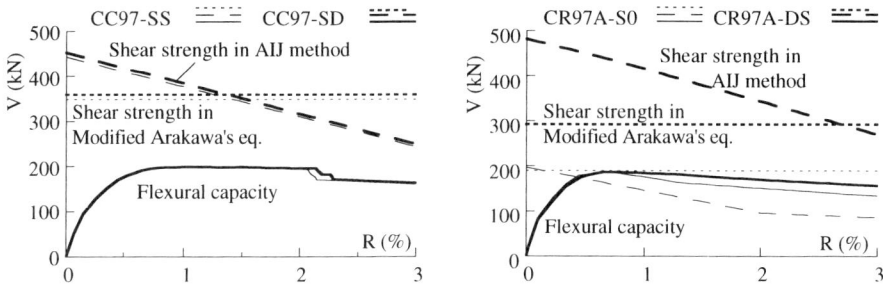

Figure 8 Comparison of shear and flexural behaviour of columns

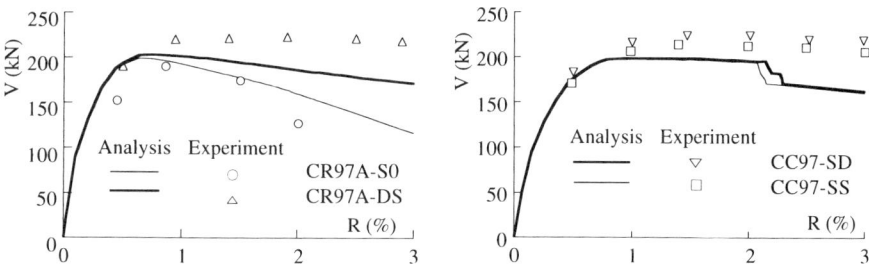

Figure 9 Comparison of theoretical flexural behaviour and experimental results

Proper fibre sheet content in CFRP tube confined column with or without hoops is examined as shown in Figure 11. Confined by one layer or two, column should be of flexural failure but ductility could not be satisfied after gained its flexural strength. In this case, column may behave shear failure for its insufficiently confined core concrete. So, three layers of carbon fibre sheet are desirable to ensure both strength and ductility of column specimens confined by CFRP tube. It is necessary to note that Kawashima's stress-strain model is not perfect yet when concrete is confined doubly by CFRP sheet and steel hoops, so flexural behaviour curve for three layers indicate lower ductility than one for 2 layers. This contradiction could also be found in Figures 7 and 10.

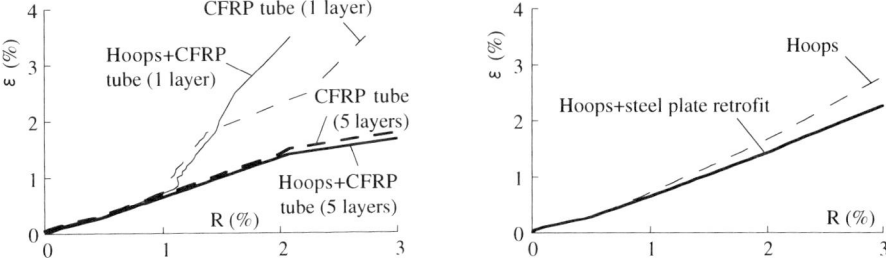

Figure 10 Analytical results of extreme fibre compressive strain versus the drift angle

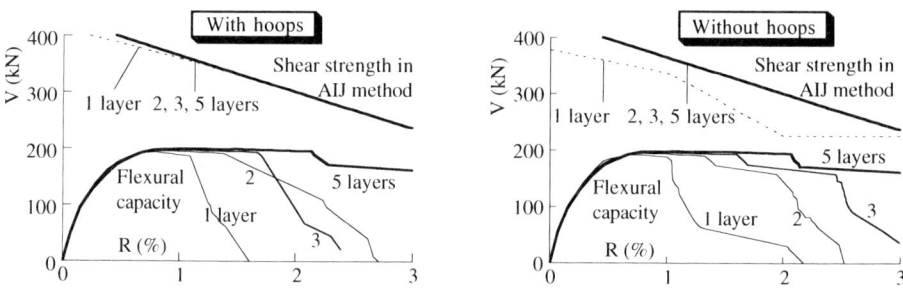

Figure 11 Analytical results of layers of carbon fibre sheet and shear strength of columns

CONCLUSIONS

Hybrid columns with CFRP tube performing dual functions of stay-in-place formwork and transverse reinforcement were loaded under lateral cyclic load while subjected to constant axial force ($0.35 f_c' A_g$). Due to lateral pressure by fresh concrete, for about 2-3 mm deflection as same as thickness of wall plate in CFRP tube occurred during concrete casting. Transversely confined by CFRP tube, the hybrid column specimens CC97-SS and CC97-SD illustrated almost perfect shear force-lateral displacement hysteresis loops in spite of very slight bond slip in comparison to that of steel plate retrofitting one.

Circular and square tubes employing carbon or aramid fibre sheets should be developed. Simultaneously, constitutive law of concrete confined by those fibres also could be studied, and relevant bending shear loading experiment is necessary to be conducted. Seismic design on hybrid column utilizing carbon or aramid fibre reinforcing plastic sheets should be created through the basis of experiments and analytical investigations.

ACKNOWLEDGMENTS

The experimental study presented herein was made possible by technical supports provided by Nippon Steel Chemical Co., Ltd. and Shimizu Corporation. The authors are deeply indebted to Mr T Tokashiki, graduate student of University of the Ryukyus, for his considerable assistance in the experiment and preparation of this paper.

REFERENCES

1. MANDER, J B, PRIESTLEY, M J N, AND, PARK, R. Theoretical stress-strain model for concrete. ASCE Journal of Structural Engineering, Vol. 144, No. 8, 1988, pp 1804-1826.

2. YAMAKAWA, T, HAO, H T, AND, MURANAKA, K. Elastoplastic behaviour of doubly confined R/C columns in steel tube and hoops. Journal of Structural and Construction Engineering of Architectural Institute of Japan, AIJ, No. 500, Oct. 1997, pp 83-89.

3. HOSOTANI, M, KAWASHIMA, K, AND, HOSHIKUMA, J. Strain-stress relation of reinforced concrete columns by carbon fibre sheet. TIT/EERG 96-2, Aug. 1996 (in Japanese).

4. ARCHITECTURAL INSTITUTE OF JAPAN. Design guidelines for earthquake resistant reinforced concrete buildings based on ultimate strength concept. Nov. 1990, pp 106-121 (in Japanese).

INDEX OF AUTHORS

Akman, M S	69-78	Mercier, S	153-162
Al-Robaidi, A	389-394	Morris, J	185-198
Austin, S A	141-152	Mutlu, M	69-78
	317-330	Nakamura, M	395-404
Baker, G	121-130	Neeley, B	239-256
Beaupre, D	153-162	Nuruddin, M F	111-120
Beckett, D	289-302	Ohno, M	311-316
Bishop, J W	317-330	O'Neill, M L	331-342
Blanck, J	121-130	Pedraza, M	219-226
Bosiljkov, V	433-442	Puri, U C	131-140
Bosiljkov, V B	433-442	Quenuedec, M	199-208
Bouguerra, A	199-208	Resh`eidat, M R	389-394
Browne, T M	269-280	Ribay, E	31-40
Cabrillac, R	31-40	Robins, P J	141-152
	209-218		317-330
Cerný, R	103-110	Rodriguez, A	219-226
Cheng, Y R	405-412	Rovnaníková, P	103-110
Collins, T J	257-268	Rynhart, A D	331-342
Constantiner, D	219-226	Sakai, H	395-404
Dalhuisen, D H	405-412	Sato, Y	395-404
De Rose, L	185-198		413-424
Dennis, A	355-364	Scott, R H	365-376
Diah, A B	111-120	Shui, Z	405-412
Douzanet, O	199-208	Sopko, S J	303-310
Dumais, N	153-162	Stroeven, P	377-388
Fukai, K	311-316		405-412
Garshol, K F	163-172	Sumi, T	49-56
Goodier, C I	141-152	Takahashi, Y	413-424
Gostic, S	433-442	Tassios, T P	79-92
Goto, Y	443-454	Teichert, P	93-102
Gouvenot, D	31-40	Thooft, H	343-354
Haimoni, A M	11-30	t'Kint de Roodenbecke, A	199-208
Hata, C	413-424	Toumbakari, E E	79-92
Hockings, R	121-130	Tseng, C-C	57-68
Hrstka, O	103-110	Ueda, K	395-404
Huang, W-H	57-68	Umehara, H	49-56
Iisaka, T	49-56	Uomoto, T	131-140
Joh, O	443-454	van der Pot, B J G	1-10
Johnson, D	41-48	van Gemert, D	79-92
Jolin, M	153-162	Visagie, M	173-184
Kearsley, E P	173-184	West, R P	331-342
	227-238	Yamakawa, T	455-464
Kiyohara, C	395-404	Yunus, S A M	111-120
Koshiishi, N	425-432	Zarnic, R	433-442
Lacombe, P	153-162	Zhong, P	455-464
Luciano, J	219-226		
Maeda, T	413-424		
Malou, Z	209-218		
Marmoret, L	199-208		
McDonald, J E	239-256		
McLennan, L	281-288		

SUBJECT INDEX

This index has been compiled from the keywords assigned to the papers, edited and extended as appropriate. The page references are to the first page of the relevant paper.

Accelerating agents 131
Accelerator 121, 163
Adjoining sash 443
Admixtures 163
Aerated concrete 173, 227
Air-void size distribution 173
Alkalifree 163
Anchors 239
Anisotropy 209, 377
Antiwashout
       admixtures 239
       concrete 281
Aramid FRP rod 413
Arrival time 49
Assessment 269

Bending 433
Bentonite 69
Bingham model 131
Bonded concrete overlays 303
Bonding strength 111
Boundary layer 377
Build 141

Carbon fibre 405 425
       reinforced concrete (CFRC) 395
       reinforced plastic (CFRP) plate 433
       sheet 443
Carbon FRP sheet 413
Cellular concrete 199, 209, 219

Cement 41, 69
       grout 11, 31, 57
CFRP tube 455
Chemical grouts 11
Chloride ingress 111
Circular cross-section 443
Closed wrapping 443
Comparison of design methods 289
Composite 365, 389, 405
       hydraulic binders 79
Concrete 269, 377, 389
       cracks 49
       slabs 331

Confinement 455
Construction remainder soil 311
Cost optimization 219
Crack density 425
Crack opening mouth 425
Creep 31, 395
       limit 31
Curing 163, 227

Damp proof membrane (DPM) 355
Defined performance 1
Deflection ability 413
Delamination 303
       repair 303
DEM parameters 131
Detailing 355
Dispersion 377
Diving 257, 269
Dosing 163
Drying out 331
Drying shrinkage 317, 395
       stress 395
Dry-mix sprayed concrete 93
Ductility 153, 227, 425, 443

Early-life 317
Elastoplastic behaviour 455
Elongation 389
Engineer-diver 257, 269
Engineering properties 57
Epoxy 49, 355
Equipment 163
Eurocode 2 343
External bonding 433
Failure mechanism 433
Fibre 153, 227, 389
       efficiency 377
Fibre reinforced cementitious composite 425
Fibre reinforced plastic (FRP) 365
Finishing time 49
Flexural
       reinforcement 413
Flexural strength 425, 433
Fly ash 185
Foamed concrete 173, 185, 219, 227

Formwork 93
Freeze-thaw resistance 153
GGBS 111
Glassfibre mesh 93
Ground floor slabs 317
Ground treatment 11
Grout 11, 41
Grouted sand 31
Grouting 69
    techniques 11
Gypsum 79
    screed 331

Homogeneization 209
Hybrid 455
Hydrated lime 79
Hydration heat 103
Hygrometer 331
Hygrothermal stress 103

Image analysis 173
In situ strength 121
Injection grouts 79
In-situ strain measurement 317
Inspection 269
Insulation 185
Interior elements 93

Jobsite-produced dry mix 93

Lightweight
    cellular concrete 185
    concrete 173, 227
Local strength 377
Long-term performance 121

Masonry 79
Mechanical characteristics 209
Mechanical properties 405
Mechanical properties of CFRC 395
Micro-properties 173
Mineral grout 31
Mix proportion 111
Mixture optimization 219
Moisture and thermal transfer 199
Moisture condition 331
Morphology 389
Mortars 141

Natural pozzolans 79
NDT 269
Nozzleman certification 153

Optimization 209
Ordinary portland cement 311
Ordinary portland cement (OPC) 79
Orientation 389
Osmosis 355
Overlay construction 303

Partial cement replacement material 111
Particle packing 1
Phenomenological relationship 31
Pile supported ground floors 343
Polyethylene terephthalate (PET) 355
Polypropylene 389
    fibre 57
Porosity 209
Precast concrete 239
Prefabricated steel 239
Pre-formed foam 185
Prestressed concrete 365
Properties 93
Pulverized-fuel ash (PFA) 227

RC column 455
Reinforced concrete 365
    column 443
    structure 433
Reinforcement 365, 389
Relative humidity 331
Repair 239
    materials 49
Rheology 1, 69, 131, 141
RHPC 185
Road surface pavement 311
Robot 163
Rock 69

Sand 31
Scaling 153
Seismic rehabilitation 443
Self-compacting 1
Self-levelling 1
Separated wrapping 443
Serviceability limit state 343
Shear strength 227
Shear strength 443
Shotcrete 111, 121, 131, 153, 163
Shrinkage 153
    cracking 395
Silica fume 425
Silicate grout 31
Site instrumentation 317
Slump 141
Slurry 69
Soil cement concrete 311
Solidification agent 311
Spacing factor 153
Sprayed concrete 103, 111, 141, 153
Standard 41
Stay-in-place formwork 455
Steel fibre 163, 377, 405
Steel fibre reinforced shotcrete (SFRS) 121
Steel plate 433
Stereology 377
Stiffness 433
Strengthening 443

Structural frame   93
Structure   405
Sub 500kg/m³ density   185
Substrate preparation   355
Sulfate resistance   79
Superplasticizer   57, 69
Surface preparation   239
Surface-sculpting with hatchets   93

Temporary high air content   153
Tensile strength   389
Tension test   425
Test data   289
Theories   289
Thermal characteristics   209
Thickness design   289
Thin shells   93
Transverse reinforcement   455
Tremie   281, 239
Triaxial tests   31
Two-point test   141

U-jacket of CFRP sheet   413
Ultimate limit state   343
Ultimate load   289
Ultra fine powder cement   49
Ultrasonic mixing procedure   79
Underwater   257, 269
      concrete   239
      construction   257
      inspection   257
      repair   257
Upcycling   355

Vapour emission test   331

Washout resistance   281
Waste containment   57
Westergaard et al   289
Wetmix   163
Wet-process   141
Workability   141

Yield line design method   343